A Volume in Bioinformatics in the Post-Genomic Era Series

PHYSIOLOGY AND MEDICINE

BIOINFORMATICS IN THE POST-GENOMIC ERA

Physiology and Medicine
Ivan Yu. Torshin
2007. 978-1-60021-752-4

Bioinformatics in the Post-Genomic Era: The Role of Biophysics
Ivan Yu. Torshin
2006. ISBN 1-60021-048-1

A Volume in Bioinformatics in the Post-Genomic Era Series

PHYSIOLOGY AND MEDICINE

IVAN YU. TORSHIN

Nova Biomedical Books
New York

For permission to use material from this book please contact us:
Telephone 631-231-7269; Fax 631-231-8175
Web Site: http://www.novapublishers.com

LIBRARY OF CONGRESS CATALOGING-IN-PUBLICATION DATA
Torshin, Ivan Y.
Physiology and medicine : bioinformatics in the post-genomic era / Ivan Yu. Torshin.
p. ; cm.
Includes bibliographical references and index.
ISBN-13: 978-1-60021-752-4 (hardcover)
ISBN-10: 1-60021-752-4 (hardcover)
1. Bioinformatics. I. Title.
[DNLM: 1. Computational Biology. 2. Physiological Processes. 3. Genomics. QT 26.5 T698p 2007]
QH324.2.T67 2007
572.80285--dc22 2007017504

Published by Nova Science Publishers, Inc. ✦ New York

CONTENTS

DISCLAIMER

In this book, we have attempted to illustrate how various methods of present-day bioinformatics can be used for the purposes of clinical and biomedical research. The examples presented are based on the data we deem to be reliable. However, neither the examples nor the suggested algorithms are supposed to be used in any biomedical application without additional research. Hence, we make no guarantee, warranty or representation regarding the suitability of information presented in this book for use in any specific applications. None of the data or algorithms presented implies a direct usage by a reader in applications of a critical nature where the safety of life or property is at risk. Neither the author nor the publisher will be responsible for a commercial outcome and suitability for any particular purpose even if the author gave a particular advice concerning a practical application and gave a detailed example thereof. Under no conditions will the author nor the publisher be responsible for any losses arising from the use or failure of any application. Neither author nor the publisher will be liable for any kind of direct or indirect damage that can arise as a result of an improper application of the examples, data and other information presented in this book. The reader or any other user of the book assumes full liability for the use of the information in any such application and for the respective outcomes. Computational methods mentioned in this publication are being patented. Under no circumstances shall any user of this publication be conveyed any license or right to ownership or any use of these patents. The readers are also strongly cautioned against any attempts of reproduction of any part of the material provided by the author in the present book. Reproduction or reverse engineering of any part of the material and publication in academic journals (or in any other form of publication) without written permit of the author will be considered as plagiarism, perpetrators will face legal action according to US and international anti-plagiarism laws. Brief quotes or use of citations do not constitute plagiarism.

FOREWORD

The author goes well beyond the ordinary definition of the field of the bioinformatics as restricted merely to compilation of databases or routine sequence analyses. Defining bioinformatics as information management of biomedical research, this quite unusual volume emphasizes the importance of balancing the technological advancements in the post-genomic research with adequate use of the physiological expertise. The author reiterates that proper informational management of post-genomic research should be based on adequate physiological expertise and this expertise is paramount for the successful medicinal applications of the post-genomic technologies. The book concentrates not so much on particular computational techniques in post-genomic research but rather on the analyses and syntheses of the vast array of already published data. Illustrating a consistent application of the integrative approach to the problems of physiology and medicine, the volume often features a careful analysis of assumptions inherent in the techniques and the methods discussed. The volume is written in an involving style making it suitable for a wide readership, both professional and non-professional. The author apparently presents an alternative view of the field of bioinformatics which the author calls "purpose-oriented bioinformatics". Apparently, this view aims at providing a framework in which clinical scientists and other biomedical researchers can participate in the interdisciplinary effort to acquire multidimensional knowledge of the human genome. Caution: the book seems to have little tolerance for skimpy readers.

PREFACE

Several years have passed after the sequencing of the human genome and what might be called as "the post-genomic era" has begun. Of course, there are many different genomes and the term "post-genomic" does not necessarily imply the human genome. However, it is the data encoded in the *human* genome that hold the promise to be of practical importance in a wide range of biomedical applications. The sequencing and preliminary annotations of the human genome provided an incredible amount of the raw, largely unprocessed information. Coupled with the millions of publications on human physiology already available in public databases, it is clear that certain informational strategies should be adopted for the retrieval, analysis and representation of these data.

Among biological sciences, bioinformatics is a specific branch that deals with managing complexities in the biological information. However, the bioinformatics is in no way restricted to the compilation of large databases or elaboration of sophisticated software. The methods of bioinformatics can greatly assist the generation of productive hypotheses that allow subsequent experimental testing followed by confirmation or disproval. The main idea behind the present volume is not worrying about the steadily growing amounts of biomedical information or about the relative quality of it. This volume, as well as the entire book series, is based on the purpose-oriented attitude: *how* to make a good *use* of this information in particular research projects. Time and again, we emphasize that integration of data, essential for post-genomic research, is not merely lumping together various sorts of information into one particular database. Apparently, this would be an extraordinary naïve interpretation of what integration of data really is in biomedicine or in any other science. In simpler terms, "collecting facts" and "making sense of the facts" should never be confused.

Reading this book requires not so much an extensive specialist background as an open-minded and sincere posture of the reader. This book can be profitable for a researcher working in clinics as well as in biotech: that is, the people actually interested in practical applications. These include a doctor interested in well-being of his or her patients; a clinician who wants to further one's education; a geneticist, interested in the study of the genetic predispositions; a bioinformatician interested in clinical science. Academic researchers can considerably profit from this volume if they are able to suspend the typical tendency for academic debunking of almost anything. Journal editors and peer reviewers can find the book quite helpful in their daily work. And, surely, any general reader interested in the question of how the data on the human genome can contribute to the public health care is most welcome.

The general analysis and discussion of difficult, controversial and sometimes even absurd points in contemporary biomedicine is supplemented and illustrated by concrete examples we call "Case Studies". Dispersed throughout the volume, these Case Studies often indicate how bioinformatics can assist biomedical research in post-genomic era. The respected readers are encouraged not to take particular conclusions and statements of the author at the face value. Rather, the author expects that the readers will make at least some research of their own (in terms of searching abstract databases, for instance) in order to appreciate the extent of *already available* information pertaining to human physiology and genetics.

INTRODUCTION

Publication of the nucleotide sequence of the human genome caused considerable amount of interest in general public and produced quite many expectations among specialists. Various discussions about personalized pharmacopoeia, possibility of "superdrugs", finding "true" genetic susceptibilities to multifactorial diseases, extention of the life span, gene therapy, gene testing, *etc* continue to fill the pages of numerous professional, semi-professional and popular publications. Often, these publications produce an impression that numerous breakthroughs in human biology and biomedicine are at hand and one just needs to extend the hand to catch them by merely having the nucleotide sequence of the genome. Coupled with tantalizing phrases such as "systems biology", "artificial intelligence", "post-genomic medicine" and often mixed with other catchwords like "stem cell", "cloning", these activities of the mass media do stimulate imagination of many.

Behind the terms we mentioned above, there are wonderfully cheerful and optimistic scenarios. We, humans, are often being conditioned to respond to certain optimistic scenarios and tend to dislike any speaker who points out at the related difficulties and inconsistencies. It cannot be denied that optimistic scenarios do have a value in the hurly-burly of the human life. However, the *scientific* value of any optimistic scenario is neither how much it is spoken about in mass media nor how much of the public or private funding is spent on it *per* year nor how blatantly optimistic that scenario is. The value of a research scenario is how much it is *scientifically* justified and to what extent it can be (will be) *actually* fulfilled. Apparently, to generate and to select scenarios that are able to produce tangible results in a reasonable amount of time is a good practice both for fundamental scientific research and for business. In the present book series, we hold that purpose-oriented bioinformatics can prove invaluable logistic support for the biomedical studies. The term "logistics" implies a detailed planning and coordination of an operation, so generation of the viable research scenarios is an essential part of this activity.

While it is quite possible to generate an optimistic scenario with little or no scientific expertise using only excessive optimism and what might be called as "personal charisma", it is virtually impossible to generate a realistic research scenario that would lead to successful scientific research using only these two components. Without realistic assessment of the current situation in the area, a seemingly optimistic scenario is nothing short of a delusion. Given the informational overload in modern human physiology (Chapter III), it will be very difficult to describe the current situation in the most comprehensive manner. We can be more objective in our assessment if we focus on a certain topic of wide biomedical importance.

For example, the genetic association studies, discussed extensively through out this volume, are one of the main research scenarios for the application of the information on the human genome. The genetic association studies aim at linking certain medical conditions directly to the presence of certain polymorphisms in the human genome and thus can be of immediate practical use in various areas of medicine including medical diagnostics, individualized drug/lifestyle prescriptions *etc*. Therefore, the proper planning, conduction and analysis of the association studies are paramount in the post-genomic era.

At present, the results of tens of thousands of the association studies are already available in professional medical literature and most of them are referenced in the public abstract databases. Theses studies differ much in their scope, quality and the level of professionalism (Chapters III, IV and V). Quite many of these studies were made in the preceding years, before the highly acclaimed "Human Genome Project".

The overwhelming majority of the available association studies that link genetic polymorphisms with such multifactorial disorders as cardiovascular disease or cancer operate with only several genes and a few polymorphisms. On the contrary, the association studies performed in a post-genomic manner (that is, those including high-throughput genotyping of thousands of genetic polymorphisms, large groups of patients) are still very rare and amount to less than a few dozen (year 2007). If we follow the expectations we referred to in the first paragraph of this Introduction, we might expect these post-genomic studies will be outstanding pieces of modern biological research that unequivocally link genetic factors with susceptibilities to multifactorial diseases and report numerous breakthroughs in human physiology and medicine.

Still, let's be a bit cautious and briefly consider a few actual examples. Within the case-control study design, 210 polymorphisms in 111 candidate genes were genotyped in 352 Caucasians with coronary heart disease (CHD). A random population sample of 418 individuals was used as a control group. Out of 111 genes, polymorphisms in about 13 genes shown statistically significant associations with CHD, ten of these 13 genes were known previously to be CHD-associated [1]. It's a well performed and an interesting study that worth taking a look. Still, a more careful observer might ask the following question: why associations with polymorphisms in only 13 out of 111 genes were found while at least 50 of the genes analyzed in this study were previously known to be associated with CHD in numerous studies? For this observer, asking those impudent questions of the kind, we can only reply that the paper [1] is one of the best examples of the large post-genomic studies.

The best example, really? But where did the promised breakthroughs go? Surely, since there are "few dozen" of the post-genomic studies, as the author of this book claims, there must be better examples? OK, let's take a few more examples. Since the author of this book is not particularly fond of pointing finger at poorly made studies, here and throughout the volume such studies will be given anonymously, without providing the references to the actual publications.

A study investigated over 10,000 polymorphisms established in the course of the Human Genome Project. These polymorphisms are contained within thousands of genes. The study aimed at pin-pointing the common polymorphic loci associated with certain types of cancer.

The result? Associations with polymorphisms in only 3 (*three*) genes were reported. One of these genes was known to be associated with cancer at least for decade and the other two belong to the same functional group as the first gene.

Perhaps, the author is biased towards seeing or citing (albeit anonymously) only negative reports. Let's try another example: a post-genomic study involving, again, analysis of about 11,000 polymorphisms in almost 7,000 of various genes. The study aimed to find possible associations with the myocardial infarction (MI). The result? Polymorphisms in only *4 (four)* genes were reproducibly associated (three independent case-control studies) with MI. What is even more interesting is the physiological "explanations" the researchers provided to each of the associations they found. It's for certain that not only an expert in human physiology would be quite annoyed by that sort of explanations, but, perhaps, any reader with an average background in medical science... We might continue with citing these examples. But, however much the author of this book would like to bring to the reader's attention an example of a post-genomic study in which at least 1/4th of the genes, *already known* to be associated with the disease in question, would be confirmed, it was impossible to find one. We consider the details of these and other large-scale studies in the Volume II of the present book series.

This seemingly "pessimistic" outlook on the problems of the post-genomic studies and application of the post-genomic technologies is far from being simply a matter of personal opinion of a particular author. Indeed, phrases like "...no significant associations were found..." or "...association was not confirmed..." occur too often in the professional literature dealing with genetic associations and hardly can be ignored by anyone who studies extensive amounts of the relevant biomedical literature with an open and an inquisitive mind.

For instance, a recent review attempting to analyze the features of the "good" association studies [2] reiterates that one of the major sources of frustration and confusion in the area of genetic association studies is the common publication pattern: when initial positive findings for a gene or a polymorphism do not get confirmed or even get disproved in most of the subsequent studies. Lack of reproducibility of the genetic association studies is a source of considerable consternation of the medical scientists and editors of the professional journals (see Chapter V, section "Para-analysis"). Some literature reviews, albeit being based on somewhat fragmentary an evidence, suggest that 75-95% of the reported genetic associations were not confirmed in the subsequent studies [3]. The figures of 75%-95% are, apparently, somewhat higher than they should be (at least, according to our analyses of the thousands of the association studies registered in the *medSNP* database, Chapter IV). Nevertheless, according to the analyses of *medSNP*, made by the author in the present book series (see Chapter IV and the Volume II), more than 50% of the published association studies do feature serious violations of the relevant standards of biomedical and scientific research (Chapters III, V).

Given this kind of situation with the currently available studies and recalling the fact that the post-genomic high-throughput approach *per se* is not a magic wand that makes much difference in the actual biomedical research, how, then, modern biomedicine will face the new challenges of the post-genomic era? What would do, for example, the researchers involved in the Biobank project [4] (http://www.ukbiobank.ac.uk/) and similar projects that aim at collection of the detailed medical information, including, probably, the genetic data on hundreds of thousands of individuals? The author of the news item [4] does make the right point by using the word "gamble" in his article. Apart from the exploits of James Bond we know from the movies, everyone knows that gambling is hardly a *source* of funds ... If the largest of the currently available association studies (that, by the way, do not include more than 10,000 individuals) report very bleak associations at best (see Volume II of the present book series), what sort of result can be expected from a study that includes *hundreds* of thousands?

The major message of the present volume, however, has nothing to do with either "criticism", "pessimism" or "optimism". Clearly, the whole endeavor of scientific research is not based on trivial human emotions such as these. On the contrary, the science is based on the adequate and timely use of the logical and other capacities of the human mind and not on superficial emotionality that characterizes mass media. Since the science of physiology did not appear yesterday (Chapters I, II), the major message of the present volume is that we need first learn how to extract and apply what is *already known*.

Human mind always tried to deal with the apparent complexity of the human body, the real extent of which physiology only begins to unravel today. For millennia, the medicine was an epitome for comprehensive *knowledge* and it was always implied that the student of medicine should have a wide-open mind and be intellectually capable of managing vast amounts of *information*. The essential difference between "knowledge" and "information" was very clearly understood and not less clearly explained by the outstanding medical authorities of the centuries past. In the past (at least, according to the books and other documents we can study today), there was little tolerance for "experts in the field" who confuse knowledge with information or who are narrowly specialized in a particular field of medicine to the neglect of all the others. In the last couple of centuries, despite the apparent explosion in physiological and medical *information*, there is a distinct trend of ultra-narrow specialization of the most of the medical practitioners. This trend is a matter of urgent concern not only for medical educators who constantly need to adjust the teaching course to include new and new discoveries. This trend is also of considerable concern to the researchers who aim at making real science and real discoveries and who are not so much retarded as to remain merely at the stage of extorting more and more grant money from the governmental and private institutions.

Albeit there is nothing wrong with the growing specialization into different areas of medicine (no one would go to a dentist for operation on appendix, apparently), it is the *ultra-narrow* specialization of the medical researchers that holds considerable danger for the medicine as a scientific enterprise. At the same time, ultra-narrow specialization of some medical scientists often couples with its opposite: almost complete lack of the specific medical expertise in the case of scientists from the allied fields of research such as genetics, molecular biology, biochemistry, biophysics or bioinformatics. The lack of specific medical expertise plagues, in particular, the overwhelming majority of the published genetic association studies. How much longer will the application of the post-genomic technologies such as DNA biochips or functional genomics (Chapter VI) systematically produce results which are, at best, simply irrelevant to clinical applications? How many more thousands of half-baked association studies will we read in journals before it becomes apparent that the proper informational management is essential to produce an adequate association study? This book, of course, cannot provide answers to these high-flying questions. What we can and will do here is to indicate the features and peculiarities of the *adequate* studies. Throughout the volume, the numerous examples of proper informational management will be presented. The author's hope is that this book will allow to any reader interested to perceive the beauty and simplicity of a truly scientific study in biomedicine, should the reader find it in an academic journal or elsewhere.

Prior to developing such a capacity for adequate perception, we must be aware that in modern biological sciences there are numerous assumptions, contentions and even entire "methodologies" which are based on nothing else than statements of the kind "I believe that there is life on Mars". Using a medical analogy, we might, perhaps, say that the current biology and biomedicine are ill with what we call here as "the disease of improperly managed research". The four main symptoms of this disease we allegorically refer to here are the where-the-fly-has-gone symptom, the doctrine-of-signatures symptom, the elephant-in-the-dark syndrome and the non-falsifiability syndrome. In the rest of this our allegorical introduction, let's consider these.

Biological systems are clearly much more than a mere sum of their parts. Far from being a mere *bon mot*, this statement is an essential principle of which modern biologists should be keenly aware at all times and under all conditions. There is an old story of a smart boy who was well versed in Latin. Once upon a time, he saw and caught a fly. Marveled by the fact that this tiny creature flies so adroitly and behaves so intricately, the boy decided to investigate the nature of the "fly phenomenon". Seizing the poor insect, the boy methodically began to separate the *parts* such as wings, legs, the head, the body in order to see *how* the fly operates. To the great dismay of the boy, when he finally managed to split the fly into the major anatomical parts and label each with a specific Latin word, it turned out that the fly "disappeared". The fly flew. The parts were not. Moreover, they were not even moving. This kind situation and whatever follows from it is what we call as the **'where-the-fly-has-gone symptom'**.

Couching the story in the scientific terms, we can say that there is fascinating complexity of integration of the biological systems, especially systems as complex as the human body. This inherent complexity effectively precludes dissection into some "key parts" and considerably hardens the analysis of the functional blocks, at any level: from the physiological level of function (Chapter I) down to the biophysical function of individual molecules (Volume V). The living body of a human can be conceived as a dozen of interlocked organ systems, a hundred of tissue types, thousands of molecular cascades, tens of thousands of proteins or as millions of protein-protein interactions. However, it still remains the singular entity: the living human body. Accordingly, the reductionist attitude of finding only some few "key" residues in the protein (Volumes IV,V), a few "key" disease-related genes in the genome (Volume II) or a few "key" cellular mechanisms (Volume III) is bound to fail. In the view of the "where-the-fly-has-gone symptom" hyperbole, we might also ponder why a genetic association study of the post-genomic kind that operates with thousands of genes tends to come out with a few "significant" ones...

Physiology is supposed to promote integrative research and teaching, which are foundations and unique strengths of physiology. Unfortunately, integrative thinking and synthesis often seems to get lost in the drive to understand individual proteins at deeper and deeper levels [5]. In the post-genomic era, the introduction of the concept of systems biology is assumed to be a sufficient description of the study of living systems from a holistic perspective [5,6]. However, avoiding the "naïve reductionism" [5] should not, apparently, lead to the other extreme: the "naïve integrativism", shall we say. In other words, despite the overly "integrative" approach that is claimed, for instance, in the systems biology, there is already a significant danger of neo-reductionism in the area (Chapters VI-VIII). An individual affected by this kind of the attitude only *pretends* to have an open mind and to exercise an "integrative approach" but, in fact, operates with the same centuries-old reductionist assumptions in mind (perfectly described, for instance, in Goethe's "Faust", T.1, S.4).

Some of these assumptions can be collected under a rubric we call here as the **'doctrine of signatures'**. The term has several meanings, not all of them are entirely nonsensical. Some of them are. In the medieval European occultism, the "doctrine of signatures" states that mere arrangement of certain magical signatures will bring certain physical effect--- through magic, of course. In medieval medicine, "doctrine of signatures" referred to an idea that in order to produce a necessary medical effect, one has to act upon like with a like. A statement like this is extremely susceptible to be misinterpreted in the occultist sense as above. These sorts of the "doctrine of signatures" are much ridiculed by contemporary scientists. And yet, paradoxically, this doctrine still thrives in contemporary biology disguised as hidden curriculums of various sorts.

Sometimes, reading the professional literature on the post-genomic biology, one does get a fleeting feeling of some medievalism that brings to mind those witchy parts of "Dr. Faust" and "Macbeth". The authors, writing in highly acclaimed academic journals, seem to assume that by merely reiterating endlessly phrases like "systems biology", "post-genomic", "significant achievements" and other magic words of power, it will be possible to make some real scientific discoveries through indoctrination by signatures. There is also another common manifestation of the same magic of "signatures" which can be stated as: "If a phenomenon *I* observe looks to *me* like a phenomenon *I* know, then it *must be* the phenomenon *I* know. And *not* any other ".

Let's step back to association studies, one of the main subjects of the present volume. If, for instance, a biomedical researcher, a peer reviewer or an editor "observes" a paper that only *looks* like an association study made by an outstanding "expert in the field" or published in a journal with a high impact factor, then it *must* be an important association study. The fact that this "study" presents insufficient data, contains ridiculous assumptions or conclusions about human physiology, features severe data distortions or even an outright forgery is of no concern. What matters is how this "study" conforms to some arbitrary biases of that particular researcher/editor/peer reviewer/*etc*. Among the most trivial, those biases might be a particular surname of a particular expert in the field ("...Ah, I know that guy, *everything* he does is reliable..." or "...*everything* he does is unreliable..." or "I do not know this guy and don't want to" *etc*), what would be the "impact factor" of this publication [7] or how intense the publication advertises the alleged results [8]. Often, no objective criteria of study assessment are of any interest to such an individual or an individual in such a peculiar state of mind. Thus, the numerous details that clearly indicate multiple inconsistencies of the data are happily neglected and the individual "observer" continues to live in the bliss of the ongoing ignorance.

Whether this astounding phenomenon is a manifestation of basic laziness or of something like that is hard to tell. What can be told from an extensive review of current biomedical literature using scientific criteria (see Chapter III) is that biomedical studies that only *look like* valid scientific studies are not simply numerous. In fact, there is a distinct possibility that they might even prevail in the current biomedical literature [3,9]. And, marvelously, hardly anyone seems to notice this state of affairs while continuing to bubble conditioned slogans like "significant achievements in post-genomic research", "breakthrough of the systems biology" and many, many other (being, apparently, at the level of comprehension similar to the "doctrine of signatures" of the medieval occultism).

Of course, a consistent repetition of some principle is not necessarily indicative of magism or obsession. For instance, in the present book we refer the respected reader, time and again, to the Diagram of the Levels of Function (see Figure 1-5 in the Chapter I), merely in order to underline the essential need to abandon reductionist assumptions in the post-genomic era. However, I am not sure that phrases alike to "significant achievements in post-genomic research" that occur all too often in the current academic publications, have anything to do with modern scientific thinking. Rather, they remind us of the medieval magical thinking.

Strict adherence to the medieval assumptions of this kind, coupled with the social heterogeneity of the participants, inevitably results in multiple self-contradicting and mutually excluding observations and associated ideas. We call this phenomenon as the **'elephant-in-the-dark'** syndrome. The phrase, apparently, refers to the millennia old story of a group of blind men and the elephant. This archetypal story goes largely the same in various versions of the tale throughout various cultures: men in darkness (or blind men) touch an elephant to learn what it's like. Later, comparing their observations they find that their opinions are in complete disagreement since they touched different parts of this admirable animal. One of the countless variations of the tale was provided by an American poet J.G. Saxe in his opus:

"...Rail on in utter ignorance
Of what each other mean
And prate about an elephant
Not one of them has seen!..."

Apart from numerous philosophical and metaphysical implications, we can also apply this fable to illustrate a phenomenon that can often be detected while taking a good look at the wide range of contemporary biomedical studies (including the genetic association studies). For instance, take the matter of "contradicting evidence". To put it in the simplest way, some researchers persistently report one result (say, a positive association of a genetic polymorphism with certain medical condition) while some others persistently report the opposite conclusion (absence of any kind of such association). The application of the meta-, ortho- and para-analyses of the published biomedical evidence (Chapter V) does show that contradictions of the kind are abound in the present-day professional literature. In terms of the general science, a presence of numerous apparent contradictions (or, in other terms, absence of even qualitative correlations) strongly suggests that some important or "confounding" factors were not taken into account either during the experiments or during data interpretation. In the terms of informatics, this sort of situation often corresponds to a very low signal-to-background ratio which considerably obscures the biological significance of the results and makes impossible to make any substantiated conclusions. The above-mentioned techniques of the data analyses (meta-, ortho- and para-analyses) can serve as effective instruments to indicate the presence of the elephant-in-the-dark syndrome in the vast array of the published biomedical evidence. Application of these theoretical techniques also indicates the ways of tackling the problem.

The elephant-in-the-dark syndrome is also closely related to the **non-falsifiability syndrome.** Thus, given the abundance of the mutually contradictive statements, some biomedical researchers do not want to risk saying anything contradictory and instead only produce non-falsifiable statements. Statements of the kind are often abound in the publications in journals spezializing on bioinformatics. Albeit the words like "falsifiability" and "non-falsifiability" tend to associate with a sour science teacher in the middle school (or with a more flowery university course on the philosophy of science), in no way this term implies that scientific discoveries are made by generating falsifiable hypotheses first and then and only then by testing those hypotheses with appropriate experiments. Alexander Fleming did not discover what turned out to be penicillin in such a highly tedious manner. Neither did Prof. I. Pavlov and his team ever worked in such a highly non-productive manner, at least according to the detailed reports of the laboratory discussions (see reference in the Chapter I). If some of the present-day popular science writers equate the *descriptive* term "falsifiability" with the *manner* of research, it only indicates a mere infantile rationalization that ensues from confusing causes and effects.

In fact, falsifiability of a particular contention or a statement has nothing to do with the manner of the scientific research. Rather, falsifiability refers to an important *quality* of any currently established scientific *fact*: namely, that this fact can be demonstrated by a clear-cut experiment. A contention like "human body is made of cells" can be directly corroborated by using a microscope. A contention like "substances consist of atoms" can be corroborated through the X-ray diffraction studies of various substances. Similarly, if someone would say "all cells are made exclusively of iron" then this statement, whatever we might think about the person producing it, still can be disproved by the contemporary experimental science.

However, a statement akin to "there are little grey extraterrestrials living on alpha-Centauri", albeit including a number of exciting details, cannot be corroborated nor disproved in any way, at least with the modern level of science and technology. Even if the SETI project would, finally, discover some auspicious signal from the direction of that particular star, this signal does not prove anything of the sort. Contrariwise, a failure to receive a SETI signal from that direction in space does not disprove the statement, either. In order to verify the statement we, using the *Star Trek* lingo, would have to take a ride on the *USS Enterprise*, find an M-class planet near the alpha-Centauri, pack up tricoders, then ask Scotty to beam us out *etc*. However, we, modern scientists, are not in the 23rd century of the *Star Trek* universe (just yet) and do not have access to the starships, tricorders and beaming technology. Therefore, application of the falsifiability principle to the statement in quotes at the beginning of this paragraph clearly indicates that this statement has nothing to do with science. This might be seemed to be "an obvious conclusion" and, of course, it is. But that is what the principle of falsifiability is about. What is much less obvious is how this relatively "simple", "obvious" or "trivial" principle seems to be almost entirely forgotten in the course of a considerable number of biological studies, including the studies dealing with biomedicine, bioinformatics and, in particular, genetic association studies (Chapters III, VI, VII and Volume II).

Generally, all of these four symptoms we just considered might seem "obvious" or "boring" to some. In this case, there is not great benefit to read this book any further and we should part at this point. The above consideration of these symptoms might also be considered as "unnecessary". In the same way, however, it is unnecessary for a savage to question the belief that fire is sent by some sort of a thundergod. Indeed, he's just a savage and exceeding questioning might do him much harm in the society he lives. For the reader who takes the risk of reading the rest of this book, however deficient and incomplete it might be, the author would like to offer an ancient saying: "...*Sapere aude*" (Horace, ep. II).

To round up with the Introduction, we can add that the bioinformatic and biomedical ideas we discuss in this book are illustrated by a number of Case Studies, which, often, were the actual case studies from the consulting experience of the author. The examples used in the Case Studies and throughout the book were crafted according to the levels of structural organization of biological systems (Chapter I) and can either be molecules (say, proteins of blood coagulation cascade), molecular cascades (blood coagulation cascade), organ systems (say, cardiovascular) and diseases (CVD). Our Case Study, however, is not simply an example of an object (molecule, molecular cascade *etc)*. It is an example of an *approach to study* the object. Any study includes definite research strategies and the strategies are, generally, of the primary interest for bioinformatics.

REFERENCES

[1] McCarthy JJ. Large scale association analysis for identification of genes underlying premature coronary heart disease: cumulative perspective from analysis of 111 candidate genes. *J Med Genet.* 2004;41(5):334-341.

[2] Hattersley AT. What makes a good genetic association study? *Lancet.* 2005;366(9493):1315-1323.

[3] Herrington D. Eliminating the improbable: Sherlock Holmes and standards of evidence in the genomic age. *Circulation.* 2005;112(14):2081-2084.

[4] Coghlan A. One million people, one medical gamble New Scientist 20 Jan 2006 (see the news item at http://www.newscientist.com/article.ns?id=mg18925353.800)

[5] Strange K. The end of "naive reductionism": rise of systems biology or renaissance of physiology? *Am J Physiol Cell Physiol.* 2005;288(5):C968-C974.

[6] Wang M. Metabolomics in the context of systems biology: bridging traditional Chinese medicine and molecular pharmacology. *Phytother Res.* 2005;19(3):173-182.

[7] Smith R. Unscientific practice flourishes in science. *BMJ* 1998;316(7137):1036.

[8] Fahey T. Evidence based purchasing: understanding results of clinical trials and systematic reviews. *BMJ* 1995;311(7012):1056-9;
[9] Ioannidis JP. Why most published research findings are false. *PLoS Med* 2005;2(8):e124.

Chapter I

SYSTEMS WITHIN THE SYSTEMS WITHIN THE SYSTEMS

A HISTORICAL PERSPECTIVE

> *"A mind is like a wee empty attic... a fool takes in all the lumber of every sort that comes across so that the knowledge which might be useful to him gets crowded..."*

As the reader, perhaps, already deduced, the above citation is from Sir A. Doyle's "Study In Scarlet". The citation, as well as the continuation of it which we'll recall a bit later, might well fit in with the approach, scope and ideas of the present volume. In the books of this series, we define bioinformatics as management of informational complexity of biological systems. Apparently, this definition is not restricted only to some efficient computer algorithms for ultra-narrow technical tasks or to cramming huge databases. This definition of bioinformatics also implies certain mental skills that need to be developed and practiced, especially in the post-genomic era of biological science.

In accordance with this definition of bioinformatics, what is generally known as a "scientific theory" can also be considered as a tool for management of the scientific information. Consequently, considering the informational effectiveness of the scientific theories (or, at least, of hypotheses) can be quite useful for the post-genomic research. To get a better feel for it, let's start this book with looking into a wider historical perspective rather than with droning on some "exceptional importance of the post-genomic technologies".

This volume primarily deals with the analyses of the information on the human physiology. Human physiology studies the functions and activities of the living human bodies and their components. Although differentiated into a separate area of biomedical science somewhat recently (19th century Europe), the basics of physiology as well as the major objects of study and the practical applications have much to do with the entire history of medicine.

The first lessons in medicine came to the primitive man by injuries, accidents, bites of animals and other impacts which little by little crystallized into practical knowledge. Magic represented the essential attitude of the primordial man to nature, and, at that time, it was a sort of a primitive "science". The primordial man had no idea of fundamental laws but, as Dr Osler once said, "...regarded the world as changeable and fickle like himself..." [1]. Despite what we said about magic in the Introduction to this Volume, the modern allopathic medicine has made quite a way from those inflexible magic rites through the simple schematics of the four humors and\or Chinese "meridians" well into the present-day scientific views of the health and disease.

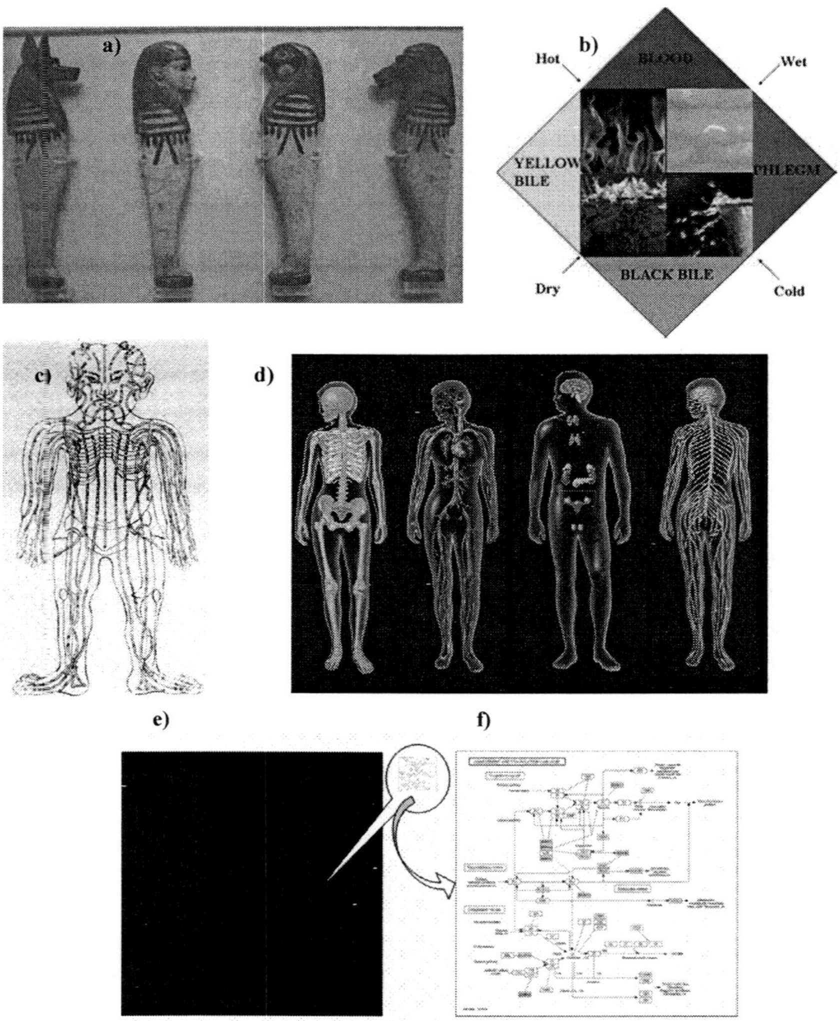

Figure 1-1. Modern physiology in a historical and cultural perspective. a) The theory of the four humors in ancient Egypt (so-called "four sons of Horus"); b) A modern diagram from a natural system of medicine known as "Unani"; c) Meridians of Chinese acupuncture; d) The skeletal, cardiovascular, endocrine and neural systems in modern anatomy and physiology; e) a modern, post-genomic state of the physiological theories...; f) ...and a detailed explanation of a tiny part of it (the blood coagulation cascade).

Let's look at a pictogram that shows some flashes from this historical perspective (Figure 1-1). The first two slides illustrate one of the most important medical theories in the millennia past: **the theory of the four humors.** Since this is one of the oldest theories in medicine and physiology, it seems appropriate to consider the issue at least in brief. The briefest consideration is provided by the Chaucer's verse *"The fyr, the eyr, the water, and the lond, in certeyn boundes that they may nat flee"*. This verse well summarizes the central doctrine of the ancient medicinal systems from all over the globe.

Rather than belonging to the astrological area or being something of a nostrum, the humoral theory was the basis of the medical diagnosis and treatment during millennia of the human history. In this ancient theoretical view that was known already in ancient Egypt and which still survives in a number of its modern adaptations, the human bodies are thought to be "made" primarily of the four "elements" conventionally called as 'earth', 'water', 'air' and 'fire'. These four baffling "elements" correspond to the four "humors" that can be extracted from the human body: 'black bile', 'phlegm', 'blood' and 'yellow bile'. According to the theory, each dietary component, medicine, climatic environment, internal or external activity can modify the humors of the body thus affecting the entire balance of those "humors". In the framework of the theory of the four humors, it is interplay of these factors that can restore health from sickness or cause the sickness to worsen. Health, then, issues from a proper equilibrium of the four "humors" in the patient's body while it is their imbalance that creates a disease. The medical practice based on this theoretical view also came to recognizing that a person's temperament often plays a significant role in determining how the person would be affected by certain diseases and how they would react to certain medicines. This "temperament" is considered to be an inborn factor, thus being a crude description of what is now known as "heredity".

This theoretical view is, actually, not something that totally died out after 19th century AD. In fact, still throughout the world we live in, various forms of the humoral theory are being practiced in the native medicinal systems such as Traditional Chinese Medicine, Unani, Tibetan and a number of others systems of medicine. The experimental studies of Prof. I. Pavlov suggested that the peculiar four-partite classification of the humoral conjecture is far from being a figment of someone's imagination or mere dogma but appears to reflect some physiological reality applicable not only to humans [2,3]. Today, WHO encourages the promotion of safe, effective and affordable systems of traditional medicine (http://www.who.int/mediacentre/ factsheets/fs134/en/) and most of these systems are invariably based on some or other variation of the millennia-old humoral theory. We might also notice that the primordial concepts of equilibrium and imbalance can remind us of the present-day concept of *homeostasis* (see the following section).

As we stated in the Introduction to this Volume, we do not stand for return of medievalism in our approach to modern medicine. By giving some attention to the ancient medical systems, we would like to consider the informational possibilities of those theoretical constructs. Let's recall that the present volume is focused on how the modern view of health and disease can be *practically* assisted by the genomic and post-genomic studies. Apparently, some informational structure within a research scenario is essential to link the clinical observations with the data from the post-genomic experiments.

What might be interesting in the relation to the present volume is that the past and current practitioners of the theory of humors seem to be able to link together the results of the inspection of the patient (in modern terms, "symptoms and questionnaire"), the results of the simple "lab tests" (like pulse and urine analysis) with the inherent predispositions to certain diseases. Then, they consider the ways in which the patient will respond to certain drugs and treatment. In the framework for the humoral theory, assessment of a "humoral predisposition" to a disease is not less important than the immediate causes of the disease such as bacteria or viruses known to the modern medicine. For an experienced practitioner of the sort, the concept of the four humors also has a number of implications for the multifactorial disorders such as cancer or cardiovascular disease. In a certain sense, this relatively simple theoretical construct allowed to formulate the medicines and regimes of treatment which were highly personalized. Modern biomedicine, too, aims at personalized drugs, though considerably more complex compound of genetics and physiology will serve this purpose.

Let's consider another theoretical construct from ancient medicinal history. From informational point of view, **the meridians of the Chinese acupuncture** provide considerably more complex and comprehensive view of the human health, disease and medical treatment than the concept of the four humors alone. Strictly speaking, the theory of meridians, unlike the theory of the four humors, is not part of the history of our Western medicine. We should take into account that the meridian theory comes from an ancient Eastern culture which is very different from the relatively recent cultures of the Europe. In the European medicine, the medieval studies (especially, the work of Dr. Vesalius) have provided a detailed anatomical picture of the human body. These anatomical studies are invaluable for carrying out surgeries unimaginable in any other culture. Nevertheless, anatomy *per se* provides quite crude (if any) means to investigate *how* these anatomically observed systems function or, being more exact, how they live.

In contrast to anatomical dissection, the view of the Chinese medicine is straightforwardly integrative concerning the mechanisms: the meridian theory maintains that the physiological organs are linked together by some sort of "vital streams", or meridians, running throughout the entire body [4]. Modern practitioners often refer to the meridians as "the key power lines" running through the body. The acupuncture points are then likened to the "switches" while the ***Qi***, or "The Force" flows through the meridians keeping life in the body.

The theory and practice of acupuncture holds a prominent place in the modern medicine: at present, WHO officially recognizes more than 400 of the acupuncture points with established medical value (see the relevant documents at http://www.who.int/medicines /library/trm/acupuncture/acupdocs.shtml). However, there are still no scientifically acceptable explanations on how the meridians and acupunctural practices do work, at least within the framework of the contemporary Western system of medicine. It is true that those 'meridians' follow aspects of the four physiological systems (the skeletal/muscular, cardiovascular, endocrine and neural systems, Figure 1-1d) known in anatomy and physiology. However, these 'meridians' are quite elusive, from our Western point of view, since they do not spear to correspond directly to the anatomical subdivisions such as nerves or blood circulation pathways. Acting, apparently, largely through the central nervous system, the acupuncture is thought to alter brain chemistry by changing the release of neurotransmitters and regulating various physiological functions such as immune reactions, blood pressure and body temperature.

Having mentioned these millennia-old theoretical views, embodied into a number of the medical systems throughout the world, we are coming closer to considering what the arising post-genomic biomedicine might offer. The previous discussion was intended, in particular, to provide a comparative scale to look at the modern state of the art. Let's briefly reconsider the historical and cultural perspective, presented in the Figure 1-1. It all looks nice and picturesque until we come to what the author of this book claims to be "post-genomic state of the Western physiology".

At first glance, the general outlook of the modern state of the Western physiology is somewhat stunning (Figure 1-1e). However, rather than being an example of suprematist art, this picture, clearly, represents the concept of a "black box", which is not only a fundamental term in cybernetics and informatics term but is also a widespread cultural word. Though it looks the least comprehensible, it is, perhaps, the most adequate. If we consider the next slide (Figure 1-1f), we might also infer that the blackness of the square is because of too much information being encoded in the diagram in the Figure 1-1e. If we enlarge some peculiar tiniest dot on this perfectly black square, we'll see a scheme of the blood coagulation cascade (generated, actually, from the data in the KEGG database, http://www.genome.jp/kegg/). Blood coagulation cascade is only one of the thousands of molecular cascades (Chapter VII) that sustain the life and the healthy operation of the human body.

In modern biomedicine, such a tiny location is often also a lifetime working place for a particular researcher, a laboratory or even a consortium of laboratories. It is also worth noting that the large majority of the molecular cascades in humans are not yet known. Therefore, the question that a practicing clinician would inevitably ask is: how, then, all this your stuff can be useful for the practical medicine? How biomedical, lifestyle, dietary and other factors will be *practically* taken into account in this theoretical framework to benefit the practical biomedicine and, hopefully, the public health? The present Volume, as well as the Volume II of the book series, focuses on these and other questions relevant to the practical biomedicine and shows how post-genomic approach to bioinformatics can provide some helpful answers. For now, let's return once more to the two ancient medical systems we have briefly mentioned.

It would be natural thing to ask: why it might be important to consider these very old theories in a book on post-genomic bioinformatics? The answer is simple: in this book, we are interested in the informational aspects of biomedicine: the informational components, the ways of structuring, compacting and retrieving the medical information. It is the compound of these elements is what generally known as "knowledge". We must be aware that both the humoral and the meridian theory resulted from a huge array of 'clinical data': that is, from the actual observations of millions of patients made by many thousands of doctors during the millennia of the human history. The post-genomic bioinformatics that aims at practical biomedical applications definitely faces quite a similar task of structuring the highly heterogeneous information. Let's consider how the two ancient physiological theories, the humoral and the meridian theory, look from the point of view of the informational management.

In the case of the **four humors** (five humors in the Chinese system, three in Tibetan), the particular patients are characterized by prevalence of one of the four humors. According to the theory, each dietary component, drug, environmental factor, climate, internal or external activity modifies the balance of humors of the body. Thus, disease treatment is considered as a patient-specific manipulation of the humors by means of the above and many other factors. Accordingly, what can be considered as a symptom in one patient might be a normal physiological peculiarity of another.

Let's consider a few examples from the empirical four-humour typology. *Phlegmatic* (roughly translatable as "not easily excited"), or patient with "phlegm" humor normally augmented, can be often characterized by forgetfulness, runny nose, dullness as well as by a few dozen of other peculiar characteristics. None of these characteristics (say, a slightly runny nose) can be considered as being definitely a symptom (in this patient), since they are often peculiarities of this particular type of the physical makeup. At the same time, these phenomena in a *sanguinic* ("cheerful") can indeed be the significant symptoms. Sanguinics, normally, are characterized by augmented veins, reddish skin, they can sometimes experience shortness of breath or pricking pains in their temples *etc* so none of these would be a significant symptom for a sanguinic. Not so for other types: the reddish skin or augmented veins in the case of a *melancholic* ("indifference to pleasure") or phlegmatic might be an important indication of a medical problem. Similarly, pricking pains in the head might be relatively normal for a *choleric* ("easily excited") or a sanguinic, but this phenomenon will attract more attention of the practitioner in the case of a phlegmatic patient.

As we can see, the concept of the four humors provides a flexible informational framework for a medical doctor to work. Furthermore, a four-humor typology classifies the peculiar life-style habits, likely diseases each of the types might have as well as the recommended food stuffs. For example, according to the empirical four-humor typologies, a patient determined to be a "phlegmatic" is often a deliberate eater and would stay long at the table, while a "sanguinic" patient can be expected to consume food in large chunks and abuse mild stimulants such as sugar or coffee. The characteristic diseases in a *phlegmatic* patient include thinning of the hair, loose teeth, weakness in hearing and vision while a *sanguinic* is pre-disposed to yeast infections, renal gravel and hypertension. Consequently, the *phlegmatics* are recommended to avoid so-called "phlegm-producing" diets (such as excess of wheat and dairy products) and are highly recommended to "warming" foods such as, for instance, lamb, liver or chicken. *Sanguinics* are recommended to avoid sugary foods and consume more of the vegetable food such as, say, parsley, chives and dark green lettuce. It might also be added that recommendations of the kind are far from being culture-specific and the author's comparative analysis of the four-humor approaches adopted in quite distant cultures (such as Egyptian, Greek, Arabic, Chinese and Tibetan) provides a remarkably consistent picture of the medical diagnostics and treatment in the framework of the humoral theory. This volume still does not seem to be a proper opportunity to discuss quite interesting results of this cross-cultural analysis. For interested readers, more medical details on the humoral theory can be found in the original medical treatises of Galen (such as *De temperamentis)* and also in the works of Ibn Sina and el-Razi.

While the humoral theory provides a comprehensive and very compact view of the health and disease, which can, generally, be explained in a couple of paragraphs, the **meridian theory** provides a comprehensive but not as compact an outlook. Rather, for each condition there is a definite "switch" (an acupuncture point) or combination of such switches that are to be treated in a specific way, by needles or other methods, in accordance with numerous other factors. The meridian theory is also a compound of empirically collected information resulting from numerous experimental observations which allowed identification of hundreds of such "switches". There are "switches" to be acted upon in treating hypertension. There are "switches" for treating specifically nasal congestion, hypertension and so on (one might take a look at a quick reference at http://www.geocities.com/jrh_iii/acupressure/acupoints.html).

The meridian conjecture states that these specific switches are organized into "meridians" or "channels" through which flows some mysterious "force" or ***Qi*** (which does resemble the concept of "The Force" from the well-known movie serial, at least in its elusiveness). In the meridian theory, each of the 8 (or 12) major "meridians" is named after one of the major organ systems and is not limited to a specific anatomic allocation. Rather, a meridian implies a broad range of the corresponding physiological and psychological activities [5,6]. The *Qi* energy, whatever it may mean, is described as "circulating" through the meridians, from one to another, in continuous loops or circuits. The 12 most important meridians are those which pass through the vital organs: the liver, the gallbladder, the heart, the small intestine, the spleen, the stomach, the lung, the large intestine, the kidney, the bladder, pericardium and an integrated entity called "triple burner" in the Traditional Chinese Medicine. In the framework of the meridian theory, there are many external factors that can affect the "flow of *Qi*" through the meridians, such as weather, emotions (anger, joy, grief, fear, worry, and anxiety), diet, exercise, rest and particular lifestyle habits. Acupunctural practices are used to stimulate the "flow of *Qi*" through some meridians, and to inhibit the "flow of *Qi*" in other channels to restore the health.

To sample the kind of information operated within the framework of the theory, let's consider the heart and the kidney meridians in a general way. The actual, physiological heart and kidney are important organs in the cardiovascular systems and in the etiology of cardiovascular disease (Chapter II). The *heart meridian* is thought to be intrinsically related to the blood and to influence the thyroid and thymus glands. The system of the "heart meridian" can be disturbed by strong emotions. The stress from overwhelming emotions such as a sudden shock, too much sorrow or too much excitement might lead to a heart attack or a stroke. If "*Qi* energy" of the blood is "insufficient", this will manifest as an abnormal blood pressure, or as a shortness of breath, as cold extremities, skin discoloration of the fingers or lips, or as forgetfulness and insomnia. Some specific foods such as wheat or beans are assumed to influence the heart meridian to a greater extent than others. Within the working framework of the meridian theory, a patient diagnosed with a "blood *Qi* deficiency" can improve the blood circulation by increasing, for example, consumption of black soybean.

As for the *kidney meridian*, in the framework of the theory of meridians it is considered to be one of the most important of the organ systems involved into the aging-related loss of vitality and decline. The kidney meridian is also thought to produce some specific "energy" for the brain and to influence the skeletal system. For example, brittle bones, pain or inflammation of the joints and progressive decay of teeth are taken to be symptoms of a disorder that would be named as a "kidney *Qi* deficiency". Some foods, thought to be peculiar for the kidney meridian (duck, mutton, walnuts, sesame, prawns *etc*) can be used by the practitioner to regulate the meridian, alone with acupuncture, drugs and other procedures.

Finally, it's also worth noting that the meridian theory, actually, is not entirely a culture-specific phenomenon. By no means is this theory restricted to the Chinese culture or to the recent Western adaptations. Definite indications of the acupunctural practices were already known in the ancient Egypt and in other Middle Eastern cultures. For example, the famous 'Ebers papyrus' (dated 1550 BC) appears to describe some peculiar system of channels and vessels in the human body which is considerably closer to the system of the Chinese meridians rather than to the anatomic outlines of the cardiovascular or the nervous systems. Whether due to peculiar ways of the cultural transmission or because of being the result of independent medical studies (in the case of largely isolated cultures, for instance), the various theories of the four humors and the meridian theory are, undoubtedly, a significant part of the medical heritage of the entire humanity.

What can we learn from the above discussion for the purposes of post-genomic studies in bioinformatics? A careful reader, perhaps, was able to notice the informational components as well as the peculiar ways of structuring, compacting and retrieving the medical information in these ancient medical systems. Speaking generally, these systems are down-to-earth practical and seem to allow making direct links between the clinical observations, the "lab tests" (such as simplistic urine analyses), the peculiar "reason" for disease (say, a "phlegm imbalance") and then to come to the practical prescriptions for the particular patient. Apparently, these systems operate exclusively on the base of the expert knowledge and incorporate the clear notions of "predispositions" to disease as well as more direct "causes" of the diseases.

It is not less important that these systems routinely take into account a considerable number of external factors such as diet, behavior traits and life style not only for adjusting diagnosis but also suggesting means of treatment, along with the herbal drugs and other pharmacy. Apparently, *any* system of medical knowledge, especially the one incorporating the enormous amount of experimental data from the physiological and post-genomic studies (as illustrated pictorially in the 1-1ef), should be able to account for these basic factors. Otherwise, such a system, a theoretical view (or just a particular) will be considerably more primitive than the ancient systems we just mentioned. Moreover, such a primitive and over-simplified system will be entirely impractical from the point of view of the actual clinical science. In addition, a valid system of medical knowledge should not overestimate the influence of either heredity or external factors (as, unfortunately, many of the modern genetic association studies seems to do, Chapters III-V).

The integrative perspective on the human physiology also allows us to reconsider some of the aphorisms and maxims that the founders of the modern physiology have given a century ago. For instance, it's all very well to remember the maxim "when entering a laboratory one should leave theories in the cloakroom" phrased by Claude Bernard, an outstanding French physiologist who was granted with the discovery of the internal secretions of organs. At present, however, given the extent of the technological development and the steadily growing amount of information (Chapter III), such a simplistic "check-it-all" approach will be greatly misleading. It would simply mean that a biomedical scientist will never leave the laboratory for the whole life while doing what amounts simply to verifying, again and again, the statement similar to "the blood coagulates". Though verification of such a basic fact, by means of various technological novelties, is a laudable perspective for a studious person, it is quite a bleak perspective for the development biomedical science and, especially, for the *novel* biotech applications that do work as advertised.

We should also remember that aphorisms of the kind were produced more than 100 years ago: that is, in an epoch when the fundamental physiological information was limited to a few basic facts and aberrated vestiges of the older theories. At present, however, the main problem of physiology is not the lack of information, as it was a hundred years ago. It is, rather, the need for a common and comprehensive theoretical basis that will allow to take on and to digest this information. The exponentially growing amounts of information in the post-genomic era necessitate the proper management of these data, while not forgetting the close connexion between physiology and genetics, physiology and cell biology, physiology and biochemistry. The essential need of the biomedicine in the post-genomic era is the creative integrating of the data from all of these science branches rather than "leaving theories in the cloakroom". The *purpose-oriented approach* to post-genomic bioinformatics that features in the present book might be one of the practically sound solutions.

Now, let's recall the rest of the citation from the "Study in Scarlet" with which we started the present section. Here it goes: "*...the skillful worker is very careful indeed... He will have nothing but the tools which may help him in doing his work...*". Hopefully, the meaning of this statement in relation to the present discussion is clear. That is, bioinformatics as a branch of biological science will not necessarily inform the clinician or any other experimental researcher what the patient should eat or, say, which muscles a patient should train to be healthy. Nevertheless, an experimental researcher in biomedicine can save much time and, literally, much of the sheer muscle effort by rationally applying bioinformatic methods and models to clarify which experiments should be made, when, how and in what order.

In order to be of real assistance to biomedical science, a bioinformatician in the post-genomic era is required to have a clear understanding of the fundamentals pertaining to physiological systems as well as to the modern view of the etiology and diagnostics of disease. We discuss these issues further in the present Chapter. An efficient bioinformatician also needs to be versed well enough in the relevant area of biomedical research (for example, cardiovascular disease, Chapter II).

The Levels of the Physiological Systems

In modern physiology, the human body is an intricately woven system of trillions of cells involved in incalculable electrochemical processes sustaining the dynamic equilibrium we generally call "life". In order to make sense of what's going on in the human body, modern physiology offers several fundamental points of view which we'll briefly consider in the present section. These theoretical views include the context of physiology among other biological sciences, the levels of structural organization of the physiological systems, the informational flows and the notion of homeostasis.

The term "physiology" can be translated into English as "knowledge of nature". The word itself predates Aristotle who later used the word to refer to the functioning of the living bodies. Hippocrates, perhaps, was first to coin the word in relation to medicine and he used it to denote "a healing power of nature". At present, the science of physiology studies functioning of the living organisms at many levels: from cells and their constituents towards the organisms and organism populations [7]. The integrative relationship between the physiology and other research fields of modern biomedicine can be explicated in at least two different ways: in relation to the levels of structural organization of the living organisms and in relation to the main informational flows in organisms. Although these two classifications appear to include the same research fields, they are not identical and provide two distinct points of view on the science of physiology. In both cases, physiology appears to be one of the most comprehensive fields of biomedical research.

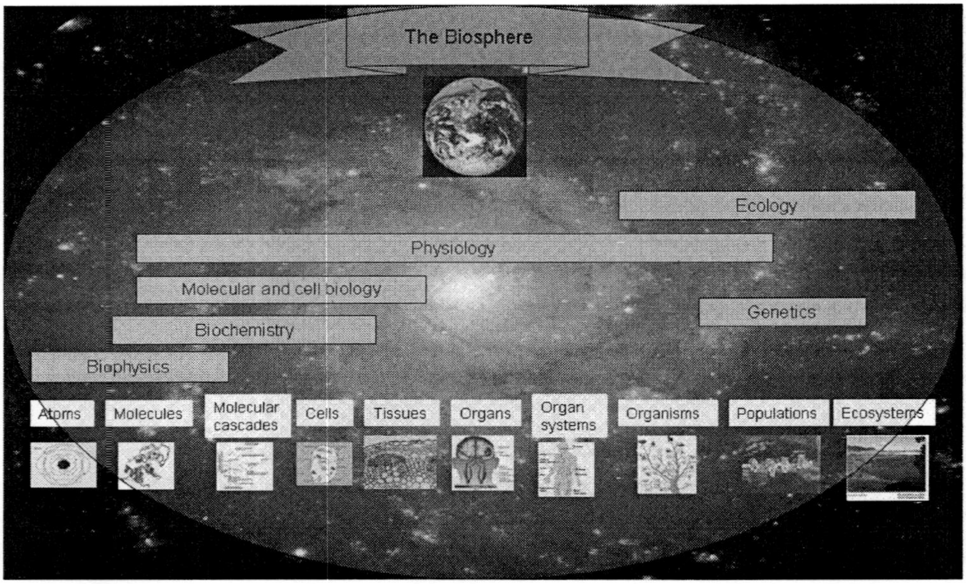

Figure 1-2. The levels of structural organization of the living organisms and the relevant branches of bioscience. As the figure shows, among biosciences physiology has the widest overview and studies objects across many levels: from molecular cascades to the entire organisms. From the point of view of astronomy, though, the objects studied by physiology as well the entire biosphere are exceedingly small…

According to the **levels of structural organization** known to the modern biological sciences (Figure 1-2), cells are the basic units of life as we know it. Each viable cell in a multicellular organism is capable of performing the basic functions such as energy and substance exchange that maintain its own life or "metabolism". In the long run, any properly functioning cell contributes to the continuation of life of the entire organism. The run across the levels of structural organization is, literally, quite long (Figure 1-2): the cells are made from the atoms that form molecules that are involved in molecular cascades *etc*. Then, the cells are progressively organized into tissues, organ systems and, finally, the entire organisms. Though, no doubt, the natural environment of the organism does determine the physiological processes in the organism, the scope of research interest in physiology is largely restricted to the level of organisms and their most immediate environment.

The primary interests of the modern physiology include, more often than not, the studies of the molecular cascades, of the tissues, of the organs and of the organ systems. The **molecular cascades**, whether inside the cells or freely flowing in the physiological fluids are the smallest functional units in physiology. In a single organism, there are thousands of the important types of these "nanotech" molecular cascades that are involved in all aspects of the biological function at the subsequent levels of the organism. There are four primary types of the **tissues**: the *muscle tissue* that is involved in motion, the *nervous tissue* participating in the signal transmission circuits throughout the entire organism, the *epithelial tissue* that forms the protective shell, deals with the exchange of the energy and matter and the *connective tissue* that interconnects various elements of the body. Blood can also be considered as a tissue, albeit a very special one. The **organs** often consist of many different tissue types and are designed to perform specific functions within the entire body. Several dozens of the organs are organized into a dozen of the organ systems that interact to accomplish a required activity. The major **organ systems** of the body include the central and the peripheral nervous system, the circulatory system (that includes cardiovascular and lymphatic systems as well as the blood), the endocrine system, the respiratory system, the immune system, the muscular system, the urinary and digestive systems, and the reproductive system. In physiology, the entire **organism** is the interconnected whole of the interlocking organ systems. We'll return to the relationship between the systems of the body later, when considering the notion of homeostasis.

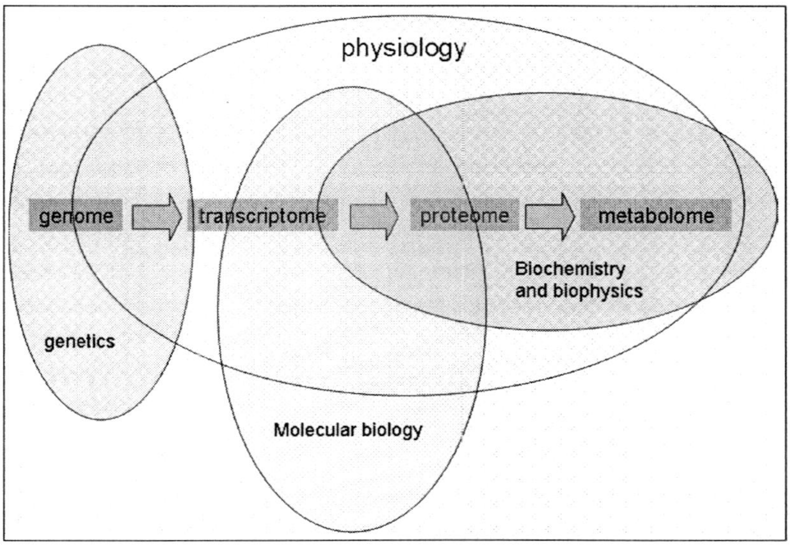

Figure 1-3. The post-genomic tetrad and the relevant areas of bioscience.

The peculiar **informational flows in organisms** occur at each of the levels of structural organization mentioned above. The brain sends signal through nerves and also releases the chemicals into the blood stream. The blood carries nutrients to all the cells of the body and takes the wastes off. All the tissues and organs of the body are in constant communication with each other and there are as many information flows as there are organs, organ systems and interactions between them.

Cells, however, are the basic units of any living organism and the most fundamental informational flow occurs inside the cells and by means of the cells. Informational flows in the cells are considered in greater detail in the 3rd volume of the book series that deals with molecular cell biology. The main informational flow is the one directed from genome to the metabolome through the transcriptome and proteome, the well-known tetrad of the post-genomic studies. This is mostly top-down (rather than left-to-right, as in the Figure 1-3) hierarchy. That is to say, information encoded in *genome* DNA is first decoded into *transcriptome* mRNA. Then, the proteins are synthesized on the base of the data in mRNA and the proteins of the *proteome* support the anabolism and catabolism of the metabolites as well as an extraordinary number of protein-protein interactions. The more commonly spoken triad of "genome-transcriptome-proteome" should certainly be complemented by the *metabolome*, which is the entire array of what is called as 'small molecules' or metabolites peculiar for the organism. Transcriptome, proteome and metabolome are also characterized by the concentration and dynamics of their individual components: mRNAs, proteins, metabolites.

According to the classical model of the molecular biology, this information flow is unidirectional: from genome strictly to metabolome and not vice versa. The growing field of epigenetics suggests that there are more intricate informational links between the genome and the rest of the hierarchy and the genome might be specifically altered by means of the cellular biochemistry represented by the nucleic acids, proteins and certain metabolites (Volume II).

Considering the main flow of biological information in a cell, we arrive at somewhat different point of view at the major research areas in modern bioscience. Again, physiology includes all of the tetrad and is the most comprehensive field (Figure 1-3). Indeed, medical problems are not restricted to genome variations or metabolic abnormalities but can be rooted at any of the four levels. However, physiology on its own would lack the precise technical details which are studied in the allied fields of genetics, molecular and cell biology, biochemistry and biophysics. Integration between the studies in these research fields is paramount for the success of the post-genomic studies. In distinction from what we might call as "pre-genomic" studies, where many researchers worked with little or no coordination and on the problems of individual (very often, individualistic) interest, interdisciplinary coordination and cooperation is the must for the post-genomic era. The value of the post-genomic knowledge is in its unremitting level-to-level continuity: from the physiological flows to molecular cascades and then to the detailed functional and structural characterization of the relevant proteins. According to the main information flow, each physiological process can be assigned a basic level of function (genetic, transcriptional, proteomic and metabolic).

Having considered the context of the physiology among biosciences, it'll be easier to comprehend the cornerstone of the modern physiology: **the concept of homeostasis**. From the point of view of the ancient medical systems, the modern term "homeostasis" has some similarity to the ancient term "equilibrium" or "life". Using modern terminology, a physiological system is alive if it has metabolism, definite borders and actively exchanges energy/substance with the environment while preserving its individuality in the course of time. The notion of homeostasis, first introduced by Dr. W. Cannon in early 1930s, states that every organ in a living system has a functional dependence on any other, either in a direct or an indirect way [8].

In the modern physiology, the concept of homeostasis is formulated at the level of cells and the mechanisms of homeostasis are held to maintain constancy of cell environment in the physiologic fluids. The basic functions of the cells include import of nutrients and oxygen, conversion of the nutrients to energy, elimination of wastes, control of the exchange of material, component synthesis (proteins, DNA, lipids), sensing changes in environment and reproduction. Cells are highly specialized and are subsequently organized into specific tissues and organs. According to the reformulated concept of homeostasis, most of the body cells in multicellular organism cells contribute to survival of organism as a whole and cannot survive without other cells. Each cell is in contact with its specific internal environment and an intricately woven communication is essential for the survival. One of the keys to this communication is the aqueous inner environment that includes the blood plasma. This environment facilitates cell-cell communication and physiological function. When the natural cell environment is disturbed, abnormal function and pathology follow.

The notion of homeostasis at the cellular level, though sounding truly fundamental, remains too abstract when taken out of the context of the particular organs and organ systems. At the level of cells, the homeostatic mechanisms are as many as there are molecular cascades, which often seem to act separately with little apparent coordination. When we view from a higher level, the level of the organ systems, we can get an integrated picture and what we call as the Black Square of the Modern Physiology (Figure 1-1e) becomes much more colorful and understandable (see the Figure 1-4). Given the dozen of the organ systems, the whole of the homeostasis of the organism can be envisioned and comprehended in terms of the pairwise and higher-order interactions between the organ systems.

The organ systems that control and maintain homeostasis are generically classified into *extrinsic* (regulatory mechanisms outside an organ) and *intrinsic* (inherent to an organ) controls. The outermost extrinsic controls are the nervous and the endocrine systems. Using cybernetic terminology, the homeostatic equilibrium is maintained by interplay of negative and positive feedbacks provided by both extrinsic and intrisic controls. *Negative feedback* loops are most common in homeostasis for the maintenance of the parameters within the specific ranges. *Positive feedback* is much less common a pattern of regulation since it often implies an avalanche-type of processes which are often employed for the amplification of the weak signals. Alternatively, positive feedback can be involved in various hysteretic switches. By the way, the blood coagulation cascade, which often features on the pages of the present book series, makes an important example of a positive feedback loop. In the framework of the concept of homeostasis, the cause of a disease is a disruption in some of certain feedback loops of the homeostasis that leads to failures in one or several of the organ systems of the body.

Thus, the levels of structural organization of the living organisms represent the basic anatomy of the living systems: from atoms to molecules to organisms and indicated specific spatial location; the basic informational flow is represented by the genome-transcriptome-proteome-metabolome tetrad and the notion of homeostasis underlies the integrity of the living matter. Rather than being just of a general academic interest, these three points of view are important principles which can often be of a practical value for the post-genomic bioinformatics. An impatient reader might well ask "How could this be?". Well, the present volume was intended to detail some of the answers…

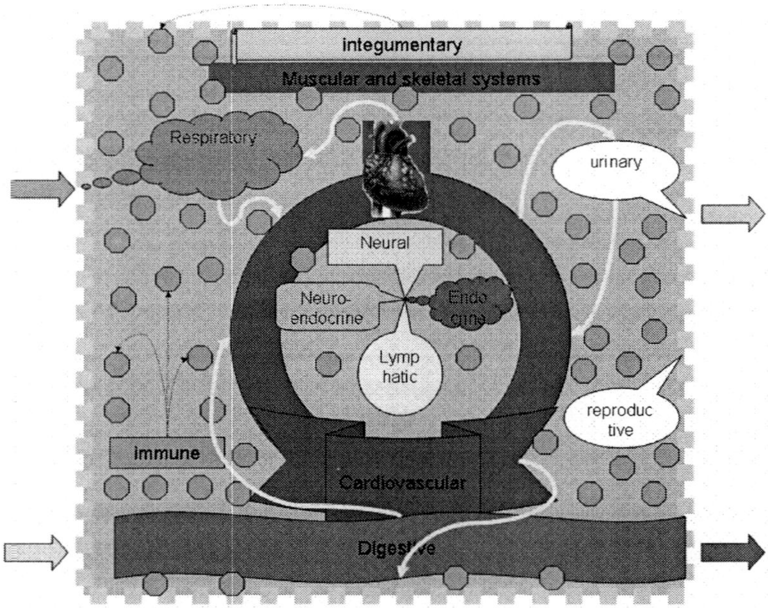

Figure 1-4. The organ systems of the body interlocked in homeostasis (after [7]). Actually, the organ systems are highly interlocked and share some of their functions so the short annotations below emphasize rather the *main* functions and the *main* organs of the systems. The **integumentary system** (skin, hair, nails) is involved in protection, temperature regulation, sensory reception, biochemical synthesis and absorption. The primary function of the **immune system** (blood, thymus, spleen, lymph nodes) is the defense against alien particles such as microorganisms. The **muscular and skeletal** system (skeletal muscles, bones) deals with physical support of the body and the movement of the limbs. The primary functions of the **digestive system** (stomach, intestines, liver, pancreas) include the conversion of the food stuffs into the units that can be taken up by the cells of the body as well as the removal of solid wastes. The **urinary system** (kidneys, bladder) maintains the water-solute balance in the internal environment and also removes the liquid wastes. The **reproductive system** (ovaries/testes) is involved in perpetuation of the species. The **respiratory system** (lungs, bronchi, airways) exchanges the carbon dioxide and other gaseous wastes from internal the environment for the oxygen of the external environment. The **cardiovascular system** (heart, blood vessels, blood) transports the materials and the heat between the organs, tissues and cells of the body. The **lymphatic system** (spleen) and the cardiovascular system are closely related structures joined by a capillary system. The lymphatic system filters out organisms that cause disease, produces white blood cells, generates antibodies and is important for functioning of cardiovascular and immune systems. The **nervous system** (brain, spinal cord, periphery) is involved in the coordination of the body functioning through the electrical signals while the **endocrine system** (thyroid and adrenal glands) coordinates the body functioning through the release of specific regulatory molecules. The **neuroendocrine system** (pituitary) regulates the entire homeostasis through hormone secretion from the pituitary (see also http://wwws.innerbody.com/ and http://www.emc.maricopa.edu/faculty/farabee/BIOBK/BioBookTOC.html).

First of all, the above outline of the human physiology apparently suggests a definite structuring of the information. Thus, disorders can be classified according to the *locations* in the body which correspond to malfunctioning and vice versa: peculiar body systems are characterized by peculiar disorders. Going further and deeper, to the *functioning* of the organs, of the tissues, of the cells and then to particular molecular cascades suggests an efficient packing of the relevant information on any imaginable disorder. The principle of the *integrity* of the living systems, incorporated into the concept of homeostasis, suggests, among other things, that clinical symptoms might be not always what they seem (see the next section).

The brief overview presented in this section also allows us not to loose the sense of the whole and to see the relative value of the newly emerging post-genomic technologies. For example, the functional genomics (Chapters VI,VII) endeavors to address biomedical problems exclusively at the cellular level of function, often ignoring the hierarchy of the levels of function and almost invariably ignoring the extrinsic controls such as those provided by the nervous and endocrine systems. The "structural genomics" (structural proteomics, rather, see Volume V) disregards the molecular cascades and even protein-protein interactions thus operating exclusively at the level of structure of individual protein molecules. The quest for "genetic markers" for common human diseases (nucleotide polymorphisms, Volume II) sometimes is done at the expense of the sane physiology so physiological value of these "getic markers" becomes very obscure (Chapters III, IV and V). These and other areas of post-genomic research are, nevertheless, of definite value. Their value will considerably grow if a physiologically plausible integration of the data from these fields of research is achieved through bioinformatics.

Generally, the human genome research, physiology and all the biomedical sciences are supposed to have the human health as their primary aim. This aim can be achieved when disease diagnostics, treatment and prevention are adequate. And this, in turn, becomes possible only when etiology of a disease is clearly understood from the three points of view considered above.

Disease Etiology and Medical Diagnostics

"...Strange details, far from making the case more difficult, have really had the effect of making it less so..." (A. Doyle, A Study in Scarlet).

In accordance with both the levels of structural organization of the living organisms (Figure 1-2) and the basic informational flow in organisms (Figure 1-3), we can compile a simplified diagram. This diagram (Figure 1-5) is quite general a scheme, since many of the important intermediary levels were not shown. Nevertheless, the diagram can help us to illustrate the inherent complexity and interrelation of seemingly distant phenomena in human physiology and medicine. Also, using diagrams such as this more often, it is possible to reduce the deleterious effects of "naïve reductionism" (see Introduction) on the mind of the researchers...

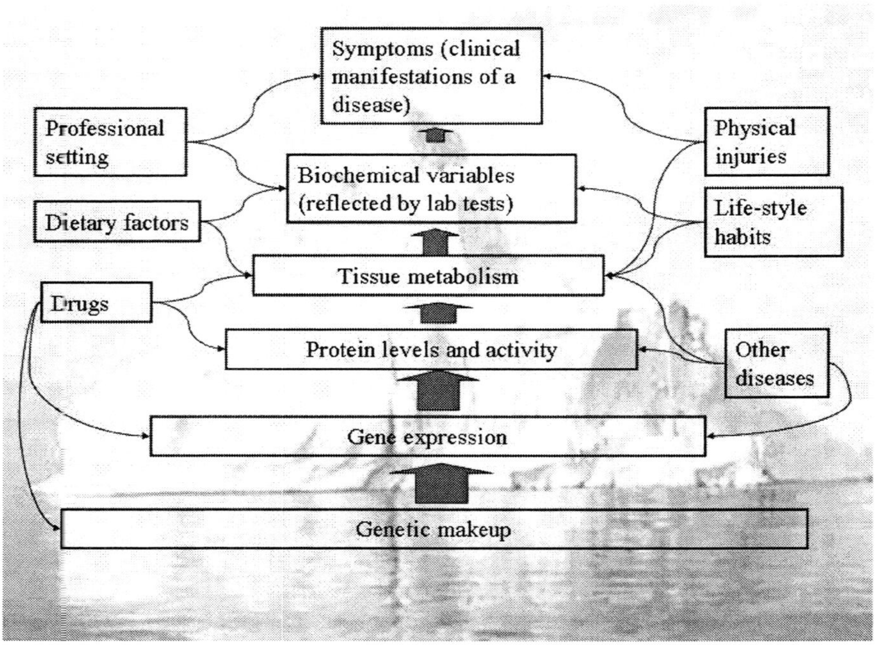

Figure 1-5. The levels of function in biomedical science. The diagram shows only few of the external factors that influence the clinical manifestations. Unlike the rest of the levels, there are no direct methods to observe the tissue metabolism *in vivo*, albeit it might be indirectly observed using MRI, X-ray, ultrasound *etc*.

As we can infer from the Figure 1-5, the clinical manifestations of a disease are, literally, just the tip of the iceberg. Even a brief consideration of this simple diagram and of the three physiological points of view from the previous section suggests numerous ways of how a disease might arise in a functioning human body. To name a few, the body is always in some external environment, with which the body should properly exchange the substances and energy. If this exchange is somehow impeded, a disease will arise. As the body needs definite arrays of nutrients and energy and also needs to avoid improper substances, the patient's diet should neither overload the digestive system nor poison the body. The natural biofeedback, provided by the extrinsic controls of the neural system should not be ignored by the patient. The activities of the neural and endocrinal systems should not throw into disarray otherwise harmonious function of the entire system of the body. The lifestyle habits should not lead to rapid degradation of the physiological systems of the body and, consequently, to an untimely demise *et cetera*. Therefore, a predisposition for a multifactor disorder, such as a cardiovascular disease, can be encoded at a number of different levels of structural organization and function. The clinical science inherently deals mostly with the higher levels of the physiological hierarchy (from clinical manifestations down to the protein levels) while post-genomic studies inherently deal with the lower levels (from the protein levels to the genetic makeup, Figure 1-5).

This section presents some particular examples the author had encountered in his consulting practice. The examples illustrate some of the clinical and physiological sophistications of which a researcher attempting to bring post-genomic technologies into the modern medicine should be keenly aware. Otherwise, if the integrity of the biomedical data is left out of view, straight-forward application of the post-genomic type of approach (such as functional genomics or high-throughput genotyping) can lead grossly astray and, inevitably, result only in a huge waste of public and private funds. Although some of the following examples might seem to be 'untypical' (to some) or 'trivial' (to some), they were selected to illustrate the three important principles related to disease etiology: primary *vs.* secondary origin of disease, the role of the completeness of medical diagnostics and the importance of understanding at least the basic physiology behind the disease in question.

Primary *vs.* Secondary Origin of Disease

Concerning the analysis at the higher levels of the physiological hierarchies, it is worth noting that a disease in a patient might be of primary or secondary origin. For example, CHD ("coronary heart disease", also known as IHD "ischemic heart disease") is a popular phenotype for the phenotype-genotype association studies. However, CHD is not always a primary disease (caused, say, by some unfavorable combination of improper lifestyle, diet and genetic predisposition). Secondary origin implies that the disease occurs as a side effect of another disease or influnce. For instance, hypertension might be a result of poisoning with heavy metal ions. Sometimes, CHD might have secondary origin: as it is the case of when there is a congenital heart disorder in the patient. Although this kind of disorders is formally called as "congenital", the clinical practice strongly suggests that it is, more than often, caused by parental smoking and/or alcohol intake, rather than by some obscure genetic predispositions. In this case, analysis of the known genetic factors of predisposition to CHD will not work as expected and our *Case of the IHD200* somewhat details the issue (Chapter III).

The Role of Complete Diagnostics

> *"If you are poor, they will prescribe you aspirin. If you are rich, they will take out your appendix"* (overheard at a conference of medical students)

Usual diagnostic procedure in the clinical practice goes as follows. A patient comes to a doctor with a few major complaints. The doctor listens, watches for the range of symptoms and then suggests additional diagnostic procedures (X-ray, ultrasound *etc*) as well as certain lab tests. Then, the patient is recommended to consult more narrowly specialized clinicians. Most of the patients do find the entire procedure cumbersome: it takes lot of considerable amount of talk, takes much time, not to mention expenditures. What is considerably more important is that the results obtained are often disjointed and the entirety of the clinical observation often gets lost in lots of paperwork. Hence, the entire procedure becomes exceedingly prone to misdiagnosis, mistreatment and, consequently, can sometimes even result in a worsened case of a chronic disease.

These statements are far from being a personal opinion but the result of the author's extensive discussions with clinicians who practice for decades. Moreover, a report from the Institute of Medicine of the National Academies (USA) based on the findings of one major study indicated that medical errors kill some 44,000 people in USA hospitals each year [9]. Another relevant study in the report puts the number much higher, at 98,000. Even using the lower estimate, more people die from medical mistakes each year than from highway accidents, breast cancer, or AIDS. The authors of the report concur that the majority of medical errors do not result from individual recklessness, as the report puts it, but from basic flaws in the way the information management in the health system is organized. For example, when a patient is treated by several practitioners, the complete medical information on the patient becomes highly distributed and specialists do not have complete diagnostic and procedural information (be it information about the medicines prescribed or simply the complete list of the patient's illnesses). The highly degree of the scattering of information results in errors of diagnostics and treatment [9].

Generally, clinical diagnostics is notoriously difficult (you might want to take a look at the Box 1, for a change). In particular, there is dearth of large studies dealing with the value of particular diagnostic tests, let alone diagnostic value of the test combinations or diagnostic procedures. Estimates of sensitivity and specificity of diagnostic accuracy are usually derived from small studies and are often imprecise since they are characterized by wide confidence intervals (CI, Chapter III). This makes it difficult to assess just how informative a single test may be. Apart from the necessity of the methodological improvements that concern the sample size [10], to improve clinical diagnostics it is also necessary to measure the diagnostic accuracy of combinations of readily available tests.

Keeping in mind this image of the difficulties with the routine clinical diagnostics (Box 1), we might understand why there would be many association studies that aim at investigating the genotype-phenotype links but instead produce only another negative finding. For example, a patient with severe liver disabilities comes to a cardiologist and is diagnosed with CHD. He might well have CHD, but, unfortunately, the liver problems are often of little concern to the cardiologists. The liver problems, however, can be of great concern to the cardiovascular physiology of that particular patient. Some of such cases, including several cases of hepatitis among CHD patients, are mentioned in our *Case of IHD200* in the Chapter III.

Another example: a patient with migraine (caused, originally, by a severe poisoning by heavy metals in a professional setting) comes to a therapist and is treated as a case of essential hypertension. How reasonable, then, would it be to search for any genetic component of "migraine" or "hypertension" in this case when a simple analysis of the nails or hair for metal poisoning would suffice? From statistical point of view, these examples might be seen as "exceptions". However, when combined with the above-mentioned lack of integrity of the medical information on the patients and other factors, these "exceptions" become considerably more typical and can easily skew the results of the association studies (Chapter V).

Box 1. Science and humor in medical diagnostics

Scoffers may see the following story as a derision of medicine for not being a science. However, the author of the present book finds that this story describes a pattern not uncommon in contemporary clinical science. The anecdote amply illustrates what will happen to an association study (or to any kind of biomedical research, for that matter) if the diagnosis was incorrect. It also indicates some of the problems with current state of common medical diagnostics (see details in the text).

A man was limping and wincing with pain as he walked down a street. A doctor stopped him and said: "If I were you, I'd get yourself seen to a doctor as you definitely need a surgery on your appendix". So the patient had his appendix out.

Now, the patient went to another doctor, claiming that he still had the same trouble. So he was put on a course of tranquillizers. This procedure wasn't very effective so he went to a hospital, where he was prescribed a specific diet and physical exercises.

Some time later he was strolling in the park and met one of his many doctors. "Glad to see you are better" said the physician, "and that I could be of service". "Bloody service, indeed!" said the patient, "both the pain and the limp went away the moment I took that nail out of my shoe!"

The examples dealing with microelement poisoning or, on the contrary, with microelement deficiency underline both the complexity of disease etiology and the necessity of as complete diagnostics as possible. Etiology of a disease can, indeed, be complex and unexpected. Supposing, there is a deficiency of intake of a certain microelement (say, zinc) in a patient. Objectively, the primary cause of this deficiency is the lack of the nutrient zinc in the foodstuffs regularly taken by the patient. Many enzymes require Zn^{2+} ions as a cofactor, so the activity of a number of enzymes in a number of tissues will drop. As a result, complex pathological conditions are likely to follow. Medical practice shows that a chronic zinc deficiency manifests itself in a number of ways, of which learning disabilities, short stature, late sexual maturity and, quite often, malfunctioning of prostate. As underlying physiology and biochemistry can be quite complex, the symptomatics of what is, actually, a microelement deficiency will be complex as well. For example, a number of symptoms of a dietary zinc deficiency are also the symptoms of diabetes, since the zinc ions are intricately involved into insulin and carbohydrate metabolism. If misdiagnosed, a zinc deficiency might be attributed to some genetic defects that lead to defects in molecular cascades, though the main point in the patient's case, is, of course, the lack of an important microelement in the regularly consumed foods. In other words, before the analysis of the molecular cascades and individual genes (that is, what post-genomic technologies aim at), the analysis at the higher levels of the physiological hierarchies should be as complete as possible and this is where the clinical diagnostics works.

Calcium deficiency is another example of the complex relationships between physiology, genetics and biochemistry. Let's take a patient with persistent dental problems from youth. On one hand, the deteriorative effects of the lack of calcium are well known to dental practitioners. It is also known among the medical practitioners that consumption of the white sugar is an efficient way to cause a severe calcium deficiency in the body, even if the Ca-ions are abundantly supplied in the patient's diet. On the other hand, there are a number of genetic markers that are relevant to the dental problems. Accordingly, one can apply state-of-the-art post-genomic technologies to analyze potential genetic determinants related to the manifested symptoms. Such a search, however, will be, largely, a fool's enterprise when the basic fact (that the patient consumes too much sugar from his early age) is left out of view. Or, in terms of biomedicine, one needs to account for all the potential confounders (Chapter III).

Accordingly, there is a need not only for the proper collection of the clinical data but also for the proper informational management of all these data. The "proper informational management' in the framework of an association study does not only imply technically trivial issues such as single database entry and data security. It implies, before all, presentation and structuring of the data in such way that allows to reconstruct the entire clinical picture from the data without bothering either the patient or the doctor. Author's experience shows that the higher levels of the physiological hierarchy (Figure 1-5) can be more completely characterized if the following items of information are available: detailed clinical observation (symptoms and questionnaire), ultrasound study of the internal organs, classical cardiography, complete urine analysis, complete blood count and detailed blood work, analysis of the blood on the hormones, analysis of the blood for viral infections and spectral analysis of the hair/nails for microelements. Of course, this is a general list and each disease requires its own specific profile of diagnostic procedures.

Understanding the Basic Physiology Behind the Disease

Contrary to the current magical beliefs and the general overconfidence typical for the post-genomic community at present, the careful analysis of the available information on human physiology will be essential for the successful experimenting of the post-genomic sort. It would be naïve to expect that by simply making post-genomic experiments on cells in culture one would ever be able to unravel mysteries of the human physiology and disease. First of all, there are hundreds of cell types that comprise a living human body. Secondly, cell type should be selected in complete accordance with the research purpose of a post-genomic study. An arbitrary selection of the cells for the experiment (which usually happens when a researcher neglects basics of the physiology of the problem under study) will result in arbitrary results (see examples in the Chapters VI&VII).

Prior to attempting any kind of post-genomic experimentation, there should be an adequate outline of the basic physiology of the problem under study. In the present volume, this principle is illustrated, in particular, by *The Case of Estrogens and Venous Thrombosis*. Without having at least general outline of the basic physiology, it will not be possible to plan a scientifically productive study in functional genomics. This is actually happened when endothelial cells in culture were taken for the study of the problem, while careful analysis of the physiological background emphatically suggests that it is the liver cells that should be taken for this particular study (in the Chapter VI, we discuss the matter in greater detail).

The following example with thyroid emphasizes how the phenotype-genotype association studies can go awry if basic facts of human physiology are either deliberately suppressed or simply forgotten. This thyroid story might seem a bit long as it provides all the necessary biomedical background for a non-professional but a careful reader. The reader's task, then, is to see how the basic physiological knowledge can be used in a framework of a post-genomic study. The reader might also get a general idea of where a bluntly hacked post-genomic study will go very wrong and result in a "negative finding" and similar pitiful outcomes. This example also touches upon the general difficulties with the clinical diagnostics we mentioned before.

Thyroid gland is a two-lobed endocrine gland located at the base of the neck and secreting thyroxine ("T4") and triiodothyronine ("T3") which are the extremely important hormones (the word "hormone", by the way, comes from Greek *hormé* meaning "impulse" or "stimulus"). These two hormones act on virtually every cell in the body to alter gene transcription thus regulating the metabolic rates, the heat balance, tissue growth and development. Release of these hormones in the blood stream regulates the metabolic rate of the body [11]. For the synthesis of the thyroid hormones iodine is required. In the form of the iodide anion, iodine is quite abundant in the seafood.

When there is a thyroid dysfunction, two abnormal conditions are possible: *hyperthyroidism* (sometimes referred to as "Graves disease") characterized by higher levels of the hormones and *hypothyroidism* when too little of the two hormones is produced. Both kinds of disorder are characterized by remarkably similar sets of symptoms such as chronic fatigue, anemia of extremities, dryness of the hair and skin, susceptibility to infections, headache and emotional disorders such as depression, unspecific irritability and insomnia. Hypertension is another disorder associated both with hypothyroidism and with hyperthyroidism. Thus, the complex links between the emotional disorders, functions of the nervous and endocrine systems and biochemistry are quite apparent if we keep in mind the concept of homeostasis.

From the diagram depicting the levels of function (Figure 1-5) and the physiology basics we covered in the previous section, we can deduce that hypothyroidism disorder may follow, for example, from a lack of iodine in the diet, from genetic factors, from a failure of the pituitary gland to produce sufficient TSH (thyroid stimulating hormone) or because of a primary failure of the thyroid due to abnormal tissue metabolism. *Dietary* iodine deficiency can often be the most likely explanation, since, for example, even a decade ago nearly 30% of the world population was calculated to be at risk of the iodine deficiency (according to WHO). As the symptoms of both hyper- and hypothyroidism are remarkably similar, biochemical tests are required which might actually measure if not the function of the thyroid but, at least, the hormone levels.

As for potential *genetic factors*, analyzing the known metabolic pathways of the hormones' synthesis, we can come up with thyroid peroxidase, TSH receptor, thyrotropin-beta as one of the most obvious gene candidates. Less obvious are the relations between MTHFR (methylenetetrahydrofolate reductase, a highly pleiotropic gene), the genes of the homocysteine metabolism and the thyroid dysfunction. For example, plasma homocysteine levels are increased in hypothyroidism and decreased in hyperthyroidism [12] and the well-studied MTHFR 677C/T polymorphism was associated with the thyroid status [13]. Anyway, to clarify the genetic factors in the hypothyroidism or hyperthyroidism in a scope of a phenotype-genotype association studies, a biochemical assay for thyroid function is important.

Biochemical assays of circulating thyroid hormone concentrations are quite common and are usually referred to as 'thyroid function tests' (TFT). Here is a snag, though. For example, for almost 30 years the 'protein bound iodine' test was the cornerstone of the thyroid disease diagnosis but was decommissioned by mid 1970s as unreliable. Later, the lab tests for T3, T4 and TSH became available, but these tests only measure circulating levels of these hormones and say nothing about the amounts of T3 and T4 available to the cells. Indeed, the practice in endocrinology shows that these tests are also not entirely completely reliable for making a diagnosis.

Looking through the thyroid example again, we might slowly begin to realize the difficulties that will *routinely* arise during the large-scale phenotype-genotype association studies. Without the proper biochemical characterization and the detailed medical data on the patient, such post-genomic studies will, effectively, run into the sand without producing any useful results whatsoever. And the cause for it will not be necessarily lack of scientific integrity, lack of formal medical education or insufficient diligence of the particular researchers. The failure will be because of the informational mismanagement of the studies and failure to apply the relevant basics of physiology. In the author's experience, such mismanagement often arises because of neglecting the essential requirement of the post-genomic bioinformatics: the integrity of the levels of biomedical function.

How an expert bioinformatician may help in such a situation? The help often comes through analyzing the delicate physiological points as well as by carefully digging through the published data. Returning to the basics of physiology and keeping in mind the general diagram we've drawn to illustrate the concept of homeostasis (Figure 1-4), we might remember the fact that the thyroid gland consists of *two* lobes. These lobes produce hormones largely independently. Therefore, whatever blood TFT would be used, it will measure only the *average* of the released thyroid hormones without giving details on the activities of the individual lobes.

In fact, many times thyroid under-activity is not detected in conventional TFTs, the fact known to general practitioners quite a while ago [14]. Therefore, the above analysis of the integrity of the levels of function clearly suggests that for a phenotype-genotype study of the thyroid function to be truly scientific, another method that measures tissue metabolism of the *individual* thyroid lobes will be essential. This method may be more careful clinical observation that would include, for instance, an analysis of the asymmetry in the bodily symptoms. Alternatively, it could be some technological advancement (such as, say, using MRI, thermal scan or whatever method modern technology can come up with) to map the individual lobes. Whatever the source of the experimental data would be, it's not up to the bioinformatics consultant. What an expert bioinformatician will say without any doubt is that it is utter waste of resources, money and time even to attempt to make such a phenotype-genotype study (in the present example, a study of the thyroid function) without such an additional method.

Hopefully, the few examples above illustrated the necessity of the intense specialization of the bioinformaticians and post-genomic researchers into the various areas of human physiology, especially in the post-genomic era. At the same time, this specialization should not become an ultra-narrow one. The examples also show the necessity of the proper information management (which is prerogative of bioinformatics). Knowing and being able to operate the knowledge related to a particular disease is a must for post-genomic researchers, if they do not want to be lost under the tons of data while hectically writing grant reports to waste even more millions of the state funds.

Another point that we would like to make here is that the very nature and the complexity of etiology of a disease suggests that some permanent type of structuring information (such as a strictly defined database format) won't be very helpful when analyzing a disorder of interest and will be even less helpful in the case of a particular patient. Therefore, in the framework of the present book series, post-genomic bioinformatics is not interested so much in general academic-type or database-type structuring of information. For a clinically significant post-genomic study, it rather would be necessary to make if not an entirely new but, at least, an updated and adjusted structure of medical knowledge for *each* particular research purpose. This restructuring of the information is to be done by making the most of the currently available information. We call this approach as *purpose-oriented bioinformatic research*. Examples of this information structuring are available in the Case Studies throughout the present volume.

The purpose-oriented research requires specialization into physiology of the problem under study as well as an expertise in analyzing and selecting the valid biomedical studies. The next two chapters cover both of these issues. In this volume, we approach post-genomic studies using cardiovascular disease as the major example of the multifactor disease. Accordingly, CVD is the main problem under study and the Chapter II deals with the basics of the cardiovascular physiology and genetics. The Chapter III discusses the criteria to assess the quality of the available biomedical. The Chapter III also presents an example that illustrates peculiarities of conducting an actual association study (*The Case of the IHD200).*

CHAPTER SUMMARY

Physiology studies the functions and activities of the organisms and their components. **Bioinformatics** is a branch of biological science that specifically deals with management of informational complexity of biological systems using **scientific theories** or theoretical constructs as informational tools. A bioinformatician in the post-genomic era is required, before all, to have clear understanding of the **fundamentals** pertaining to physiological systems and to the modern view of the disease. These fundamentals include the context of physiology among biological sciences, the **levels of structural organization** of the physiological systems, the **informational flows**, the notion of **homeostasis** and, most important, the hierarchy of the **levels of function**. Clear understanding of the **etiology of the disease** within the framework of the modern physiology and the valid medical **diagnostics** are essential for adequate planning of the large-scale studies of the post-genomic kind. The **purpose-oriented research** in bioinformatics requires intense specialization into physiology of the problem under study as well as an expertise in analyzing and selecting the valid biomedical studies.

REFERENCES

[1] W. Osler. The evolution of modern medicine (lectures at Yale University, 1913), available at http://etext.lib.virginia.edu/toc/modeng/public/OslEvol.html

[2] Pavlov I.P. *Pavlovskie sredy (in 3 volumes)*, volume 3. Izd Akademii Nauk SSSR, 1949, Moscow-Leningrad.

[3] Pavlov I.P. *Complete works (in 6 volumes)*. Izd Akademii Nauk SSSR, 1952, Moscow-Leningrad.

[4] *The Yellow Emperor's Classic of Internal Medicine*. Huang Ti (The Emperor of China), Ni Maoshing (Tr.), Shambhala Pub, Boston, MA, 1995

[5] Reid, Daniel J., *Chinese Herbal Medicine*. Shambhala Pub, Boston, MA, 1993

[6] H. Beinfeld, E. Korngold. *A Guide to Chinese Medicine Ballantine Books*, NY, 1991

[7] D Unglaub, P. Silverthorn. *Human Physiology: An Integrated Approach (1st ed)*, Prentice Hall, 1997

[8] Cannon WB *The wisdom of the body*. Norton Pubs, 1932.

[9] Committee on Quality of Health Care in America, L Kohn, J Corrigan, M Donaldson (eds). To Err Is Human: Building a Safer Health System. Institute of Medicine (2000), http://books.nap.edu/books/0309068l/html/index.html

[10] Bachmann LM, Puhan MA, ter Riet G, Bossuyt PM. Sample sizes of studies on diagnostic accuracy: literature survey. *BMJ*. 2006;332(7550):1127-1129.

[11] S. Nussey, S. Whitehead. *Endocrinology: An integrated approach*, BIOS Scientific publishers, 2001.

[12] Diekman MJ, van der Put NM, Blom HJ, Tijssen JG, Wiersinga WM. Determinants of changes in plasma homocysteine in hyperthyroidism and hypothyroidism. *Clin Endocrinol*. 2001;54(2):197-204.

[13] Hustad S, Nedrebo BG, Ueland PM, Schneede J, Vollset SE, Ulvik A, Lien EA. Phenotypic expression of the methylenetetrahydrofolate reductase 677C-->T polymorphism and flavin cofactor availability in thyroid dysfunction. *Am J Clin Nutr* 2004;80:1050-1057.

[14] B. Barnes. *Hypothyroidism: the unsuspected illness*, Harper-Collins, USA, 1976.

Chapter II

Some Basics of Cardiovascular Physiology, Pathology and Genetics

CVD: Cardiovascular Disease

Recent WHO reports indicate that at least one-third of total global deaths (~17 million) result from various forms of cardiovascular disease (CVD). The coronary heart disease and stroke appear to be the most common disorders that contribute to the cardiovascular mortality (Figure 2-1). As cardiovascular diseases serve as the major example in this book series, let's consider the major types of CVD.

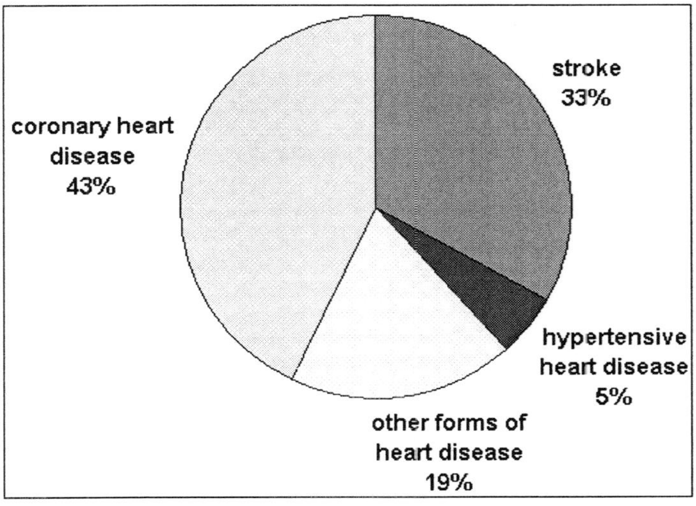

Figure 2-1. Cardiovascular mortality over the globe (WHO, 2002); more detailed and recent data are available in Global Cardiovascular Infobase (http://www.cvdinfobase.ca/).

What we call as "CVD" in this book include multifactorial disorders described below. Monogenic disorders are, generally, not included into the notion of CVD and are specifically referred to as 'F2 deficiency', 'dislipidproteinemias', 'hyperlipidaemias' *etc*. The major types of the cardiovascular disease are delineated in the Figure 2-2 and include (in alphabetic order) atherosclerosis, cerebrovascular disease and stroke, congenital heart disease, coronary heart disease (CHD), deep venous thrombosis (DVT), hypertension, peripheral arterial disease and rheumatic heart disease.

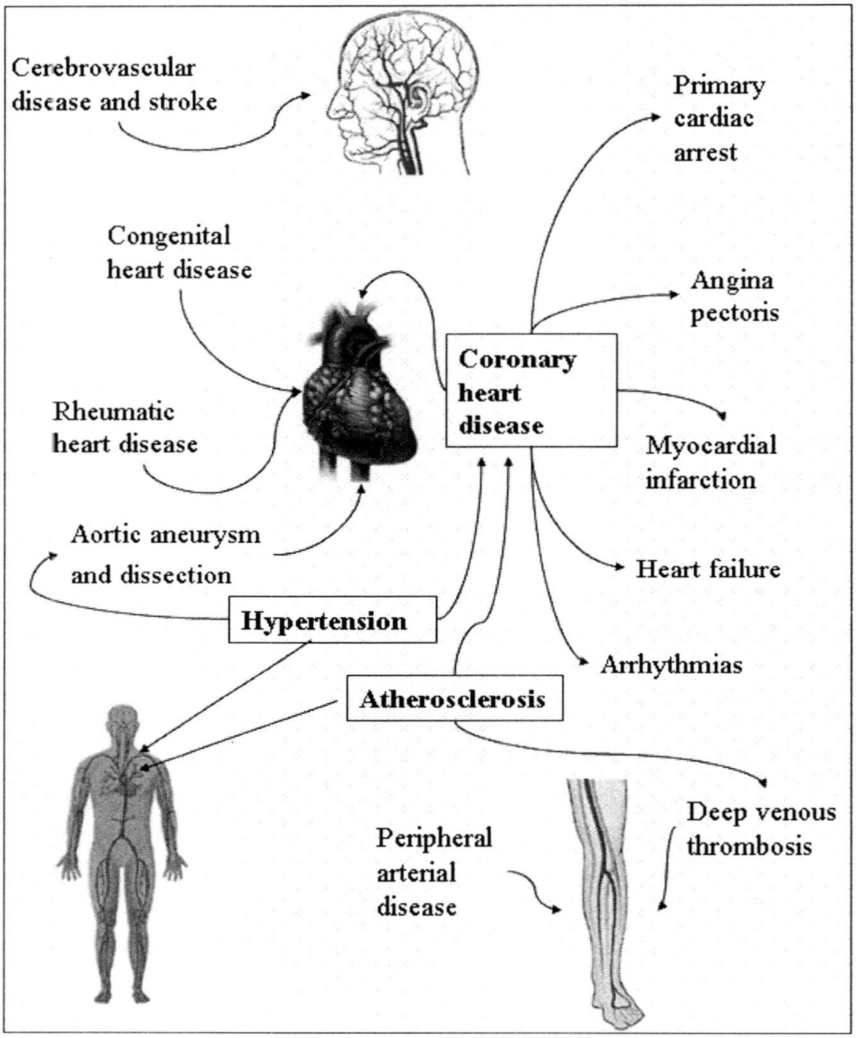

Figure 2-2. The major types of the cardiovascular disease and the related physiological systems. **Aortic aneurysm and dissection:** dilatation and rupture of the aorta. **Atherosclerosis**: abnormal thickening and loss of elasticity in the arterial walls. **Cerebrovascular disease and stroke**: disruption of the blood supply to the brain. **Congenital heart disease**: inborn malformations of heart structures. **Coronary heart disease (CHD)**: severe imbalance between coronary blood flow and myocardial requirements. **Deep venous thrombosis (DVT)**: blood clots in the leg veins. **Hypertension:** a chronic arterial condition characterized by elevation of the blood pressure. **Peripheral arterial disease:** disease of the arteries supplying the arms and legs. **Rheumatic heart disease**: damage to the heart muscle and heart valves from rheumatic fever caused by streptococcal bacteria. (after WHO's Atlas of Heart Disease And Stroke, http://www.who.int/cardiovascular_diseases/).

Atherosclerosis is an abnormal thickening and loss of elasticity in the arterial walls. In a common form of arteriosclerosis fatty substances form a deposit or 'plaque' on the inner lining of arterial walls. Atherosclerosis affects the entire cardiovascular system and is a major risk factor for all of the CVD forms (especially CHD, stroke and DVT). The common risk factors include high blood cholesterol, unhealthy lifestyle and advancing age.

In **cerebrovascular disease and stroke**, there is a disruption of the blood supply to the brain. In ischemic stroke, this disruption is the result of the blockage of the lumen while in hemorrhagic stroke disruption occurs from rupture of a blood vessel in the brain. The risk factors include hypertension, atrial fibrillation, atherosclerosis, tobacco, unhealthy diet, lack of physical inactivity, diabetes and advancing age.

Congenital heart disease is characterized by inborn malformations of heart structures (such as abnormal valves, abnormal heart chambers) that occur, in most of the cases, because of the adverse exposures during gestation. The known risks are consanguinity and gestation factors: maternal alcohol use, infections, drugs (warfarin, in particular) and poor maternal nutrition.

In **coronary heart disease (CHD)**, a severe imbalance between coronary blood flow and the myocardial requirements is caused by changes in the coronary circulation. The terms "coronary heart disease", "ischemic heart disease" and "coronary artery disease" are synonymous. CHD often arises because of the constricted heart arteries. In accordance with the WHO classification (Figure 2-2), CHD is of five types: primary cardiac arrest (a sudden event, presumably due to electric instability of the heart), *angina pectoris* (transient episodes of chest pain precipitated by exercise, by emotional stress or by other situations), myocardial infarction (a sudden insufficiency of oxygen supply to the heart that results in heart muscle damage), heart failure (complication of acute or previous myocardial infarction, also may be precipitated by angina episodes or arrhythmias) and arrhythmias. The common CHD risks include hypertension, atherosclerosis, tobacco use, unhealthy diet, physical inactivity, diabetes, advancing age, inherited disposition, inflammation, blood clotting disorders and emotional factors (depression *etc*).

Deep venous thrombosis (DVT) involves blood clots persistent in the leg veins. Being carried to the heart and lungs by the blood stream DVT can result in pulmonary embolism. The risks include recent surgery, obesity, recent childbirth, use of oral contraceptives and hormone replacement therapy, long periods of immobility, air travel, high homocysteine levels in the blood and genetic predispositions.

Hypertension is a chronic arterial condition characterized by elevation of the blood pressure. Hypertension affects the entire cardiovascular system and is a major risk for other types of CVD. One of the most important risks for hypertension is emotional stress.

What we can learn from the material presented above? As we have seen, the most common risk factors include improper diet, damaging habits and disabling lifestyle. WHO maintains that many of the CVD incidences are preventable by action on the major primary risk factors such as unhealthy diet, physical inactivity, and smoking. It is also likely that at least 50% of the CVD deaths and CVD disability can be cut by joint effort of the doctor and the patient to reduce major risk factors such as high blood pressure, high cholesterol, obesity and smoking. According to WHO (and, especially, when taking into account the amount of information *actually* available through the abstract databases, see Chapter III), there is compelling evidence that such simple dietary measures as substitution of nonhydrogenated unsaturated dietary fats for saturated fats; diet high in fruits, vegetables, nuts and whole grains; avoidance of the excess salty and sugary foods, when combined with regular physical activity and abstinence from smoking can considerably reduce CVD incidence and mortality. Moreover, the multivariate *vs.* univariate analyses of the data in the genetic association studies (Chapter V) often indicate that the major risks are, indeed, associated with these external factors.

It's true that the worldwide rise in CVD during the recent decades reflects a significant change in diet habits, physical activity levels and tobacco consumption as a result of industrialization, urbanization, economic development and globalization of the food market. However, the inherent predispositions to CVD cannot be ignored. Thus, atherosclerosis (which is often a precondition for most of the CVDs) is apparently related to the metabolism of fats. How well the individual is adapted for the metabolism of the "risk" fats is determined by the individual's genetic makeup. Genetically inherited dispositions to CHD, which are apparent from numerous clinical studies, can hardly be denied. The defects in the genes of the blood coagulation cascade are likely to result in thrombotic predispositions (see, for instance, Chapter V, where we present and discuss results of the relevant meta-analyses). The effects of most of the genetic factors are amplified by other factors such as diet, smoking *etc.* Accordingly, a factor of CVD risk represents a greater risk for one individual (or population) and lesser risk for another individual (or population). In the next section, we consider the known and potential risk factors for CVD in greater detail.

Factors of Cardiovascular Risk

The matter of the factors predisposing to certain diseases is as old as the medicine itself. For millennia, the practicing doctors observed that some patients are likely to be more susceptible to certain diseases and that this susceptibility (sometimes inherited) is, in some peculiar manner, also related to their diet, life style habits and other factors. At present, according to WHO, there are over 300 risk factors that were associated with coronary heart disease and stroke. The risk factors established by WHO as "major" meet three criteria: a high prevalence in many populations; a significant independent impact on the risk of CHD/stroke and direct evidence that control of these factors during treatment results in a reduced risk. The following four categories of the CVD risks are, at present, significant in all populations: *major modifiable* risk factors (high blood pressure, abnormal blood lipids, smoking/tobacco, physical inactivity, obesity, unhealthy diets), *other modifiable* risks (low socioeconomic status, mental illness, emotional stress, alcohol, certain medication), *non-modifiable* risks (advancing age, heredity, gender, ethnicity) and *"novel"* risk factors (hyperhomocysteinemia, inflammation, abnormal blood coagulation). Apparently, post-genomic physiology of CVD would be most interested in the non-modifiable risks (heredity/genetics) as well in those "novel" risks.

To illustrate some of these factors we compiled the diagram (Figure 2-3) and detail the diagram in the rest of this section. This diagram is compiled for CVD in general, without making much distinction among the individual types of CVD (Figure 2-2). The purpose of this diagram is to provide a general but medicinally adequate background against which we can look at the place of the genetic predispositions to CVD. It also illustrates, albeit indirectly, the actual complexity of the physiological mechanisms behind the disease. Looking at this picture, it becomes clear that in order to assess the relative significance of each factor, there must be established some sort of a hierarchy of factors that reflects the hierarchy of the levels of function (Figure 1-5). However, the order and the contents of the levels in such a hierarchy are quite disputable and there is no unanimous agreement among the biomedical scientists as, for example, whether alcohol (in so-called 'small doses') is always beneficial to the cardiovascular patients or not. Cardiovascular effects of alcohol intake can be significantly attenuated depending on the intake of drugs, genetic makeup and other factors.

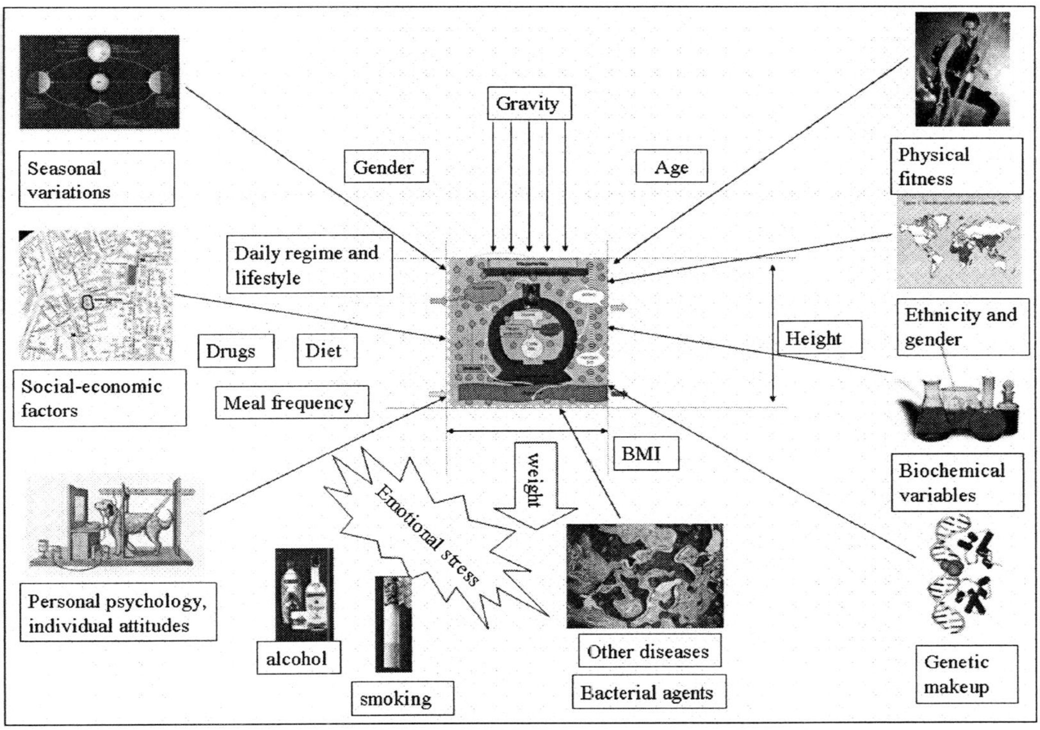

Figure 2-3. Common factors of cardiovascular risk. The factors are shown to act upon all the organs systems interlocked in homeostasis (Figure 1-4).

Accordingly, the diagram depicts most general kinds of risks relevant to the etiology of various cardiovascular diseases. Considering the diagram for a while, it also becomes quite apparent that, for instance, the common craze of "no-cholesterol!!" is no more than hyping a particular factor of cardiovascular risk for commercial purposes. Apparently, one-sided guruisms of the kind "no cholesterol and you'll have no CVD at all" have quite a bleak scientific perspective, even in the view of the pretty simplified diagram we have in the Figure 2-3. The same consideration concerns the genetic makeup: genetic makeup is only one of the factors and is not always a major one, although it is a non-modifiable, life-long and latent risk factor.

For the sake of clarity, the diagram does not show interconnections between the individual risk factors but merely represents a general context of the risk factors. The peculiarities of each factor need a separate description. For example, some of the genetic polymorphisms are specific only for certain human populations and are entirely absent in other human populations. A genetic polymorphism can be considered as a "common risk" only if it is a common polymorphism in the given population. Another example: some of the genetic predispositions are bound to the some external factors of risk and would hardly correlate with the clinics if the corresponding risk factors are absent. This concerns, for instance, the polymorphisms in the genes involved in detoxification and drug metabolism (glutathione transferases, cytochromes, alcohol dehydrogenase *etc*) or polymorphisms in the genes related to the fat metabolism (apolipoproteins, lipases and the related genes). Accordingly, if the patient does not smoke, does not take certain drugs and consumes the right kind of diet, then such polymorphisms won't be significant risks for this particular patient.

Although the diagram in the Figure 2-3 is a simplification of the data on the several hundred of the known CVD risks, it reflects, nevertheless, the variety of risk factors for CVD. Remembering about this variety of heterogeneous factors is important while analyzing and planning association studies (Chapters III&V, Volume II). The following discussion not so much details the diagram as underlines the importance of the risk factors which are routinely neglected in most the association studies dealing with genetic predispositions to CVD. In author's experience, these factors sometimes can make a difference during the data analysis of association studies. In this section, we briefly consider environmental, dietary, physical exercise, microbiological and psychological factors while biochemical variables and the related risks are discussed in the following sections. In most of the cases below, it is quite apparent how these neglected factors can routinely be accounted for within an adequately planned association study.

Environmental Factors

It cannot be denied that all of the human patients we might be aware of live on the planet Earth. Any human is constantly under the influence of the planet's gravitation which is, apparently, the most common factor from which none can escape. Gravitational stress was proposed to be one of the risk factors for hypertension [1]. The author's arguments include the results of the spaceflight physiology studies and the calculations of the gravitational stress. Spaceflight studies have demonstrated that adaptation to increased gravitational stress after prolonged microgravity includes sympathetic activation, water retention, and arterial pressure increase (*i.e.*, a set of responses very similar to essential hypertension). From this point of view, hypertension looks much like an advanced stage of adaptation to a further increase in gravitational stress. Calculations of the gravitational stress on the cardiovascular system in an upright position [1] suggest that regular prolonged sitting, typical of modern life, should cause a significant increase in gravitational stress on the cardiovascular system. This gravitational stress will require an advanced "anti-gravitational" response with sympathetic hyperactivity, vasoconstriction, volume overload, and, therefore, arterial hypertension.

The idea that gravitation stress is a CVD risk correlates with the results of the studies linking individual's height and CVD risks. Height is not always recognized as a risk factor for CVD. However, shorter people (that is, individuals under somewhat reduced gravitational stress) appear to have substantially lower rates of CHD mortality and moderately lower levels of stroke mortality [2]. For example, there are data indicating that shorter southern Europeans had about half the CHD mortality rate of northern Europeans. Among ethnic groups living in one locality, shorter *vs* taller groups tend to have substantially lower mortality rates [2].

The gravitational stress is far from being the only common environmental factor that can adjust cardiovascular risk (Figure 2-3) and other environmental factors are most general risks of multifactorial diseases. That is why in genetic association studies the careful researchers attempt to select subjects (both patients and controls) from one locality and, preferably, during the same period of time: in order to reduce the effects of otherwise "unknown" environmental confounders.

For example, the occurrence of adverse cardiovascular events, such as myocardial infarction, stroke and pulmonary embolism, is not randomly distributed over time, but shows definite seasonal patterns. The same consideration appears to be valid for deep vein thrombosis [3]. Physiologically, even a mild surface cooling has been shown to increase both red blood cell mass resulting in higher hematocrit and platelet count thereby increasing the propensity for spontaneous thrombosis. This hypercoagulable state could be further aggravated by elevated fibrinogen levels, which also appear to show consireable seasonal variations increasing up to 23% during the colder months. Limited physical activity during the colder months also can stimulate thrombotic events [3].

The intricacy of the interactions of the time factors and genetic factors may be illustrated by the circadian effects of the 4G/5G polymorphism in PAI-1. This gene (plasminogen activator inhibitor 1) is, apparently, involved in the inhibition of fibrinolysis and the 4G promoter variant is known to increase the gene expression thus increasing the thrombotic propensities and the cardiovascular risks. It was established that protein heterodimers BMAL/CLOCK (involved in circadian clock) cause almost 2-fold greater activation of the 4G-promoters than of the 5G-promoters [4]. Accordingly, the deleterious effects of the 4G variant may surface not necessarily as a higher level of PAI-1 but, rather, as the extent of the variation of the PAI1 level at certain times of the day thus increasing the cardiovascular risks at those times. In addition, each human has his or her own unique circadian rhythm so the peak levels of the PAI1 in blood will, actually, occur at various hours of the day.

Among more specific environmental risks, smoking is a major modifiable risk for CVD. What is less widely known is that passive smoking is likely to be as important risk as the smoking *per se*. Generally, cardiovascular risk can increase because of the fine particulates involved in the air pollution (http://content.nejm.org/cgi/content/full/356/5/511?query=TOC). Specifically, many studies have reported that passive smoking is also associated with increased risk of CHD. In comparison to mere questionnaire data, biomarkers of passive exposure to smoking can provide a more objective measure of exposure from all the sources of the tobacco smoke [5]. Prospective study of cardiovascular disease in middle-aged men using measurement of one of such biomarkers (cotinine) has shown dose-dependent associations with this biomarker and CHD hazards. High overall exposure to passive smoking suggests that the effects of passive smoking might have been underestimated in earlier studies [5].

Dietary Factors

It is common knowledge among medical practitioners that the dietary components of so-called "junk foods" engender considerable risks for CVD and other diseases (see, for instance, an enjoyable web-site http://www.yourdiseaserisk.harvard.edu). At the same time, more natural dietary styles alleviate CVD risk. "Mediterranean diet" is, perhaps, the best known and the most popular of the natural diets.

Although, strictly speaking, different regions in the Mediterranean basin have their own very peculiar diets, from a medical point of view it is appropriate to consider these variants under the single rubric of a generalized Mediterranean diet. The term "generalized Mediterranean diet" implies a high intake of vegetables, legumes, fruits, unrefined cereals, a low intake of saturated lipids but a high intake of unsaturated lipids (olive oil, in particular); a low to moderate intake of cheese and yogurt; a low intake of meat but average to high intake of seafood (including fish) [6,7]. Olive oil has a special place in the Mediterranean diet. Oleic acid, polyphenols, and other components abundant in olive oil are thought to be partly responsible for the antiatherogenic effects attributed to the Mediterranean diet. One of the positive impacts of the olive oil is the improvement of the lipoprotein spectrum [8].

A meta-analysis and review of the studies which evaluated the association between adherence to a Mediterranean diet and the occurrence of coronary heart disease outcomes suggests that the health benefits from the Mediterranean diet are significant in all studies [6]. The reduction in the risk of coronary heart disease varied from 8% to 45%, depending on the increment used by the investigators in the presentation of their results. Reviews such as this have shown that Mediterranean type diets increase longevity. However, there is a dearth of international studies that would assure compatibility among the group(s) analyzed. In a recent study of the sorts [7], over 70,000 participants without coronary heart disease or stroke at enrolment supplied extensive information about their diets and a number of potential confounding variables. Using a 10-point coring scale, these data were recalculated to an "extent of adherence to a Mediterranean diet". The results indicated that an increase in this score was significantly associated with lower mortality [7]. In particular, the study confirmed once again that polyunsaturated lipids are an acceptable substitute when monounsaturated lipids are not readily available. By the way, the factors of genetic predispositions to the metabolism of the unsaturated and saturated lipids are located in the genes of the fat metabolism (such as APOC3 and others, see *The Case of GeneScape* at the end of this Chapter).

The dietary risks and protective factors include not only the dietary substances but also the manner in which these substances are taken in by the patient. This factor is, of course, closely related to the individual's life style habits. Animal studies show that intermittent fasting (*i.e.*, reduced meal frequency) and caloric restriction extend lifespan and increase resistance to age-related diseases [9]. Both intermittent fasting and caloric restriction enhance cardiovascular and brain function, reduce blood pressure and increase insulin sensitivity.

Physical Exercise

Apart from being an indisputable protective factor, several forms of physical exercise testing are routinely used to evaluate patients with cardiovascular symptoms. However, many exercise testing laboratories and clinicians focus on electrocardiographic findings and minimize the prognostic importance of the duration of exercise [10]. Duration of the exercise, nevertheless, provides strong and independent prognostic information about the overall risk of illness and death, especially in respect to the cardiovascular disease. The physiological effects of physical exercise can surface through a number of various physiological mechanisms. It's also remarkable an observation that cellular and molecular effects of intermittent fasting on the cardiovascular system and the brain are similar to those of produced by regular physical exercise [9].

Genetic polymorphisms can adjust the effects of the physical exercise. For instance, polymorphisms in the ADRB2 gene (beta2-adrenergic receptor) were associated with a number of differences in cardiovascular function at rest and during exercise [11]. A Case Study in the Volume II details the relationship between genetic makeup and the effects of the physical training by analyzing a subset of data from the *medSNP* database. As another example, we consider here an issue which is almost never analyzed in association studies: the alteration of the acid-base balance.

Everyone knows well that one of the most apparent results of a physical exercise is that it stimulates breathing, when done even for a moderate amount of time. More intensive breathing fastens the metabolism through hyperventilation, the increased oxygen supply and also normalizes acid-base balance of the blood. Influence of the physical exercise on the breathing patterns and on the acid-base balance may be one of the important mechanisms of the beneficial action of the physical exercise at the level of the physiological systems. Both influences will certainly benefit the cardiovascular health, depending on a number of additional factors. The following examples illustrate how physiological, biochemical and genetic factors can be intertwined in the beneficial influence of what may seem to be a "mere breathing".

The breathing affects the gas balance of the blood and, before all, the oxygen levels. The oxygen levels are far from being the only parameter affected. In particular, analysis of the *acid-base balance* can be an important diagnostic tool if used more consistently. Normally, pH level of blood should be in between 7.32 and 7.46 and pH values outside this range indicate an abnormal condition. However, pH values are rarely interpreted in the clinical practice. It's is understandable since blood pH is rather general biochemical variable. What is interesting about this biochemical variable is that it reflects both the effects of the physical exercise and the diet. In the physiology, the levels and the dynamic of the blood pH and the related parameters can reflect a number of conditions and, in particular, the negative effects of the change from the diet of the hunter-gatherer to the modern, grain-based, "acid-loaded" [12].

If there is an abnormality in the acid balance (acid is generated or taken in faster than it is removed), it results in lowering of the pH. This condition is known as acidosis. There are two main types of acidosis: the *metabolic acidosis* (which occurs because of an increase in acids other than carbonic acid) and the *respiratory acidosis* (which corresponds to excessive retention of the carbon dioxide and is usually a sign of hypoventilation). In both cases, the adverse change in pH can be counteracted by hyperventilation which leads to excretion of CO_2 by the lungs and physical exercise is, apparently, the most natural means to induce hyperventilation. Thus, the frequency and other aspects of breathing can observably influence the pH levels and the gas balance of the blood.

These general considerations have direct medical applications in CVD diagnostics and treatment. For example, a recent investigates the impact of a 6-month correct breathing stereotype on minute ventilation, capillary blood gases and acid-base balance. This study was done on a small group of post-myocardial infarction patients [13]. It appears that correction of the breathing stereotype improved respiratory function and could be an additional measure in rehabilitation of post-myocardial patients [13].

Apart from these general medical factors, the respiratory acidosis can also be influenced by genetic makeup. One of the most important mechanisms to sustain the pH balance is the bicarbonate buffer mix (bicarbonate, carbon oxide and carbonic acid) which is maintained by the carbonic anhydrase(s) catalyzing the buffer reaction. Carbonic anhydrases are zinc metalloenzymes participating in a variety of biologic processes, including respiration, acid-base balance, calcification, bone resorption, and the formation of the bodily fluids such as saliva and gastric acid. In a very small study, serum concentrations of myoglobin and carbonic anhydrase III were measured in patients with acute myocardial infarction and other groups. The myoglobin-to-anhydrase ratio was higher in infarct patients indicating that this test can be useful in cardiovascular diagnostics [14]. Although there are no data from association studies dealing with genetic polymorphisms, the common polymorphisms in the carbonic anhydrases are likely to influence the gas and the acid-base balances of the blood thus affecting cardiovascular physiology of the patient. We describe the selection and analysis of polymorphisms in the human carbonic anhydrases in another Case Study of the Volume II "Genetics".

Microbiological Factors

Microbiological analyses are not commonly practiced in contemporary cardiovascular medicine. Nevertheless, atherosclerosis is an inflammatory process and infection with microorganisms is one of the most likely contributors to inflammation. For example, there were antibiotic trials that explored the role of antichlamydial agents in the treatment of cardiovascular events [15]. Any mechanical, chemical or microbiological insult acting upon the vessel wall will induce endothelial dysfunction. Such an insult triggers a cascade of inflammatory reactions, in which monocytes, macrophages, T-lymphocytes and vascular smooth muscle cells participate. Many infectious agents have been proposed to be potential triggers of the inflammatory reactions relevant to CVD and the greatest body of evidence is available for *Chlamydia pneumoniae* [16].

Chlamydial infections, however, are far from being the only microbiological CVD risks. Gingivitis, the mildest form of periodontal disease, is caused by the bacterial biofilms in the form of the dental plaques that accumulate on teeth surface adjacent to the gums. Advanced forms of periodontitis can result in loss of connective tissue and bone support and is a major cause of tooth loss in adults. What is also important is that common forms of periodontal disease have been associated with cardiovascular disease, stroke and pulmonary disease albeit causal relations were not established [17]. So far, there were a few dozen of the studies associating oral conditions and cardiovascular diseases. A recent review shows that patterns of high and low levels of eight periodontal pathogens and antibody levels against those organisms are related to clinical periodontal disease as well as appear to influence atherosclerosis and CHD in accordance with age, race and gender. The cumulative evidence clearly supports the association between periodontal infection and atherosclerotic cardiovascular disease [18]. We explore the subject of the interactions between bacterial agents and genetic predispositions in the Volume II.

Psychological Factors

Among negative effects of emotional stress is severe though reversible dysfunction of the left ventricle in patients even without coronary disease. Accordingly, such a stress in a CVD patient can induce severe condition or even result in lethal outcome. However, psychological factors are often difficult to measure on a quantitative scale without producing a large number of false positives. Still, carefully constructed questionnaires can deal with the problem of taking into account the psychological and emotional stress while analyzing cohorts of CVD patients. Thus, a recent review of about 50 studies dealing with behavioral influences on atherosclerosis and CHD suggests that despite inconsistencies and occasional conceptual and methodologic blunders, there is a considerable amount of evidence indicating that psychosocial variables play significant role in the CVD etiology. In particular, so-called "type A behaviour pattern" that has a high potential for hostility and inability or unwillingness to express anger emerged as a significant predictor in most of the studies reviewed [19].

There is growing amount of epidemiological literature suggesting that marital discord is a risk factor for cardiovascular morbidity and mortality. Couples who demonstrated consistently higher levels of hostile behaviors across both their interactions healed at 60% of the rate of low-hostile couples. High-hostility couples also produced relatively larger increases in plasma IL-6 and TNF-alpha levels the morning after a conflict than after a social support interaction. Comparisons with low-hostile couples indicate potentially higher inflammation rates in the high-hostility couples [20] which, in turn, precipitate cardiovascular disease.

There is growing evidence that cardiovascular diseases are more prevalent in subjects who feel chronic anxiety. In particular, a panic-like anxiety can induce hypercoagulable states. A study of over 600 subjects has shown that the level of D-dimer (a biochemical marker of the coagulation) was significantly different across the 4 scores of panic feelings.

The findings also suggest increased fibrin turnover associated with sudden onset of the feelings of panic [21].

Results of the studies that take into account psychological factors also indicate that the psychological capabilities of the subjects to tolerate stress, to adapt to stressful situations and to exert specific forms of what is generally known as "self-control" can reduce CVD risk. For example, many individuals who have apparent atherosclerotic changes in their arteries may never develop more severe forms of cardiovascular disease. A study that uses serial color word test (a semi-empirical technique that can be used for assessment of individual's capacity to adapt in a stressful situation) delineated four distinct patterns of behavior among CVD patients [22]. In this study, atherosclerosis was a significant risk factor for MI and cardiovascular mortality but mostly among the subjects who showed maladaptive behavior. No excess risk could be established in men with an adaptive behavior pattern.

It is important to notice that general notion of "self-control" that we might have learned from movies and fiction novels appears to have nothing to do with the self-control that benefits cardiovascular health. Thus, a large study (over 5,000 participants) indicated that such major personality traits as anger control or symptoms of depression were not consistently associated with myocardial infarction, stroke, or cancer [23]. At the same time, a psychological capacity which the authors refer to as "internal locus of control over disease" was significantly associated with a decrease in risk of myocardial infarction [23]. This and other studies emphasize that introduction of the psychological profiling into questionnaires of the CVD-related studies should be carefully thought out.

As we might infer from the above presentation, there are many environmental, dietary, physical, microbiological and psychological factors that need to be accounted for while conducting association studies dealing with genetic predispositions to CVD in order to obtain non-trivial results. For most of these factors, the corresponding physiological mechanisms were not established and these are considered as being "risk factors" only according to epidemiological type of evidence. Biochemical variables which we are about to consider in the following section are more directly related both to the clinical observations and genetics, at least in the framework of the diagram of the levels of function (Figure 1-5).

Biochemical Markers and Diagnostics of CVD: A Few Examples

What we call as "biochemical markers" or "biochemical variables" in this book are the concentrations (or "levels") of certain substances in the tissues of the body. These substances can be inorganic salts, organic acids and bases, hormones and numerous proteins that include coagulation/fibrinolytic factors, inflammation-related proteins, lipoproteins *etc.* In modern medicine, there are not less than a hundred of these biochemical markers which are routinely analyzed in clinical lab tests. The terms "marker" and "variable" do differ in content: a "marker" merely reflects a physiological condition while "variable" is involved, more or less directly, in the corresponding physiological mechanisms.

A biochemical variable or a marker can be analyzed in various tissues. Most of them, however, are the levels of various substances in the fluids of the body and, primarily, in the blood. Why the blood is, perhaps, one of the most important tissues in physiology? This fluid tissue connects all of the organs into the single functioning system and thus is the most important tissue for the maintenance of the total homeostasis of the body (Figure 1-4). The blood provides nourishment for and carries off wastes from each cell in the body: it transports the oxygen, products of digestion (sugars, lipids, amino acids), ions and microelements, hormones, vitamins and many other kinds of nutrients and other substances to the organs that require them while carrying off carbon dioxide, urea and other wastes away from the organs. The blood is also a self-regenerating tissue that is also directly involved in such various physiological processes as heat distribution, immunity and coagulation. Blood is the essential component of the cardiovascular system and many disease states, not only CVD, are to considerable degree influenced by or reflected in the physiological state of the blood.

In clinical practice, the state of blood is investigated from various points of view using the lab tests that analyze numerous biochemical variables/markers. The set of these biochemical tests often depends on the condition of interest. Some tests, such as the Complete Blood Count (CBC), are routinely done for any patient irrespectively of his or her specific condition. Other common tests often include determination of the levels of coagulation factors, of the cardiac enzymes, of the inflammation markers, hormones, bilirubin *etc*. A selection of the common lab tests is presented in the Box 2. We use this selection of tests later, when analyzing associations in the IHD200 cohort (Chapter III).

There at least three essential peculiarities of the biochemical markers/variables. Firstly, a value of biochemical marker reflects, by definition, the state of the body's homeostasis and, therefore, the state of the health. Secondly, most of the biochemical markers represent only a particular state and their values change with time. Thirdly, the values of the most of the biochemical markers are dependent on the external factors. In other words, most of the biochemical variables represent *current* state of the organism (current at the moment of taking the blood sample). Some of them are not. For example, analysis of a blood group (ABO, MN *etc*) is expected to produce the same result in the same individual over the entire life span. Although analysis of, say, ABO blood group belongs to the area of clinical lab testing, it, nevertheless, directly reflects certain genetic polymorphisms (Volume II).

Given that, generally, a biochemical variable reflects the current state of the homeostasis, the lab tests are used to supplement the clinical observations in order to arrive at more precise diagnosis. The results of the lab tests are also used for the prognostic purposes. Since a number of the biochemical variables depend, to various extents, on the external factors (which is reflected by the fact that the lab tests are usually taken in the morning, with no food intake in 12 hrs *etc*), in the case of a particular patient it is not always enough to measure, say, FGB levels just once. Methodologically, rather a series of measurements should be undertaken in the course of a specific amount of time (which is usually done when monitoring response to drug treatment). Apart from monitoring certain conditions, these systematic measurements also provide more reliable background to be used in genetic association studies.

Box 2. Most common clinical lab tests

No doubt, this list is neither comprehensive nor representative. Some of these tests (hemoglobin, hematocrit) belong to CBC (Complete Blood Count, the most common blood test), ESR serves as a simple inflammation marker, enzymes AST, ALT and ALP are biochemical markers of wide significance; bilirubin levels indicate the liver function and APTT, fibrinogen and d-dimer are coagulation markers. As we'll see from the IHD200 Case Study (Chapter III), a rational analysis of these simple biochemical markers can be very useful in analyzing the apparent failures of the genotype-phenotype association studies*[1].

ALP (alkaline phosphatase) catalyzes hydrolysis of the complex ethers of phosphoric acid. The ALP plasma levels are higher in bone damage or liver damage, nutritional deficiency of calcium or phosphates and in heart failure. Lower ALP levels correspond to nutritional deficiency of Zn and Mg.

APTT (activated partial thromboplastin time) test is a measure of the functional state of the blood coagulation cascade. Longer APTT might correspond to the deficiency of the coagulation factors II, V, VIII, IX, X, XI or XII, liver disease, vitamin K deficiency, anticoagulant therapy; shorter APTT corresponds to hypercoagulation.

AST (Aspartate aminotransferase) enzyme, normally present in liver and heart cells, is released into blood when the liver or heart is damaged. AST levels heighten in MI, pulmonary thrombosis, cardiosurgery, hepatitis, severe angina pectoris, cholestasis and a number of other conditions.

ALT (alanine aminotransferase) enzyme levels are determined to check for possible liver damage. Usually, ALT is used in conjunction with AST. Normally, the AST/ALT ratio ("de Ritis ratio") is about 1.33, the ratio significantly lowers in hepatitis of various origins and is significantly higher in MI.

Bilirubin is one of the most common bile components also present in serum. Higher bilirubin may indicate B-12 deficient anemia, acute and chronic liver disease, alcohol cirrhosis and drug intoxication.

D-dimer is one of the major fibrin degradation products released into plasma upon clot degradation with plasminogen. Thus, d-dimer is detected only when a clot was formed and then degraded. Higher d-dimer levels indicate excess coagulation that can occur in DVT, pulmonary thromboembolia, surgery, inflammation, therapy and other factors.

ESR (erythrocyte sedimentation rate). The sedimentation rate depends on the erythrocyte aggregation which, in turn, is largely determined by the protein fraction of the plasma. The aggregation grows as the concentrations of the acute-phase proteins (fibrinogen, C-reactive protein, ceruloplasmin, immunoglobulins and other) grow. The higher ESR values serve as a biochemical marker of the inflammation processes, lower ESR values are observed in hyperproteinemias, erythrocytoses and hepatitis.

Fibrinogen is the key component of the blood coagulation cascade. Fibrinogen levels increase during inflammation, it is one of the proteins of the acute inflammation phase. Higher fibrinogen concentrations correspond to an increase of the cardiovascular risk; lower fibrinogen levels occur in liver disease, vitamin C and B-12 deficiencies, during anticoagulant therapy.

Hematocrit (CBC test) reflects % relationship between the volume of the erythrocytes and the total blood volume. The test is quite prone to experimentalist errors both during taking of the blood sample as well as during the test itself. Higher hematocrit suggests erythremias, dehydration of the organism; lower hematocrit corresponds to anemia and hyperhydration.

Hemoglobin (CBC test). Haemoglobin is the major protein of erythrocytes that transports oxygen and carbon dioxide. The reference ranges of the haemoglobin are gender and age-adjusted. Lower haemoglobin level strongly suggests anemia (iron- or folate-deficient), hemolysis, specific haematological disorders or severe blood loss.

*1 Additional details on these and other tests are available, for instance, at http://www.labtestsonline.org/, detailed information on the reference ranges for these and other tests can be found in Clinical Diagnosis and Management by Laboratory Methods. 20th ed. Henry JB, ed. New York, Saunders, 2001.

In modern clinical practice, there are hundreds of the lab tests and a few dozen of them can be done in the case of a particular patient. The proper informational management of this array of data is important. Common sense suggests that results of a particular test are neither to be ignored (as illustrated by a statement like "this is only one out of dozens of tests") nor to be taken as a "cornerstone" overruling all the other data ("his ESR is too high, therefore he must have his appendix out" *etc*). Undoubtedly, behind each of these tests there is a body of physiological and biochemical knowledge that delineates the range of applicability and significance of the particular test. Nevertheless, for a diagnostic or a prognostic procedure to be adequate, it is important not only to know what biochemical variable this or that test reflects but also to look into the correlations between the results of various tests.

In the rest of this section, we'll consider a few of the common biochemical markers in greater detail in order to see how interpretation of the results of the tests varies in accordance with certain conditions. The studies we'll cite throughout are just a few examples of the applications of particular biomarkers. These examples also indicate a few important features of the proper information management of the association studies. They also illustrate the usage of the complementary lab tests. The first two examples (fibrinogen and C-reactive protein) deal with analysis of a prospective cohort of patients with early myocardial infarction, the third example (d-dimer) is a meta-analysis dealing with the usage of d-dimer testing for diagnostics of the deep venous thrombosis.

Fibrinogen Test for a Long-term Prognosis

Fibrinogen is the key proteins of the blood coagulation cascade (see the *GeneScape* diagram at the end of this Chapter). In clinical practice, determination of fibrinogen levels is used to analyze the pathologies of blood coagulation and also serves as a test for inflammation and necrosis. High fibrinogen level can be seen as an indicator of cardiovascular risk in healthy donors and CHD patients [24-26]. Meta-analyses suggest that people with high levels of fibrinogen might have approximately twice as high risk of developing CHD [27]. But the clinical value of fibrinogen test is not restricted to these issues only. For instance, there were a few studies that used fibrinogen test as a prognostic marker in survivors of myocardial infarction (MI).

A prospective cohort study of middle-aged MI-survivors investigated the prognostic value of plasma fibrinogen level [28] with a 10-year follow-up. A total of 247 middle-aged CHD patients were recruited at least 3 months after the most recent MI. The primary outcome measure was total mortality, and the secondary endpoint was cardiac deaths. It was found that the top quartile of fibrinogen was a significant predictor of cardiac death in middle-aged patients who had suffered MI (odds ratio 2.2, 95% CI 1.1-4.4, P=0.03).

Albeit the group of patients has a biased male/female ratio (193 men, 54 women) and the results are thus more relevant to men, this study presents several interesting features that are have contributed to the relative success of this epidemiological study. Firstly, the patients with relatively early first MI were selected: at ages 55 (men) and 60 (women). Selecting the group of patients in this way allows to remove the statistical noise related to the effects of old age in MI etiology. Secondly, the study appears to feature quite accurate fibrinogen measurements albeit analysis was done on singular fibrinogen samples (one sample for each patient). Moreover, these blood samples were stored for almost 10 years. Despite these two points, the fibrinogen measurements appear to be quite accurate: for the data presented in the tables in the study, the values of the error intervals showed almost no overlaps. Thirdly, the authors carefully analyzed a few basic factors of cardiovascular risk in relation to fibrinogen levels. This was done on the base of the data provided in an adequately constructed questionnaire. It was found that geometric means of fibrinogen were higher among current smokers, in non-consumers of alcohol, in patients with diabetes and in patients with cerebrovascular disease. Distinct and physiologically plausible correlations were also observed for the patients taking warfarin, diuretics, ACE-inhibitors and lipid reducing drugs. Although the relative risks related to fibrinogen levels lowered after adjustments for the other prognostic factors, the risks remained significant for prediction of cardiac mortality. In particular, the study also suggested that the deleterious effects of smoking with regard to CHD may mediated, in part, through fibrinogen levels [28]. Thus, the study illustrated that the standard biochemical variable such as fibrinogen can be used as a prognostic marker.

C-reactive Protein for Long-term Prognosis

The same cohort of MI patients was used in a study of C-reactive protein as a long-term predictor of cardiac death. C-reactive protein represents entirely different physiological mechanism of cardiovascular pathology: atherosclerosis (see the *GeneScape* diagram). Atherosclerosis is known to have considerable inflammation component and C-reactive protein (CRP) is one of the most sensitive inflammation markers. Usually, CRP test is prescribed for monitoring collagen diseases, acute infections, tumors and sometimes is used to assess the risk of cardiovascular complications in patients with atherosclerosis or diabetes. CRP marker appears to be a useful prognostic factor in patients with unstable angina, acute MI or acute stroke. In the study under discussion [29] it was found that CRP is also a strong predictor of cardiac death in premature MI. Thus, the relative risk of cardiac death was doubled with increasing CRP quartiles, and patients in the top quartile had six times as high risk of cardiac death as patients in the lowest quartile. CRP is a strong predictor of mortality in patients with premature MI. As the paper was done by the same research group, it includes the features we mentioned above. In addition, the following features of the study design worth noting:

- No exclusion criteria other than age and ethnicity were applied in the present study. Although quite apparent a point, many association studies, unfortunately, do not properly account even for these two basic factors.
- Diagnosis of myocardial infarction was done in accordance with the ‘gold standard’: the WHO definition (see Box 5 in the Chapter III).
- Although the population sample in the study is relatively small (n=247 patients, mortality n=44), the study is strengthened by the fact that there was a distinct semi-quantitative correlation between the CRP quartiles and the increase in the relative risk (P=0.0001). Observing such a semi-quantitative correlation is an example of how self-consistency of the data can be used to verify the scientific significance of the results apart from the established methods to ensure statistical significance (Chapter V).

D-dimer and DVT Diagnostics

The d-dimer is a degradation product of fibrin generated during fibrinolysis. In particular, the test is routinely applied for diagnostics of thrombotic conditions, deep venous thrombosis (DVT) and pulmonary embolism. It is also used for monitoring pregnancy and/or thrombolitic therapy. A raised level of the d-dimer indicates the presence of an abnormally high level of fibrin degradation products. Despite of the limited specificity of the test and the influence of a number of other facts on the d-dimer level, the advantage of the d-dimer testing is that d-dimer is formed only when both coagulation and fibrinolysis take place. In short, there is nothing unusual in using d-dimer testing for diagnostics of the thrombotic conditions.

The point here is *in what way* and *for what purpose* the results of the d-dimer testing are to be interpreted. For example, it is known that roughly 75% of the patients that might present clinical symptoms of DVT (see Box 5) do not have the disease. More precise DVT diagnostics requires more complex physical methods such as ultrasound scans. Before using this kind of modern diagnostic artillery, it would be nice to use a few lab tests to look at the biochemical variables to elucidate the preliminary diagnosis. The d-dimer is one of such tests. A meta-analysis [30] summarizes the evidence supporting the use of rapid d-dimer testing combined with estimation of clinical probability to exclude the diagnosis of deep venous thrombosis among outpatients.

We discuss the technique of meta-analysis and the related problems of informational management elsewhere (Chapter V). Here, we just notice that the careful selection of the studies to be included in meta-analysis is as important as the careful clinical characterization of each patient in an association study. Thus, the authors selected only 12 out of 84 reviewed studies. Sneering at the figures, an orthodox statistician might invoke the magic phrase "using selective evidence". An actual clinical scientist, however, would understand that a clinical diagnosis requires an independent confirmation and not all of the published studies might present the best evidence. The authors of [30] clearly describe about 10 of the study inclusion criteria, among them are: estimation of the risk of deep vein thrombosis by using a validated clinical probability tool; prospective study of consecutive outpatients presenting with features of deep vein thrombosis; evaluation of outpatient data separately if inpatients were included; evaluation of deep vein thrombosis data separately if patients with pulmonary embolism were included; follow up of all patients; confirmation of DVT using venous compression ultrasound, venography or impedance plethysmography; presentation of data that allows calculation of the sensitivity and specificity of the d-dimer assay. Accordingly, the results of the meta-analysis [30] were clear-cut and strongly statistically significant: a normal value of a highly sensitive d-dimer test reliably rules out DVT in patients at low or moderate risk thus indicating ways to reduce the need for ultrasound scans.

Taking a wider look at the human physiology, the analysis of the biochemical variables might not always be straight forward as in the three examples we just cited. For example, it is possible that risk of CHD will be associated with a particular polymorphism in a gene but without any observable differences in the lipid levels [31]. The reason for this effect would be much alike to the problem of the measurement of thyroid function we considered in the previous Chapter. The paper [31] studied the effects of the "R219K" polymorphism in a lipid transporter gene ABCA1on the lipid levels and the CHD risk. There were no apparent associations between the serum triglycerides and HDL-c but rather complex changes in the functional dependencies (see the Figures 1&2 of the paper). The authors surmise that polymorphism-associated changes in transport activity would change the net flux of cholesterol from the vessel wall towards the liver, without necessarily altering plasma lipid levels. The possible explanations (apart from the points we consider in the Chapter V) in this case include the magnitude of the changes registered by the lipid measurements (only larger changes in efflux result in measurable changes in plasma lipid levels, smaller changes might still directly impact cholesterol accumulation within the vessel wall), variety of the environmental stimuli as well as changes in the plaque lipid composition [31]. All these effects can considerably obscure otherwise straight-forward analysis of biochemical variables and need to be specifically tackled already at the stage of the study planning.

Having considered these few examples of the clinical applications of the biochemical tests, we might be able to observe that the principle of "circumstances alter the cases" is very much appropriate when dealing with the biochemical markers. The most reliable clinical diagnosis and the most adequate interpretation of the results of the lab tests will occur only when the results of various tests are analyzed together and, of course, without leaving the clinical background out of view. Loosing the view of the clinical background is, unfortunately, often the case with the genotype-phenotype association studies. In the following sections, we'll discuss the general results of the genotype-phenotype association studies in CVD and will attempt to outline CVD genetics in a Case Study at the end of this chapter.

Approaching CVD Genetics

In the previous three sections, we have covered the major types of the CVD, considered the variety of the common (and uncommon) risk factors for CVD and mentioned a few points related to modern diagnostics of CVD. The most common risk factors include improper diet, damaging habits and disabling lifestyle. In medical practice, CVD appears to be largely preventable by decisive action on these primary risk factors. In addition, the management of these risk factors by the joint effort of the patient and the medical practitioner can reduce the risk of cardiovascular mortality and disability. Apart from personal experiences of thousands of doctors, these statements were corroborated by a considerable number of independently made epidemiologic studies and clinical trials.

These external risk factors represent, however, only one side of the coin. The other side should not be neglected: the external risk factors are often specific for particular individuals or populations and are definitely linked to genetics. That there is a distinct inherited component in etiology of CVD that also affects the course, treatment and outcome of the disease is hardly a matter of discussion. What can be (and usually is) under discussion is *what* genes are involved and what would be the *extent* of the influence of the genetic factors on CVD. Apart from the rare monogenic disorders (which comprise an entirely different area of research in clinical genetics), there are no unequivocal epidemiological data in regard of the *common* genetic factors. The extent of the influence seems to vary not only from one "candidate gene" to another "candidate gene" and from one common genetic polymorphism to another polymorphism. Various authors, in accordance with the variance in their biases, will tend to use only one set of facts and not the other, either diminishing or hyping the extent of the contribution of the genetics to disease etiology.

For example, an analysis of a well-known cohort of white Northern Americans states that "genetic effects of apoE in disease etiology cannot exceed 1%". In the same publication, this particular finding for a particular gene is then somehow gets over-generalized to the point of "genetic factors cannot be important for CVD". Another example: studies of apo(a) polymorphisms in Caucasians suggest that Kringle-IV polymorphism can "account for 90% of inter-individual variation of Lp (a) concentration". Lp(a) levels are considered to be an important factor adjusting CVD risk. And yet another estimate of the sorts: contribution of genetics to etiology of arterial thrombosis "ranges from 20% to 80%" [32]. These and many other examples of the findings are not, necessarily, biased or unreliable. However, they are often abstracted out of the entire context of physiology and then are cited to suit particular interests of particular authors. In short, the question of the extent of the involvement of the common genetic factors in the disease etiology sometimes does involve as much of heated debate, flawed arguments and sheer speculation as the famous question of "Is there life on Mars?" (thus reminding us of the falsifiability principle, see Introduction).

This kind of situation is hardly something that is totally unexpected. The genotype-phenotype association studies (which are, effectively, studies of epidiomiological sort) are the primary source of any evidence concerning the extent of involvement of the common genetic factors in CVD. In terms of the diagram of the levels of function (Figure 1-5), the genotype-phenotype association studies can *a priori* be expected to have one of the lowest possible signal-to-background ratios among other types of the association studies. An investigation of associations between, say, smoking and CVD (the top levels of the function in the Figure 1-5) or an investigation of the association between the genetic polymorphism and the gene expression (the lower levels of the function) would, apparently, include much less of the statistical noise since the levels of functions involved are much closer. It is obvious that the genotype-phenotype association goes across the entire stack of the levels of functions and additional factors are likely to blur the picture. Therefore, there are many association studies that indicate very low effects related to genetics or even report a negative finding altogether. There is also a possibility that a statistically significant positive finding might be a result of a statistical fluke (Volume II).

We consider some of the most prominent failures of the genetic association studies and potential causes for those failures in the following chapters (Chapters III, V) and in the 2nd Volume of the present book series. In brief, the genotype-phenotype relationships are difficult to examine, and predictions concerning them are seldom certain, especially when several pathways of different pathophysiologies are implicated in a multifactor disease. The statistical nature of an association study directly suggests that most of the evidence of such associations deals much less with clear-cut logical argumentation than with rather vague statistical probabilities [33]. This does not mean that such relationships cannot be studied or are unpredictable, but rather that the reproducibility is not straightforward. First of all, the reproducibility of the genetic associations is hampered by the plethora of additional factors that should be taken into account (Figures 1-5 and 2-3). Secondly, the effect of the genetic factors can be considerably amplified (or, on the contrary, significantly weakened) by interaction with the external factors. Hence, in order to obtain meaningful results in a genotype-phenotype association study, not only a careful analysis of the clinical observations and these additional factors (such as those we considered in the previous sections) is required. Not less important is to have at least an overview of the most likely physiological mechanisms involved in the pathology of the disease under study as well as an overview of the roles of particular genes in these mechanisms.

Although in the present book we do not consider monogenic disorders as being "CVD", the case studies of the CVD-related monogenic disorders were often crucial for obtaining most of the presently available information on the genes and the related physiological mechanisms of CVD. This matter was considered in detail in an excellent review [34]. For instance, LDL is the major cholesterol-carrying lipoprotein in plasma and is considered to be a causal agent in CHD etiology. The studies of the relevant monogenic diseases allowed the researchers to outline some of the basic components of *cholesterol synthesis and excretion*. Familial hypercholesterolemia was the first monogenic disorder shown to cause elevated plasma cholesterol levels. The primary genetic defect in familial hypercholesterolemia is a deficit of functional LDLR (LDL receptor) molecules. Non-functional receptor fails to take up plasma cholesterol resulting in this monogenic phenotype. Familial ligand-defective apolipoprotein B-100 indicated APOB-100 as another important component that binds to LDLR, autosomal recessive hypercholesterolemia indicated ARH gene, while sitosterolemia (very high levels of plant sterols in the plasma resulting in an accelerated atherosclerosis) indicated ABCG5 and ABCG8, the transcription factors regulating liver cholesterol synthesis and clearance. Therefore, the studies of the monogenic disorders that disrupt LDL-receptor pathways have clarified the importance of cholesterol synthesis and excretion pathways in the liver and have highlighted molecular targets for regulating plasma cholesterol levels.

Similarly, other fundamental physiological mechanisms contributing to the etiology of CVD can be outlined. *Hypertension* imparts an increased risk of all types of CVD and many clinical trials indicated that reductions in blood pressure reduce the incidence of stroke and myocardial infarction. Investigation of the relevant monogenic disorders influencing blood-pressure indicated a number of the physiological mechanisms. Glucocorticoid-remediable aldosteronism, aldosterone synthase deficiency, 21-hydroxylase deficiency and other monogenic diseases outlined the basic molecular mechanisms involved in the salt reabsorption in kidneys and also indicated the genes involved in the regulation of the blood pressure. The mechanisms of *hemostasis* and pathophysiology of thrombosis were also outlined through the studies of the rare Mendelian defects in the genes of what is known now as "blood coagulation cascade". For example, the "Leiden"/R506Q variant of the F5 gene was initially found in a particular patient with familial thrombophilia. The same concerns the G20210A variant in prothrombin.

The studies of monogenic diseases are, of course, not the only source of the experimental information. The role of the models of the classical physiology cannot be denied. For instance, there are two fundamental physiological mechanisms contributing to the CVD etiology: hypercoagulativility of blood and closing of the vascular lumen. In particular, thrombosis and atherosclerosis are the most common physiological mechanisms related to CVD. As the research in the last century shows, pathophysiology of thrombosis involves complex interactions between the endothelial surface, platelets, and the proteins of the blood coagulation cascade. However, the basic concept of the pathophysiology of thrombosis was first formulated by Prof. R. Virchow as early as 1856. This concept, now known as the "Virchow's triad", reflects the three major physiological mechanisms that can lead to increased incidence of thrombosis: changes in the vessel wall, changes in blood flow and changes in the coagulability of blood [32,35]. We illustrate the concept in the Figure 2-4. This illustration introduces certain systematics to the common factors of cardiovascular risk we summarized in the Figure 2-3.

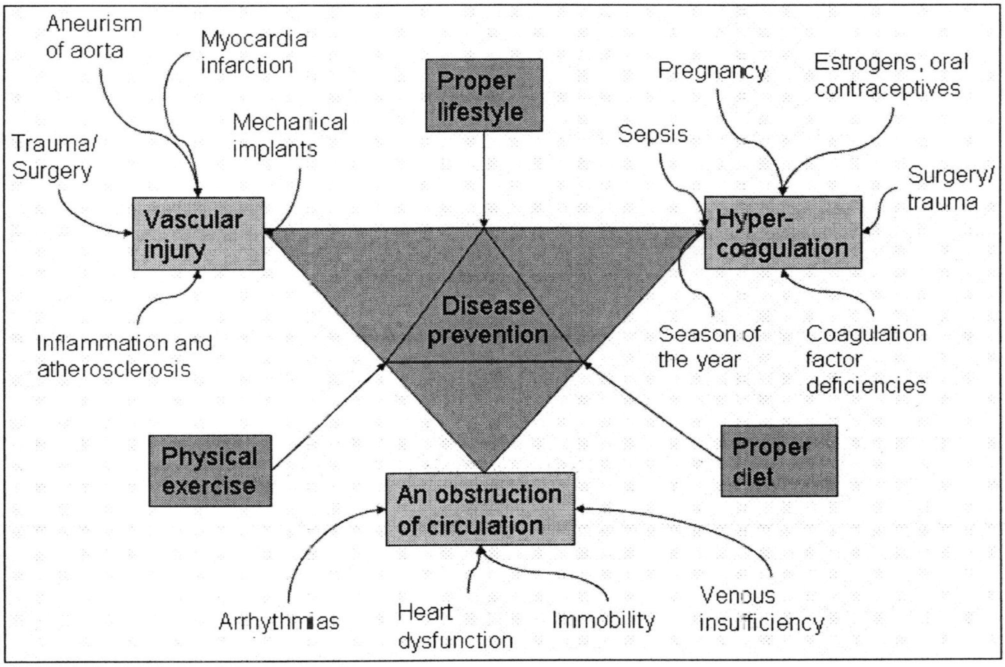

Figure 2-4. The Virchow's triad, the related factors of CVD risk (the outward triangle) and the protective factors (the inner triangle).

The Virchow's triad is an excellent example of a proper informational management of a biological complexity. This triad is still one of the most reliable and flexible concepts in the pathophysiology of CVD. No doubt, the century of research through the joint effort of clinical scientists, biochemists, molecular biologists and geneticists considerably detailed this general picture. Nevertheless, as we'll see from the results of the *GeneScape* Case Study, most of the common genetic risks act through at least one of these three general physiological mechanisms (Figure 2-5). For example, in acute coronary occlusion (which results in myocardial infarction) the thrombotic process begins with an injury of the vascular endothelium. When arterial subendothelium is disrupted, von Willebrand factor (VWF) molecules are rapidly localized to the exposed collagen, and the initial platelet contact with the wound is made through the tethering of VWF molecules to the integrin subunit GPIba. Tethering and rolling of platelets on the vessel wall causes platelet activation and up-regulates platelet integrin GPIIb/IIIa. Once platelets are activated, this integrin undergoes a conformational change to a high-affinity ligand-binding state and then platelets adhere to one another via fluid phase fibrinogen or via VWF molecules that bridge the GPIIb/IIIa integrin molecules. The result is a quick expansion of the thrombus. Platelet activation and aggregation initiates coagulation cascade by the exposure of the various blood elements to tissue factor in the vessel wall which results in the production of thrombin [32]. Thus, the mechanism of acute coronary occlusion was detailed down to the individual genes and biophysics of individual molecules. And yet, an adequate overview of this mechanism is provided by the good old triad of Prof. Virchow.

CVD Genetics and Bioinformatics

The descriptions even of a single molecular mechanism, such as the one cited above, can be considerably detailed and involve a number of other relevant genes and biochemical variables. As our Black Square of the Modern Physiology (Figure 1-1) suggests, the amount of the currently accumulated information might be so that thousands of pages in a book like this can be filled with interesting and even practically important descriptions of the molecular circuitry behind the CVD. The question here is, then, how will we manage this informational complexity, be it for the purpose of fundamental studies or for the practical applications. In other words, how can we collect, extract and systematically organize the available information on the genes, gene polymorphisms and CVD physiology within a uniform conceptual framework?

Bioinformatics helps in this endeavor. Let's begin with considering the notion of the "candidate gene". In brief, a candidate gene is any gene which is thought likely to be involved in a disease. It might be called as "candidate" either because it is located in a particular chromosome region suspected of being involved in the disease or because the known physiology of the protein product suggests such an involvement. Clearly, this definition is exceedingly broad and might include dozens, hundreds or thousands of genes, depending on the additional criteria that we'll use to elaborate the elusive term "any" in the above definition. It is also clear that these criteria should also determine what the level of evidence is taken to be reliable and what candidate genes are of the *primary* interest. From the bioinformatic point of view, we can boost efficiency of our search through three basic strategies to select candidate genes for an association study we plan:

- By using clinical data on the known monogenic diseases related to CVD
- By using data on the relevant biochemical variables
- By using published data on the previously made association studies.

Monogenic Diseases and CVD

In this Volume, we do not consider monogenic disorders under the rubric of CVD. Monogenic diseases are characterized by the classical Mendelian transmission modes (Volume II). However, current research suggests that one-genotype to one-phenotype association is an oversimplification even in the case of human monogenic diseases and there are often multiple genetic and non-genetic modifiers of the disease phenotype [33]. Nevertheless, particular genetic defects implicated in various monogenic diseases are often deleterious to the function either of the gene or of the protein and, accordingly, result in generally consistent clinics characterized by low protein levels and/or activity. On the contrary, cardiovascular disease is a multifactor disorder resulting from considerably more complex interaction of a much larger amount of environmental and genetic factors.

Both monogenic and multifactor disorders result from certain abnormal features in the DNA of certain genes. These features sometimes called as "polymorphisms", "mutations", "genetic defects" *etc.* Though these notions often overlap in everyday science talk, a "polymorphism" is more common a phenomenon, with population frequency of 1% or higher, while a "mutation" or "gene defect" is rare (population frequency <0.1%). In a historical perspective, first these rare gene defects were found (as it was the case with the basic components of cholesterol synthesis and excretion we've mentioned) and the information on monogenic disorders was collected in a number of databases such as OMIM (http://www.ncbi.nlm.nih.gov/omim/), HGMD (http://www.hgmd.org) and dozens of locus-specific databases.

Generally, the "common polymorphisms" in these genes were identified after the "mutations". The common polymorphisms, especially if they have functional value, can be used as common genetic factors of susceptibility to CVD. And they were used, in many thousands of the association studies that followed the case studies of the monogenic disorders. Therefore, instead of reinventing bicycle and formulating the candidate genes on the base of the data on monogenic disorders, we can rather analyze the published data of these association studies. By using the latter approach we, apart from listing the most likely candidate genes, will also obtain a perspective of how useful the data on monogenic disorders can actually be for determination of the common factors of the genetic risk (Volume II).

The Cardiovascular Biochemical Variables

The information on the existing lab tests and the related biochemical variables is largely independent on the study of monogenic disorders. In many cases, we can obtain a pretty good list of the potentially significant candidate genes by simply enumerating the biochemical variables of interest and then tracing the relation of each biochemical variable to the corresponding gene. For example, cardiovascular system and the kidneys play an indispensable role in maintaining the biochemical constancy of the blood and other extracellular fluids and CVD pathology often involves either direct or indirect violation of this biochemical constancy. As the hormones are the causative principles and not mere biomarkers, the polymorphisms in related genes can be important genetic markers. Let's use a common textbook on endocrinology [36] and compile a list of hormones produced by and acting on the cardiovascular system (Table 2-1).

In this Table, we traced the relation of the hormones secreted by or acting on the cardiovascular systems to the particular genes. Generally, there is no warranty that for a particular gene product related to a biochemical test there would be certain common polymorphisms that have a potential to become reliable genetic markers. Nevertheless, in this our example, most of the genes mentioned in the Table 2-1 also appear in the *CVD GeneScape* diagram (which was compiled independently, using the results of genotype-phenotype association studies in an abstract database).

Table 2-1. Hormones produced by and acting upon cardiovascular system (CVS) and the related genes (after [36])

Hormone	Physiological function	The genes
	Produced by CVS	
Atrial natriuretic peptide	Filtration rate, Na^+ reabsorption	NPPA
Endothelins	Vasoconstriction	EDN1, ECE1, EDNRA
Nitric oxide	Vasodilatation	NOS3
	Acting on CVS	
Angiotensin II	Vasoconstriction	AGT,ACE,AGTR1
Arginine vasopressin	Vasoconstriction	AVP, AVPR1A/B
Catecholamines	Vasoconstriction/dilatation in skeletal muscle	COMT, ADRB2
Bradykinin	Vasodilatation	KNG1, BDKRB1/2
Thromboxanes	Vasoconstriction	PTGIS, PTGS2

Published Data of Genotype-phenotype Association Studies

The two previously discussed strategies imply that any kind of "gene hunt", whether on the base of data for monogenic disorders or using biochemical variables can result in finding corresponding genetic associations. Accordingly, by efficiently extracting (Chapter III) and thoroughly analyzing (Chapter V) the data from the published array of literature, we can come up with a certified list of the candidate genes of primary interest (certified to various extents, of course). Several hundred of the candidate genes for CVD have been analyzed in thousands of association studies in attempts to common polymorphisms that can determine CVD susceptibility among general population. In order to extract these studies in a comprehensive and a uniform manner characteristic of the post-genome approach, we can use the methods of purpose-oriented bioinformatics. We implemented this strategy in *The Case of A GeneScape* that follows the present section.

Selection of the candidate genes is only the 1st step in a larger scheme of post-genomic biomedical research [33]. Whether this selection is done by searching literature or databases, by using microarrays or by other techniques, the compilation of the list of the candidate genes is to be followed by identification of the common genetic polymorphisms, assessment of the functional significance of these polymorphisms, characterization of the relevant biochemical variables (such as mRNA and protein levels), accurate and uniform clinical characterization of the disease phenotypes, pilot association studies in populations of interest and, finally, large-scale association studies involving multiple polymorphisms in multiple genes. Although selection of the candidate genes is only the initial step, it is also the most essential step. Careful selection of the candidate genes allows efficient performance of all the research that follows and also makes this research less time-consuming and much more cost-efficient.

THE CASE OF A GENESCAPE FOR CVD

"Not less than a thousand of the candidate genes for CVD have been proposed in various studies. Elucidation of the impact of the common polymorphisms in these genes on the susceptibility to CVD will, of course, depend on the following decades of research. What are the genes of primary interest and how the most likely of the candidate genes can be selected?"

Introduction

We already discussed the three most important sources of information that can be used for selection of the "candidate genes". To proceed from what is known to what is unknown is the only sane procedure of developing any kind of the human knowledge, including the medical one. Therefore, it can hardly be denied that proper extraction of the *already available* information is the quickest and the most reasonable way to assess the current situation with the candidate genes. It is the question, then, what, how and where to search.

We can hardly search published reviews that feature huge tables and dozens of short gene descriptions. Almost without exception, these reviews were compiled rather arbitrarily and are largely based on the lists of the hundreds of genes obtained from public databases such as OMIM (http://www.ncbi.nlm.nih.gov/entrez/query.fcgi?db=OMIM). Often, these reviews not only neglect clear physiological interpretations of how the genes they propose are related to CVD but also lack any epidemiological evidence: the actual availability (not to mention the quality) of the relevant phenotype-genotype association studies does not appear to be an important criterion in those reviews.

Also, we can hardly expect that by using the existing gene databases, we would obtain any certification of the reliability of the data these databases present. We discuss the relevance of the information in the existing databases to practical clinical research elsewhere (Volume II). OMIM is, undoubtedly, a very helpful educational resource but it will produce too many "candidates" without providing any ranking of their significance. The same two points can be said about the "Gene" database. GeneCanvas (http://genecanvas.idf.inserm.fr/), a database of "cardiovascular candidate genes and their polymorphisms" might seem to be most appropriate for the task. And yet, most of the genes and polymorphisms in this database were not investigated except in a very few studies. Moreover, in many cases, only negative findings about these genes were published. The relevance of the data collected in the GeneCanvas database to the CVD diagnostics also needs to be questioned.

Thus, as it is the case with many other research projects in the present book, the direct search of abstract databases such as PubMed and EmBase appears to be the most logical and the least arbitrary choice. Answering the question of the *common* genetic factors of predisposition to CVD and the related candidate genes, we'll extract the necessary information using PERL scripting in the framework of the purpose-oriented bioinformatics.

Methods

The *medCrawler* script allowed automated extraction of the relevant abstracts using online searches of a database of abstracts. Then, the texts of the abstracts and the keyword fields were parsed to extract the gene names and other data. The script employs several keyword search techniques which we discuss in the next Chapter. Optionally, the script can also be used to some of the criteria of polymorphism selection that we used for compilation of the *medSNP* database (Chapters IV&V). The script produces two outputs: a file with the relevant abstracts and an initial list of genes obtained after parsing of these abstracts. As it is always the case of semi-automated techniques, outputs of the script should be critically sieved by the human expert.

Results

After scripting, running the script and sieving the output data, a list of about 100 genes was obtained. We summarized these data in the *CVD GeneScape* diagram (Figure 2-5). Of course, not all of the genes in the diagram were studied extensively. For example, the diagram refers to some putative and relatively new CVD-related polymorphisms in genes (such as SAA1, SELP, LRP1 *etc*) which are, nonetheless, functional and might become gene markers for CVD. Still, for most of the genes in the diagram, a considerable number of positive associations were independently reported and these genes are the most likely candidates for the future large-scale genetic association studies addressing CVD. Brief annotations of the genes mentioned in the GeneScape diagram are presented in the Appendix I. We describe and discuss some of the results of this Case Study in the last section of the present chapter and also throughout the rest of the volume.

An Ultra-Brief Overview of the CVD Genetics

The *GeneScape* diagram (Figure 2-5) was produced as the result of *The Case of a GeneScape for CVD.* The genes selected after completion of this Case Study were classified into three general groups: **a)** cardiovascular system and other organ systems; **b)** coagulation cascade; **c)** bloodstream and the vessel wall. In the resulting diagram, the genes are grouped not so much in accordance with the known tissue expression but rather in accordance with the functional context of the gene action as related to CVD. For instance, the GNB3 gene is expressed in many kinds of tissues and cells including retina, brain, umbilical cord, colon, fibroblasts and B lymphoblasts. In relation to the CVD etiology, however, it is the influence of the polymorphism of this gene's product on vasoconstriction that is important. In the *GeneScape* diagram, therefore, this gene was attributed to the smooth muscle layer of the vessel wall. Another example: APOA1 is present in the blood plasma while it is expressed in liver, atrium, ventricle, small intestine, skin *etc.* However, it is the concentration of the APOA1 in the blood plasma that has most important cardiovascular implications. Accordingly, the label for APOA1 was placed in the vessel lumen near the HDL particles. Albeit simplistic, the diagram depicts at least general CVD-related function of the genes discussed.

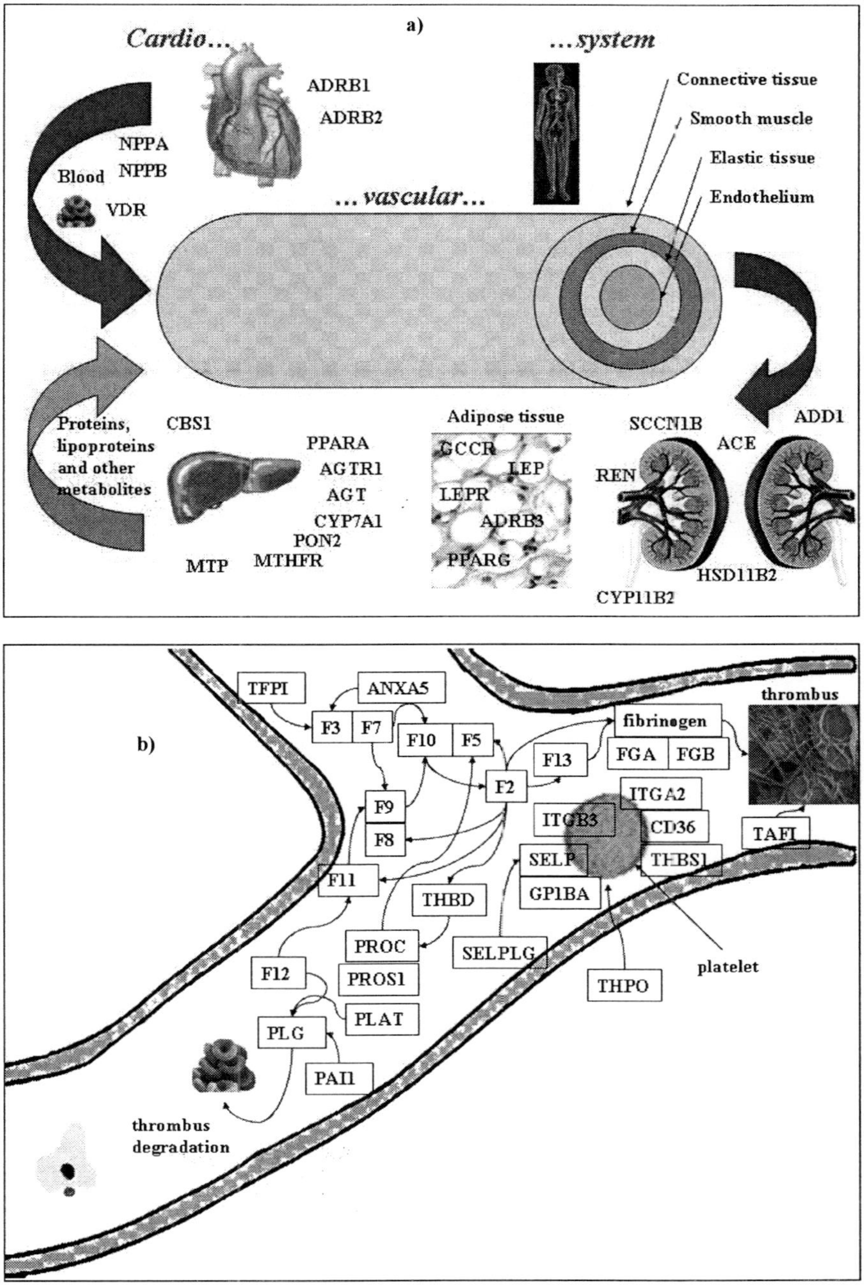

Figure 2.5. Continued

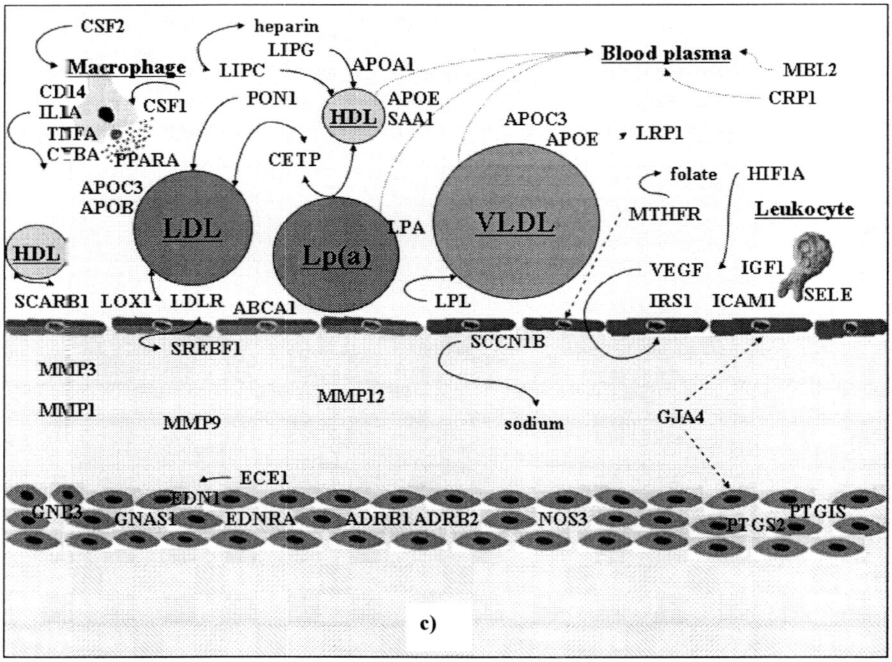

Figure 2-5. The GeneScape diagram. The gene annotations are provided in the Appendix I. The genes are grouped into three main classes (see details in the text): a) cardiovascular system and other organ systems; b) coagulation cascade; c) bloodstream and the vessel wall.

Grouping of genes into these three classes is most general so the gene annotations were further subdivided into eight more specific groups (Appendix I). Thus, the genes in the *cardiovascular system and other organ systems* (Figure 2-5a) are subdivided into the genes of the adipose tissue (fats metabolism), the genes involved in the electrolyte balance and the genes of the renin-angiotensin system. The genes in the *coagulation cascade* (Figure 2-5b) represent both clotting and fibrinolytic branches of the blood coagulation cascade. The genes in the slide illustrating *bloodstream and the vessel wall* (Figure 2-5c) are related to atherosclerosis (lipoproteins and inflammation), vasoconstriction/vasodilatation and vascular constitution/remodeling. While each gene in Appendix I has a short annotation relating it to one of these eight groups, it is clear that most of the genes in the diagram are related, either directly or indirectly, to the functional states of the blood vessels. Accordingly, we can get a compact overview of the CVD genetics by looking at the structure of a blood vessel and then recalling the associated genes.

The general structure of a blood vessel is depicted in the Figure 2-5a; a section is presented in the Figure 2-5c. Looking from outside to inside, blood vessels are composed of connective tissue (*tunica adventia* also known as 'fibrous' or 'collagenous' tissue), smooth muscle (*tunica media*), and *tunica intima* that can be subdivided into the layer of *elastic tissue* and the *endothelium*. Endothelium cells line the vessel's inner space known as *lumen*. The blood vessels differ in the relative thickness of each of the four tissue layers and in the lumen size. The apparent design of a particular type of blood vessel indicates the specific functions of that type of vessel. Thus, *capillaries* are comprised of only a single layer of endothelium. This thin layer of cells considerably alleviates the transport of the oxygen/nutrients into the tissues and the excretion of the wastes from the tissues. *Arteries* have rather thick muscular walls, in order to withstand the pressure surges which occur as the heart pumps the blood while *veins*, on the contrary, have less muscle tissue as the blood in veins is not under pressure as great as in the pressure in the arteries. Therefore, the impact of each particular gene attributed to the vessel wall will differ in accordance with these differences in the structure of particular types of the vessels. Let's briefly retrace the general vessel structure once again, from outside to inside, and mention the relevant genes.

The **connective tissue** is the outermost layer of the vessel that stabilizes and anchors the vessel as the connective tissue fibers penetrate into adjacent tissue. In veins, the connective tissue layer is often thicker than smooth muscle layer. The collagen fibers also support the entire vessel structure interconnecting the smooth muscle and elastic tissue/endothelium. The collagenases MMP1, MMP3, MMP9 and MMP12 are involved in maintaining and remodeling of the vessel structure. Collagenase activity also determines the stiffness of the vessels.

The **smooth muscle** layer consists of involuntary muscle stimulated by sympathetic nerves. The primary function of the layer is to adjust the lumen size of the vessel as required. During vasodilation the smooth muscle relaxes while during vasoconstriction the smooth muscle contracts affecting the blood pressure. Renin-angiotensin system (AGT, AGTR1, ACE and REN) is the most general mechanism for the vasopression. Apart from the activities of the renin-angitensin system, vasopression is affected by a number of other genes of endothelin pathway (EDN1, ECE1, EDNRA), cellular signal transduction (GNB3, GNAS1), adrenergetic receptors (ADRB1, ADRB2), nitric oxide synthesis (NOS3) and prostaglandin synthesis (PTGS2, PTGIS). GJA4 is a structural protein that mediates interactions smooth muscle and endothelium layers.

The **endothelium** layer consists of squamous epithelium that lines the lumen of all blood vessels. Endothelium forms the interface between the blood in the vessel lumen and the rest of the vessel wall. It plays a critical role in the mechanics of blood flow, the regulation of coagulation, leukocyte adhesion, secretion and modification of vasoactive substances. The vascular endothelium growth is stimulated by VEGF which is regulated, in part, by HIF1A. Endothelial cells contain a number of lipoprotein receptor proteins, lipases and the related proteins (LDLR, SREBF1, LOX1, SCARB1, LRP1, LPL, LIPG and ABCA1) as well as receptors and adhesion molecules implicated in inflammatory responses (PPARA, PPARG, ICAM1, SELE, IRS1). Both of these groups are involved in the etiology of atherosclerosis. Here, we can also mention genes responsible for the maintenance of the electrolyte balance of the blood (VDR, SCNN1B, ADD1, NPPA, NPPB) which is related to vasopression, albeit indirectly.

Lumen is the most important location in the vessel: it is the place of the blood flow. The volume and the size of the lumen determine the peculiarities of the blood flow. Most of the genes mentioned in the *GeneScape* diagram were, apparently, attributed to the lumen. These include the genes of apolipoproteins as well as other genes implicated in metabolism of fats (APOA1, APOC3, APOE, APOB, LPA, LIPC, CETP, PON1), the genes implicated in inflammatory responses (CYBA, CSF1, CSF2, CD14, IGF1), the genes coding the inflammatory response proteins (CRP1, SAA1, TNFA, IL1A) and, of course, the genes corresponding to the numerous proteins of the blood coagulation cascade and platelet aggregation/production (in alphabetic order, ANXA5, CD36, F2, F3, F5, F7, F10, F12, F13A1, FGA, FGB, GP1BA, ITGA2, ITGB3, PAI1, PLAT, PROC, PROS1, SELP, SELPLG, TAFI, TFPI, THBD, THBS1, THPO).

Although the above description is quite compact, it mentions almost 80% of the genes we had identified in the course of the Case Study. Interested readers can find somewhat elaborated descriptions of each of these genes in the Appendix I. In the framework of the purpose-oriented bioinformatics, we will make an extensive use of the *GeneScape* diagram throughout this book. However, for the *purpose-oriented paradigma* in bioinformatics to work, it is important to realize how much of the relevant information is actually available at the moment. Then, what is the quality of this information? How can we sieve out the useful information? What are the criteria of validity and verification of each item of the biomedical information? And, after all, why there are so many discrepancies among the individual association studies dealing with the same genes and polymorphisms in similar populations? We consider some answers to these questions in the Chapters to follow.

Chapter Summary

The purpose-oriented approach to bioinformatics requires specialization into physiology of the problem under study. In this book, our research is oriented towards **cardiovascular disease**, a leading cause of the mortality worldwide. The **coronary heart disease** and **stroke** are the most common cardiovascular disorders. At present, more than three hundred of the **risk factors** were associated with coronary heart disease and stroke and there are many **environmental**, **dietary, physical, microbiological, psychological** and **biochemical** factors that need to be accounted in a valid **genetic association study**. The genotype-phenotype relationships are difficult to examine and predictions concerning them are seldom certain since, according to the diagram of the **levels of function**, they have one of the lowest possible **signal-to-background ratios**. The genotype-phenotype association implies a complex interplay of the **biochemical variables** with a host of **genetic and external factors**. Hence, for adequate planning of the association studies it is essential to get an overview of the most likely **physiological mechanisms** involved in the pathology of the disease under study. This Chapter presents a brief **overview of the CVD genetics and physiology** (the CVD GeneScape diagram) which was based on the results of semi-automated searches of an abstract database.

References

[1] Pekarski S. A gravitational hypothesis of essential hypertension as a natural adaptation to increased gravitational stress. *Med Sci Monit*. 2004;10(6):HY27-32.

[2] Samaras T, Elrick H, Storms L. Is short height really a risk factor for coronary heart disease and stroke mortality? *Med Sci Monit*, 2004; 10(4): RA63-76

[3] Gallerani M, Boari B, de Toma D, Salmi R, Manfredini R. Seasonal variation in the occurrence of deep vein thrombosis. *Med Sci Monit*. 2004;10(5):CR191-6.

[4] Chong N, Codd V, Chan D, Samani N. Circadian clock genes cause activation of the human PAI-1 gene promoter with 4G/5G allelic preference. *FEBS Lett*. 2006;580(18):4469-72.

[5] Whincup P, Gilg J, Emberson J, Jarvis M, Feyerabend C, Bryant A, Walker M, Cook D. Passive smoking and risk of coronary heart disease and stroke: prospective study with cotinine measurement. *BMJ* 2004;329(7459):200-5.

[6] Panagiotakos DB, Pitsavos C, Polychronopoulos E, Chrysohoou C, Zampelas A, Trichopoulou A. Can a Mediterranean diet moderate the development and clinical progression of coronary heart disease? A systematic review. *Med Sci Monit*. 2004 Aug;10(8):RA193-8.

[7] Trichopoulou A, Orfanos P, Norat T, Bueno-de-Mesquita B, Ocke MC, Peeters PH, van der Schouw YT, Boeing H, Hoffmann K et al. Modified Mediterranean diet and survival: EPIC-elderly prospective cohort study. *BMJ*. 2005 Apr 30;330(7498):991-997.

[8] Habana P, Klvanova J, Zidekova E, Nagyova A. Dietary supplementation with olive oil leads to improved lipoprotein spectrum and lower n-6 PUFAs in elderly subjects. *Med Sci Monit*, 2004; 10(3): PI49-54

[9] Mattson MP, Wan R. Beneficial effects of intermittent fasting and caloric restriction on the cardiovascular and cerebrovascular systems. *J Nutr Biochem*. 2005 Mar;16(3):129-37.

[10] Kraus WE, Douglas PS. Where does fitness fit in? *N Engl J Med*. 2005;353(5):517-9.

[11] Snyder E, Beck K, Dietz N, Eisenach J, Joyner M, Turner S, Johnson B. Arg16Gly polymorphism of the beta2-adrenergic receptor is associated with differences in cardiovascular function at rest and during exercise in humans. *J Physiol*. 2006 Feb 15;571:121-130.

[12] Gennari F, Adrogue H, Galla J, Madias N (eds). *Acid-Base Disorders and Their Treatment*. Taylor&Francis, 2005.

[13] Vasiliauskas D, Jasiukeviciene L. Impact of a correct breathing stereotype on pulmonary minute ventilation, blood gases and acid-base balance in post-myocardial infarction patients. *Eur J Cardiovasc Prev Rehabil*. 2004;11(3):223-7.

[14] Vaananen H, Syrjala H, Rahkila P, Vuori J, Melamies L, Myllyla V, Takala T. Serum carbonic anhydrase III and myoglobin concentrations in acute myocardial infarction. *Clin Chem*. 1990 Apr;36(4):635-8.

[15] Kuppuswamy VC, Gupta S. Antibiotic therapy for coronary heart disease: The myth and the reality. *Drugs Today (Barc)*. 2005 Oct;41(10):677-85.

[16] Kaperonis EA, Liapis CD, Kakisis JD, Dimitroulis D, Papavassiliou VG. Inflammation and atherosclerosis. *Eur J Vasc Endovasc Surg*. 2006;31(4):386-93.

[17] Pihlstrom B, Michalowicz B, Johnson N. Periodontal diseases. *Lancet*. 2005;366(9499):1809-20.

[18] Beck JD, Offenbacher S. Systemic effects of periodontitis: epidemiology of periodontal disease and cardiovascular disease. *J Periodontol*. 2005;76(11 Suppl):2089-100.

[19] Manuck SB, Kaplan JR, Matthews KA. Behavioral antecedents of coronary heart disease and atherosclerosis. *Arteriosclerosis*. 1986;6(1):2-14.

[20] Kiecolt-Glaser J, Loving T, Stowell J, Malarkey W, Lemeshow S, Dickinson S, Glaser R. Hostile marital interactions, proinflammatory cytokine production, and wound healing. *Arch Gen Psychiatry*. 2005;62(12):1377-84.

[21] von Kanel R, Kudielka BM, Schulze R, Gander ML, Fischer JE. Hypercoagulability in working men and women with high levels of panic-like anxiety. *Psychother Psychosom*. 2004;73(6):353-60.

[22] Andre-Petersson L, Hedblad B, Janzon L, Steen G. Asymptomatic atherosclerosis and myocardial infarction: risk modified by ability to adapt to stressful situations. *Med Sci Monit*. 2004 Oct;10(10):CR549-56.

[23] Sturmer T, Hasselbach P, Amelang M. Personality, lifestyle, and risk of cardiovascular disease and cancer: follow-up of population based cohort. *BMJ*. 2006;332(7554):1359.

[24] Wilhelmsen L, Svardsudd K, Korsan-Bengtsen K, Larsson B,Welin L, Tibblin G. Fibrinogen as a risk factor for stroke and myocardial infarction. *N Engl J Med* 1984; 311:501-505.

[25] Kannel W, Wolf P, Castelli W, D'Agostino RB. Fibrinogen and risk of cardiovascular disease. *JAMA* 1987; 258:1183-1186.

[26] Woodward M, Lowe G, Rumley A, Tunstall-Pedoe H. Fibrinogen as a risk factor for coronary heart disease and mortality in middle-aged men and women. *Eur Heart J* 1998; 19:55-62.

[27] Ernst E, Resch KL. Fibrinogen as a cardiovascular risk factor: a meta-analysis and review of the literature. *Ann Intern Med* 1993; 118:956-963.

[28] Retterstol L, Kierulf P, Pedersen J, Bohn M, Bakken A, Erikssen J, Berg K. Plasma fibrinogen level and long-term prognosis in Norwegian middle-aged patients with previous myocardial infarction. A 10 year follow-up study. *J Intern Med*. 2001;249(6):511-8.

[29] Retterstol L, Eikvar L, Bohn M, Bakken A, Erikssen J, Berg K. C-reactive protein predicts death in patients with previous premature myocardial infarction--a 10 year follow-up study. *Atherosclerosis*. 2002;160(2):433-40.

[30] Fancher TL, White RH, Kravitz RL. Combined use of rapid D-dimer testing and estimation of clinical probability in the diagnosis of deep vein thrombosis: systematic review. *BMJ*. 2004;329(7470):821.

[31] Srinivasan S, Li S, Chen W, Boerwinkle E, Berenson G. R219K polymorphism of the ABCA1 gene and its modulation of the variations in serum high-density lipoprotein cholesterol and triglycerides related to age and adiposity in white versus black young adults. *Metabolism*. 2003;52(7):930-4.

[32] Williams MS, Bray PF. Genetics of arterial prothrombotic risk states. *Exp Biol Med* (Maywood). 2001;226(5):409-19.

[33] Herrmann S, Paul M. Studying genotype-phenotype relationships: cardiovascular disease as an example. *J Mol Med*. 2002;80(5):282-9.

[34] Nabel E. Cardiovascular disease. *N Engl J Med*. 2003;349(1):60-72.

[35] Schafer AI. Hypercoagulable states: Molecular genetics to clinical practice. *Lancet* 344:1739–1742, 1994.

[36] Nussey S, Whitehead S. *Endocrinology: An integrated approach*, BIOS Scientific publishers, 2001.

Chapter III

BIOMEDICAL STUDIES AND THE STUDIES OF THE STUDIES

Hypotheses non fingo
"I feign no hypotheses"
Isaac Newton

This sentiment, featuring in the Newton's *Principia Mathematica,* does not necessarily imply "I don't make up any false hypotheses". Rather, it can be taken as "I do not hold any [particular] hypothesis [to be true *a priori*]". It is a very useful attitude for any scientific researcher who needs to be as objective as possible and who should, ideally, never allow personal likes and dislikes to interfere with the sane processes of scientific evaluation. However, this principle alone is, at best, useless and even misleading if no objective criteria of evaluation of that or other hypothesis are forthcoming. The present chapter deals with some of the basic criteria of the assessment of the relevance and significance of biomedical studies in general and of the genetic association studies in particular.

The Case Study of the previous chapter allowed us to identify a few dozen of genes that are known to be implicated, in some or other way, in CVD etiology. This Case Study, which provided a list of the genes with polymorphic markers common in Caucasian populations, employed computer-assisted search of the published data. Although the *medCrawler* script did operate within certain rules and restrictions, we did not discuss neither the origins of these rules nor the significance of the evidence produced. From how many studies the references were selected? What were the reasons for their selection? And, the most important, how can we judge the reliability of the evidence for each particular association?

THE INFORMATIONAL OVERLOAD IN MODERN HUMAN PHYSIOLOGY

To begin answering these questions, we'll elaborate a bit on the Black Square of the Modern Physiology (Figure 1-1) by making a few general searches in PubMed and

EMBASE. These simple searches will also illustrate us the enormous amounts of already published data available through these databases. Our purpose here is to observe the extent and the trends of the information growth in specific areas of physiological and biomedical interest. In particular, physiological studies dealing with human genetics are of especial interest in post-genomic era. As those areas, we'll consider human physiology in general, the genetic studies relevant to human physiology and the studies dealing with estimates of genetic predispositions to common diseases. The results of these database searches are summarized in the three slides of the Figure 3-1.

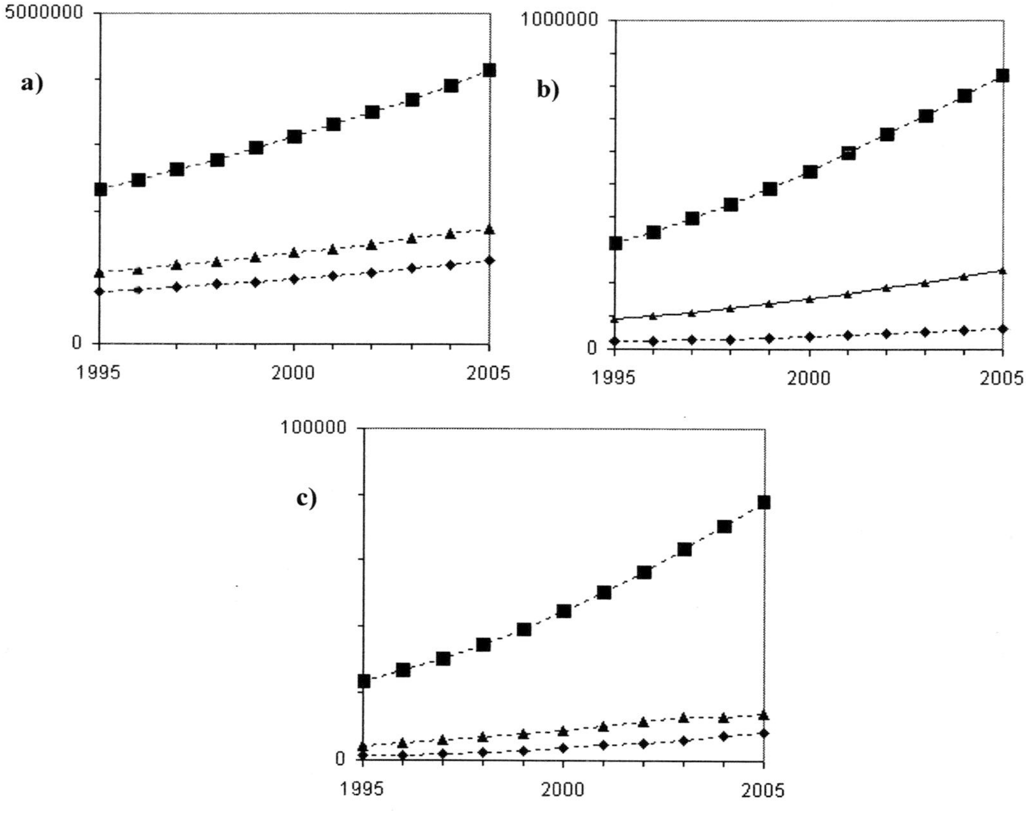

Figure 3-1. A steady growth of the information on human physiology during the last decade. The squares denote the number of publications on human physiology in general; triangles, the data on cancer; the rhombi, the data on cardiovascular disease. The keywords for PubMed searches were straight-forward: "cardiovascular disease", as in slide (a); "cardiovascular disease" AND "genetics", slide (b) *etc.* **a)** The total number of publications in the relevant areas; **b)** Publications dealing with genetic factors in human physiology, cancer and CVD; **c)** Publications, in which particular gene polymorphisms (related to human physiology in general, cancer and CVD, respectively) were studied.

Totally, there were over 16 millions of abstracts in Pubmed and about 8 million of abstracts in Embase, albeit these two databases do overlap quite a lot. The amount of papers indexed in PubMed that deal with human physiology exceeded 4 million (as of fall of 2005, Figure 3-1a, squares) and about 1 million of these papers dealt, in some or other form, with genetic issues in human physiology (Figure 3-1b). Although the relevant numbers of publications for the specific biomedical issues such as cancer and cardiovascular disease were considerably less than 4 million (1.7 mln and 1.2 mln, respectively), such an amount of publications well exceeds the abilities of the current experts in the field. Despite the common notion that a real "expert in the field" is supposed to have a comprehensive and updated knowledge, no current expert in the field of human physiology will, presumably, be able to read (not mentioning a thorough scientific scrutiny) just *millions* of papers.

Even in the case of pretty specialized sorts of information (such as studies addressing some particular genetic polymorphisms associated with particular multifactor diseases), the amounts of the related publications are measured in thousands for specific diseases (CVD, cancer) and in tens of thousands for the human physiology in general (Figure 3-1c). These polymorphism-related subsets of publications are vital for a number of biomedical and biotech applications such as gene testing, genetic diagnostics, pharmacology and pharmacogenomics (Chapter IV, Volume II).

Of course, author readily admits that the numbers cited above and illustrated by the Figure 3-1 are too general. A careful researcher definitely uses more precise search strategies. And, by the way, a *number* of publications isn't, exactly, an indicator of the amount of the *reliable* information, not to mention real scientific advancements. For example, according to a recent study, many of the published research papers in biomedicine are likely to be widely off the mark and represent somewhat misleading results: see, for instance, a news item in New Scientist, 30.08.2005 (http://www.newscientist.com/article.ns?id=dn7915) as well as the references in the Introduction to this volume. Nevertheless, the data shown in the Figure 3-1 definitely illustrate close-to-exponential growth of the amount of available information on human physiology as well as the steady growth of the pool of information on the genetic determinants of most common diseases such as cancer and CVD. Hence, the problem of managing all of this information is imminent while working on a research project in the field of human physiology.

Given these considerable amounts of information (and information of the kind which is supposed to be processed, sieved and analyzed by human experts), we might be able to realize instantly the danger of the wasteful rechewing of the thousands of facts *separately* and taking emotional stimulation in countless arguments of the academic style. Alternatively, there is also the opposite danger: of lumping together all those data without any regard for the quality of the evidence and with even less regard for the principles of the human physiology. Both sorts of these very common activities are performed, no doubt, at the expense of both the fundamental research and of the novel biomedical applications.

Clearly, in order to manage these vast and steadily growing amounts of physiological information, a correct structuring of the information is essential. At the very least, the structuring should be correct enough for a particular research purpose. To separate the gold from the dross, it is necessary to keep operating in mind not only a clear picture of some important research purpose, but also the most basic concepts of human physiology as well as the intricate relationships between physiology, genetics and biochemistry (which we have covered, in a most general manner, earlier in this book).

In this chapter, we'll walk through the world (largely a "paper-world", so to say) of searching, finding, selecting and analyzing the needed items in this overwhelming flow of information. The primary goal of this chapter is to consider the strategies for an efficient retrieval and assessment of the biomedical data. The following sections of the chapter will consider the general strategies for retrieval and assessment of the biomedical data, the major types of the cohort studies, the cohorts being studied and a number of other aspects of the study appraisal. Finally, while discussing *The Case of IHD200,* we'll consider a detailed example of an actual association study and the inherent problems that routinely arise.

SEARCH FOR AND RETRIEVAL OF STUDIES

"Where ignorance is bliss
'Tis folly to be wise"
T. Gray, 18th century

The entire issue of literature search might seem to be all too banal, especially for an experienced medical scientist. Nevertheless, elaborating and implementing adequate strategies in search of the published evidence is cornerstone of the purpose-oriented bioinformatics. In addition, what is obvious and common for a clinical scientist might be not so, say, for a geneticist and there will be a number of *caveats* for an over-confident individual who assumes this important issue to be "banal".

For example, "coronary artery disease", "ischemic heart disease" and "coronary heart disease" are largely synonyms for one and the same CVD condition (Figure 2-2). If, for example, a geneticist neglects this basic fact of the biomedical terminology, a significant proportion of the published data will be neglected. This situation is far from being a figment of someone's imagination. It becomes very prominent when one reads meta-analyses published in respectable academic journals and sees that most of these papers present no more than 1/3rd of the available evidence (Chapter V). This appalling lack of data might arise as a result of sloppy research. Alternatively, it might be the result of using an inadequate strategy of literature search.

We can illustrate the above point in numbers using another series of simplistic searches. In PubMed, by the fall of 2005, there were 2397 papers for the keywords "ischemic heart disease" and "polymorphism", 1960 papers for "coronary heart disease" and "polymorphism" as well as 1190 papers for "coronary artery disease" and "polymorphism". The three sets of the abstract do overlap and majority (88%) of the studies and can be found using the single keyword "ischemic heart disease". However, using any other single keyword will produce much less evidence (72% for "C.H.D." and 45% for "C.A.D"). If we try NOT-searches (such as "coronary heart disease NOT ischemic heart disease" or "ischemic heart disease NOT coronary heart disease" *etc*), we will also find that there are several hundreds hits for each of these NOT-searches (while, ideally, the number of these hits should be close to zero, since the three terms are synonyms). Clearly, the synonymous terms do not appear to be such according to the data indexed in PubMed. Accordingly, the most sensitive search should be inclusive: "(ischemic heart disease OR coronary heart disease OR coronary artery disease) AND polymorphism" *etc.*

The above example underlines the two major problems to be taken care of while searching published evidence: to include all the relevant information available at the moment and to exclude much of the irrelevant information. In other terms, these are the problems of *sensitivity* and *specificity*. Although even an issue as simple as this might be an important research problem in bioinformatics, it is often left out of view in medical studies. Thus, in biomedicine, there is little discussion and an exceptional dearth of studies dealing with the efficient strategies for the biomedical literature. At the same time, there are much more studies of the kind in other areas of science (say, in physics). This situation looks even more puzzling if we recall that biology and medicine feature one of the largest bibliographies among the natural sciences.

Occasionally, a few papers addressing the matter do get published. For instance, a number of important points relevant to the strategies of literature search were discussed in [1]. Among them, the authors mentioned several common psychological inhibitions that, apparently, prevent an efficient retrieval of information. One of these inhibitions is the exhilarating "90 second rule" which gives many opportunities to reflect upon the results of Pavlovian conditioning in human subjects. That is, many clinicians don't even try to find information unless they think they can do so in a minute and a half [1,2]. In other words, to reorganize various routine activities to look for the best evidence (which is already available) appears to be unthinkable or highly repugnant to most of those researchers.

Another stumbling block is that finding published evidence often appears to be "too easy" (perhaps, through the mechanism of assuming it to be "banal"). Compliance with this assumption inevitably leads to missing a significant proportion of the available evidence (as it was in the above example with IHD/CHD/CAD). Apparently, the proverbial principle of "ignorance is bliss" suggests that individuals that have similar attitudes to the search of literature are, perhaps, blissfully unaware of both that they missed at least half of the published evidence and that they are quite far from finding the *best available* evidence. These and other psychological blocks might follow from the lack of understanding that the proportion of useful information is, more than often, very small and quite a few of the studies found in a simple search would have immediate clinical relevance. Limitations in the basic skills of the scientific analysis, be they real or imagined, can also be considerable barriers to effective information retrieval [3].

At the same time, an efficient search of literature is not to be equated with a mere absence of these or other psychological inhibitions. It is also necessary to employ an appropriate information technology to search the evidence. Merely using a sort of a "free-text search" (such as the one commonly used in Internet search engines) is hardly enough for efficient extraction of the *reliable* information. In spite of a recently published small study that claims to employ Google to aid diagnostician (see [4] and the heated discussion at http://www.bmj.com/cgi/eletters/333/7579/1143#150402), there are strong reasons to assume that a free-text search made in this manner is extremely ineffective way of finding *the best* evidence. Firstly, it is a common knowledge that Internet search engines (even as excellent as the Google) do tend to offer extremely low *specificities*. In other words, there are too many false positives for a few really useful hits. Often, these useful hits are not among the 1st hundred but rather are scattered among the thousands of the hits. Secondly, the reliability of the information found often cannot be verified. So, a free-text search is not likely produce *the best* available biomedical evidence, be it papers or any other sort of data. And, thirdly, only a few of the biomedical papers can be accessed directly through Internet (since, obviously, many journals have password-protected subscriptions). Accordingly, using Internet search engines to extract biomedical evidence is a very non-efficient technique based on an assumption similar to the "90-second rule" we mentioned previously.

For the search to be efficient, it is not only necessary to know how to formulate an adequate set of search criteria in each particular abstract database and being aware that even the best text filters will not compensate for a poorly structured search strategy [1]. Often, a bioinformatician will find that it is very important to know the word usage that the papers featuring reliable evidence tend to feature. Only then, provided an adequate search strategy, will using the text filters be helpful in separating the wheat from the chaff.

Several particular examples of large-scale assessments of the most efficient strategies crafted for particular purposes can also be found in the published literature (see [5,6] and the references therein). In these studies, the authors attempted to extract scientifically strong studies using a sophisticated set of text filters or "hedges". In the paper [5], combinations of search terms reached peak sensitivities of 99% at specificity of 74%. A large-scale analysis of optimal search strategies (sensitivity, specificity, precision, and accuracy of 4862 unique terms in 18 404 combinations) [6] has shown peak sensitivities of 99% at specificity of 70%.

Studies of the kind often require a sort of a "training set" of the reliable publications on the base of which the initial keywords, the word combinations and the strategies are generated. Then, the most appropriate search terms are selected on the basis of this "training set" and then are further corrected by taking into account peculiarities of indexing in particular abstract databases. Finally, the strategies of search are formulated and evaluated. Each of these stages is required to arrive at sound search strategies. At the stage of *compilation of the training set* in the study [5], the authors searched for specific kinds of studies that deal with evaluation of the diagnostic techniques. The methodological criteria for selecting studies into such "training set" were quite elaborate: inclusion of a range of participants; use of a current clinical standard for diagnosis; participants receiving the new test and some form of the diagnostic standard; interpretation of diagnostic standard without knowledge of test result (and vice versa) as well as analysis of the results consistent with study design. At the stage of *selection of the appropriate search terms* in the study [6], the PubMed's MeSH terms and the text words related to study criteria were compiled into list. Then, this list was presented for independent assessment to clinicians, librarians and other experts for corrections. As the result, *formulation of the efficient search strategies* that will extract the most reliable clinical trials was done by using iterative analysis of the search terms and their combinations. Apparently, the actual issue of the efficient search of published evidence is not as "banal" as the frenzied adepts of the "90-second rule" would assume…

Particular strategies that were obtained in the two studies we cited above are most appropriate when, indeed, a clinician is about to search particular thing: publications of evidence-based clinical trials of reasonable scientific quality. When, however, a bioinformatician embarks on a PubMed search in the scope of purpose-oriented bioinformatics, the criteria similar to those so nicely formulated in the above publications can often become too rigid or outright irrelevant to retrieve the required studies. For example, the two studies cited above dealt with evaluation of the diagnostic procedures. Since genetic testing is a potential diagnostic procedure, someone might be tempted to apply directly the terms formulated in the studies to find "the most reliable evidence" among tens of thousands of the genotype-phenotype association studies. Every reader is cordially invited to try the strategies. In author's experience, however, directly applying the terms from [5,6] to search for reliable association studies shows low sensitivities (less than 60% of the studies of interest) and even lower specificities (less than 40%). Accordingly, one needs to beware of the improper generalizations of the search terms formulated in the evaluation studies.

Anyway, publications similar to [5,6] are quite instructive in showing that what might seem a real simple task turns out, after due consideration, to be a branch of bioinformatics research. First of all, these studies formulate the strategies to arrive at the efficient methods of literature search. Secondly, these studies allow to notice and to deal with some general problems. One of these problems is the language change known among linguists as "language evolution". Especially in biomedical area, the rates of this evolution are incredible: terminology can substantially change during a decade, not mentioning longer periods of time. The matter is further complicated by the facts that indexing terms and methods are also modified over time and that only a few changes are implemented retrospectively, even in such exceptionally well-kept databases as PubMed. In addition, indexers of the abstract databases often choose only a small number of terms for each abstract they index and the indexing is not always exhaustive as our example of the IHD/CHD/CAD already pointed out.

Clearly, a biomedical scientist usually does not need to go in all of these ramifications of selecting the most efficient strategy for literature search. It is, however, very important for bioinformaticians in the post-genomic era, when the amount of the already available information (Figure 3-1) makes adoption of the efficient search strategies a professional necessity. An important piece of experience that bioinformatician can share with the rest of biomedical scientists is that it is often very helpful to elaborate the search criteria on fly as it were. That is, during the time while one is searching for some particular information, using the previously made searches as a mental feedback. This experience, apparently, entirely abrogates the funny "rule of 90 seconds" and underlines the necessity of bio-logical thinking.

The efficient strategies of literature search were important for a number of the Case Studies as well as other bioinformatic projects we mention in this book. For example, the efficient literature searches were important in compilation of the HPRD database to find the tissue expression patterns genes of interest, to retrieve data on the known protein-protein interactions and the data on post-translational modifications [7] (we discuss HPRD and other databases of the kind in the Volume III). Similar strategies were elaborated and implemented for the Case Studies in the Chapters II, IV and VII. Among other examples available in this volume, the *medSNP* database (next chapter) was compiled after elaboration of the efficient search strategies to extract the reliable association studies for particular genes and polymorphisms. The *CVD GeneScape* diagram featuring in the previous chapter was compiled using *medCrawler* script that implements search strategies we mentioned here. A complex keyword search technique, which we jocularly called as "*in silico* gene expression analysis", was essential for the formulation of the research hypotheses linking venous thromboembolism and physiological effects of estrogens (Chapter VII).

Once studies of interest are retrieved, there comes the assessment of the actual quality of the evidence presented in those studies. There are different kinds of biomedical studies and a study of certain kind can be assessed only with an appropriate set of criteria. Since one of the major themes of the present volume is usage of the genetic polymorphisms in clinical science, this and the following chapters will contain numerous details pertaining to the assessment of the genotype-phenotype association studies.

Assessment of Biomedical Studies

As we might have noticed while discussing the efficient search strategies, retrieval and assessment of the studies seem to be interrelated. So are the social, biomedical and statistical considerations regarding significance of a particular biomedical study. From a social point of view, one can never "assess" quality of research by considering publication date, "prestige" of the author or "impact factor" of that paper or of the journal. When dealing with large-scale analyses of the published data, it becomes clear that many poorly designed and scientifically flawed studies do appear in "high-impact" journals or published under the guise of some "prestigious" individuals. And, by the way, so-called "impact factor" has nothing to do with the actual quality of the particular study. Assessment of the quality of research by using the "impact factors" is an antiscientific practice which, nevertheless, appears to flourish in biology and medicine, despite the clear warnings of the originator of the "impact factors" [8].

Moreover, it was hypothesized that poor study design, poor informational management, researcher bias and the usual difficulties in getting the research published make a significant proportion of the published research if not wrong but certainly questionable [9,10]. At the same time, the procedures of anonymous peer review have long been criticized for failing to identify flaws in research as well as for finding flaws that are not there [11]. For example, spurious criticisms of statistically sound data are common and appear to proceed mainly from reviewers with limited knowledge of statistics (such is the case of the grant reviews in biology and medicine). This most common type of the peer reviewer likes, apparently, to make unfounded statistical criticisms, especially in such inherently controversial areas as "sample size" and "multiple comparisons" [11]. Sometimes, the "anonymous" component in peer-review procedure outweighs all the others thus leading to further distortions and extortions [12]. To round up, we'll just coin another intriguing quote: "The authors affect the scientific style of the disinterested observer when in truth they are advocates. The reader-deterring style in which most scientific papers are written has evolved because they are written not to be read but to be published" [13].

If we are to abstract from these petty, but very common and certainly counterproductive social aspects while dealing with the assessment of biomedical research, we'll have to admit that not all of the authors who actually publish anything would be monomaniacs who want to see their name in print. Neither most of them are paranoiacs with uncontrolled suicidal drives to get published in a journal or to perish. Neither all of the peer reviewers are "failed scientists with personality disorders" [14]. Indeed, a sense of the rapport with the writing style of a particular study does, sometimes, indicate the general "quality" of what is presented to the reader (including indications of the psychopathological aspects we just mentioned, see [13]). A proportion of well-designed, well-made and well-analyzed biomedical studies does get published so biomedical and statistical considerations are, to say the least, not less important for the further assessment of a published study than the automatic thinking in terms of "prestige", "impact factor" *etc.* In the following sections, we consider general criteria pertaining to the assessment of the quality of biomedical studies. The topics include the use of hierarchies, questionnaires, general kinds of biomedical studies, appraisal of observational studies, study cohorts, statistical considerations pertaining to the significance of the findings as well as a brief consideration of the possibility that particular findings were forged.

STUDY ASSESSMENT: THE USE OF HIERARCHIES

"Nothing matters very much and
very few things matter at all"
A. Balfour

Both the informational overload in the modern biosciences, coupled with the social difficulties we just mentioned suggest that the principle we cite as the epigraph to this section is likely to have a place in the assessment of the significance of biomedical research. This principle, attributed to an English statesman, was, presumably, first formulated in relation to political issues. Still, it's quite applicable to the issue of the assessment of the biomedical studies. Therefore, in order to be able to recognize a valid piece of evidence, we will need certain criteria that would allow us to judge, at the very least, the relative quality of the evidence. These criteria might range from a rule-of-a-thumb to more elaborate sets of criteria peculiar for a particular kind of study.

Use of hierarchies of biomedical evidence is, perhaps, the most apparent rule-of-a-thumb. A "hierarchy" or ranking of the general types of the biomedical studies provides a general scheme for the preference of the various types of biomedical evidence. The current hierarchies were formulated and have been used almost exclusively to determine the effects of interventions [15]. A hierarchy states that systematic reviews of the published research (i.e., the publications that attempt a comprehensive analytic overview of the currently available evidence) have precedence over a particular randomized trial or a meta-analysis. Observational studies, small group studies and case reports have even lower positions in the hierarchy of the biomedical research. Depending on particular research purpose, the rankings might be somewhat adjusted thus producing different hierarchies. The major problem of using hierarchies is that the simplifications involved in creating and applying hierarchies have also led to a number of misconceptions. In particular, some of the specific criteria that were originally designed for the intervention studies have been indiscreetly applied to questions about etiology, diagnosis and prognosis [15], while, apparently, entirely different kinds of hierarchies are required for answering these questions.

Nevertheless, a hierarchy does provide some general guidance in judging the evidence available on particular subject. The type of publication known as "systematic review" is at the top of most of the hierarchies. A systematic review is deemed to be the most reliable kind of biomedical evidence. This kind of publication summarizes the medical literature on a particular point of interest and involves a critical appraisal of the individual studies as well as of the entire set of data. The major advantage of a systematic review is that it does not take in isolation any particular study but rather provides a wider overview of the published evidence. The major weakness of a systematic review is that such a review might sometimes contain some cherished conjectures of the author(s) which hardly bear any significant clinical support ("non-falsifiability syndrome", Introduction). Another problem is related to the informational overload and is quite apparent: only a tiny fraction of the hundreds or, sometimes, thousands of the relevant research papers are considered for possible inclusion in systematic reviews [15]. Often, a systematic review might not even feature any description of the criteria that were actually used to find, to select and to analyze the evidence reviewed.

It is also clear that for most of the time, a systematic review (or, at least, a meta-analysis) on the subject of interest will not be available. In this case, to answer a therapeutic question, a hierarchy would suggest looking for randomized controlled trials. If the literature search fails to identify any randomized studies, observational studies are to be looked for. For non-therapeutic questions (as it is the case of the genotype-phenotype association studies) search strategies should accommodate the need to look for observational designs that are able to answer the questions of interest pertaining to etiology or diagnostics. In the absence of observational studies, case reports on particular point of interest can be used. In the case of intervention/treatment studies, case reports are important for alerting the researcher to potential rare harms or benefits of an effective treatment. In Volume V of the present book series we consider how clinical case reports can significantly contribute to biomedicine of a particular disorder while we discuss the maps of protein structure and function.

Given the flaws in the current hierarchies of biomedical evidence, it is sometimes suggested that to deal with these flaws might require abrogation of the entire notion of the hierarchies. Instead, general principles of research conduction and assessment should be attended to so that they can be applied to appraisal of the quality and relevance of particular studies. This kind of approach is especially important in the post-genomic era when it becomes necessary to broaden the scope by which biomedical evidence is assessed [15]. Although systematic reviews will always be necessary, independently on the kind of research, clinically significant conclusions can be arrived at only when such a review involves a variety of types of research. We will touch again the issue of the systematic reviews when we'll deal with meta- and ortho-analyses of the published research on genotype-phenotype assocaitions in the Chapter V.

Study Assessment: Questionnaires

Clinical observations can be roughly divided into the two main groups: the doctor-made (symptoms, physical examination, lab tests, modern imaging methods *etc*) and the patient-made (perception by the patient of his or her state of health, most apparent complaints, self-reported life style, diet *etc*). Although the points pertaining to life-style and diet can often be deduced by the doctor from his or her observations, much of the information concerning these issues is derived from questioning the patient by the doctor or by patient's filling of a specifically designed questionnaire. In the framework of an association study, questionnaire data often provide an opportunity for cross-comparisons with other sets of biomedical data. Accordingly, a carefully designed questionnaire is an important part of a genotype-phenotype association study.

Sometimes it is said that the great popularity questionnaires have is because they provide a "quick fix for an actual research methodology" [16]. Indeed, questionnaire data may be especially important when no precise physiological mechanism for the disease is established or can be deduced. Questionnaires offer a means of collecting information about patients' life style attitudes, diet and behavior. When properly designed, questionnaires can be used as the sole research instrument (as in the case of a cross-sectional survey) within clinical trials or epidemiological studies. However, no single method has been so abused and despite a wealth of detailed guidance in the specialist literature, brutal methodological errors in questionnaire design and application are very common. The paper [16] provides several general guidelines concerning elaboration of questionnaires and analysis of the questionnaire data.

The most important point is that reaching a consensus among independent researchers from different fields is one of the most crucial issues in the questionnaire design. This is especially important when a researcher is not well versed with the clinical science of interest. It might also be helpful to use some pre-existing questionnaires, if there are such. This principle, however, can be as abused as any other if followed to the exclusion of anything else, as the HTR2A example illustrates (see below).

Assessment of the validity and suitability of a particular questionnaire for a particular study is another important point. A valid questionnaire should measure what it claims to measure. The paper [16] cites a number of examples of inadequate questionnaires. For example, a self-completion questionnaire that seeks to measure people's food intake may be invalid because it does not have cross-checks to ensure that it assesses what they actually eaten and not what they just say they've eaten. As shows almost a century-long history of psychological research (which is, 90% of time, is based solely on questionnaire data), an efficient questionnaire will contain questions that will verify and cross-reference the statements alleged by the questioned. This intricacy in questionnaire design suggests that a questionnaire developed for a different country or a culture will not be a valid measure just for any other group under study. It is also clear that in order to be precise, a questionnaire should avoid using multiordinal terminology such as "commonly", "usually", "many" *etc*.

Thus, in relation to the assessment of the study, it is important to look at the quality and the relevance of the questionnaire used for a particular study. This point is not of small significance. For instance, there are dozens of the phenotype-genotype association studies that claim to link a polymorphismin the HTR2A gene with schizophrenia. The most schizophrenic, if the readers would pardon the expression, about these studies is that diagnosis "schizophrenia" was almost exclusively made on the basis of... a questionnaire! "Schizophrenics" were, thus, defined not through clinical records, not using the standard DSM-IV manual but simply a series of questions in a self-filled questionnaire. These papers, published in respectable journals, look as if their authors deliberately shunned contacts with the mental health professionals and purposefully avoided using the standard diagnostic manual well known world-wide. Instead, they decided to use an obscure questionnaire. Moreover, these studies appear to use the *same* questionnaire which, after closer inspection and consultations with several mental health practitioners, was found to be extremely inadequate. Some of these practitioners even gone as far as filing protests to the editors of the journals where these alleged "schizophrenia studies" were published. If the geneticists who bluntly hacked these studies without any regard to psychiatry, physiology and medicine would be more aware of the problems and abuses common in the questionnaire designs [16], they surely would have consulted the actual psychopathologists prior to rushing with publishing or perishing. *Ergo*, they, probably, won't be lowering their professional prestige by publishing such low-grade papers of extremely doubtful scientific quality.

STUDY ASSESSMENT: APPRAISAL OF OBSERVATIONAL STUDIES

In the most general manner, the studies in biomedicine can be subdivided into careful studies of a few individual patients ('case studies') and the studies that deal with groups of many patients and/or healthy individuals ('cohort studies' or 'population studies'). The cohort/population studies are of the two major kinds: randomized and observational. A randomized double-blinded study is currently considered to be a "gold standard" of a population study in medicinal treatment, while a prospective observational study appears to be a "gold standard" in epidemiology.

From practical point of view, a randomized study is most suitable for testing efficiency of medical treatment, be it effects of a drug or a set of treatment procedures ("intervention study"). In a randomized study, at least one of the variables can be explicitly manipulated. In an observational study, the variables of interest are observed and not manipulated. The bulk of the published randomized and observational studies happen to be intervention studies. Therefore, it is clear beforehand that the details of assessment of these studies are quite far from the actual problems in genotype-phenotype association studies. Nevertheless, general ways of assessing the cohort studies in evidence-based medicine do have some implications for the assessment of the genotype-phenotype association studies, as we'll see further.

The observational and randomized studies represent two different kinds of the study designs which are often considered as mutually exclusive methods of clinical investigation. Often, a randomized study is considered as having higher priority (in other words, having a higher rank in a hierarchy). A few of the recent reports (reviewed in [17]) question this common dogma. In these reports, both randomized and observational studies had been used to evaluate the same clinical question. In contrast to the expectation that randomized studies would produce more reliable estimates of the treatment efficiency, comparative analyses indicated that observational studies did not overestimate the size of the treatment effect compared with the randomized counterparts. On the contrary, both kinds of studies appear to produce largely concordant results. Moreover, the concordance has been worse when the results of the large randomized trials on the same topic were compared among themselves while the respective observational studies even tend to have similar odds ratios among themselves [18]. These results, of course, do not challenge the current opinion about the hierarchy of study designs in clinical research. They do alert us, however, to the possibility that well-designed observational studies might not systematically overestimate the magnitude of the associations as it has been so often advertised.

A general comparison of these two main types of studies is presented in the Table 3-1. The foremost advantage of a randomized controlled study is that the random allocation of the treatment procedure allows explicit control of at least one major variable. This allocation by chance is supposed to mean that the groups compared are similar in terms of both measured and unmeasured baseline characteristics. In other words, a randomized study is often deemed to be "more objective" only because the random allocation of the treatment is assumed to reduce selection bias and confounding. At the same time, randomized studies are expensive, often pose problems with recruiting the participants and not infrequently tend to raise a number of questions concerning scientific integrity. Randomized studies are also relatively short-term thus the long-term effects of the treatment are neglected almost entirely.

Table 3-1. A comparison of observational and randomized studies (after [19])

Characteristic	Observational	Randomized
Populations	Populations of patients under various settings	Population 'randomly' selected on the base of a strict set of demographic and other criteria
Outcomes	Can be defined *a posteriori* and might include unexpected outcomes	Strictly defined *a priori* in relation to the risks or protective factors of interest
Follow-up	Both retrospective and long-term prospective	Only short-term follow-ups of less than a year (costs, social pressure to produce results)
Analysis	Multivariate techniques for the analyses of the 'confounder' variables	Straightforward $\chi 2$ and *t*-tests

As for observational studies, despite the number of their distinct advantages (Table 3-1), they are generally assumed to be more susceptible to the selection bias and confounding (due to possible differences in what is called baseline characteristics such as gender, age, intake of drugs *etc*). It cannot be denied that genetic association studies are much closer to the observational cohort studies than to randomized studies since the most important variable (*i.e.*, the genetic makeup) is something that can hardly be explicitly randomized (still, see our 'scrambling' exercises in the Volume II). We summarize the types of the observational phenotype-genotype studies in the Chapter IV. Here, let's consider a few important points that had to be kept in mind while dealing with any kind of an observational study.

Observational cohort studies can be thought of as natural experiments in which real outcomes are measured in real world rather than in an artificial experimental setting. A flexible design of an observational study allows evaluations of large groups of diverse individuals, long-term follow-ups and detailed characterization of a wide range of outcomes [19]. Dealing with an observational study, the foremost point of attention is on the *definitions of the groups of subjects*. A well-designed and well-written observational study should contain unequivocal definitions and the selection criteria of each of the groups involved in the study. Not only the statistical significance but also the clinical relevance of all the comparisons made must be thoroughly discussed within the study.

There are several important terms relevant to the appraisal of the observational studies: selection bias, confounding and analytical strategies. A ***selection bias*** consists in the possibility that 'intervention' ('case') and 'comparison' ('control') groups may differ in some characteristics that affect the study outcome. If these characteristics have independent effects on the observed outcome in each group, they will produce differences in outcomes between the groups, independently of the factors of the primary interest. This effect is known as *confounding* and the characteristics affecting the outcomes are *confounders* [20]. The potential sources of the selection bias and the potential confounders need to be analyzed prior to attempting any analytical strategy to deduce the results of primary interest.

Selection bias occurs when there is an inherent difference between the factors characterizing each of the groups under consideration. For example, we have seen that there is a variety of the risk factors for CVD (Chapter II) which act at different levels of function (Chapter I). If we study an association of a genetic polymorphism with a disease or some other aspect of physiological interest, many of these factors can independently influence the result of the study thus becoming a confounder of the study. Accordingly, the most reasonable strategy to detect and then to account for selection bias is to characterize each of the groups in accordance with as much of the significant factors as possible. Although the most commonly used factors are age, gender and familial history, we already mentioned quite a lot of factors deemed to be "unusual" as factors of CVD risk which can become potential confounders (Chapter II).

In statistics, ***confounding*** is a mixing of effects of two or more independent variables on the response variable. In other words, the contributions of each of these variables on the response variable cannot be distinguished from each other. When a confounding factor is recognized, adjustments have to be made to ensure that confounding does not distort the major results of the study. For a characteristic to be a *confounder* in a particular study, it must meet two criteria. Firstly, it must be known to influence the outcome of interest, either in terms of prognosis or susceptibility. This statement is applicable, for example, to the factors of CVD risk (Figure 2-3). Secondly, the distribution of this characteristic should significantly differ between the groups compared. For example, for age to be a confounder in an observational study, either the average age or the variation in age among the groups under comparison would have to be different [20, 21]. While in a randomized study all potential confounders, whether known or unknown, are supposed to be evenly distributed between the groups being compared (because of the randomized selection), observational studies are, at least formally, more prone to some unknown confounders. The problem of unknown confounders will especially plague the large-scale studies of the post-genomic era leading to false associations, negative associations and other statistical flukes (Volume II). Once again, a careful analysis and data collection on all the known factors of risk (the potential confounders) can significantly reduce the confounder problem. Statistically, the effects of the potential confounders on the outcomes are analyzed by using common significance tests such as $\chi 2$-test and t-test for dichotomous and continuous variables, respectively.

Regression and stratification are the two ***analytical strategies*** that are used to analyze both the confounder effects and the significance of the findings in observational studies. A *regression analysis* attempts to model relationships between one or more response variables ('dependent variables') and the predictors ('independent variables') using a linear or a non-linear equation. In the case of confounder analysis, regression can estimate to what extent the potential confounders biased the outcome. The validity of the results of a regression is based on specific assumptions such as constancy of the effect across the groups analyzed and identity of the baseline characteristics between the groups. *Stratification* means a process of structuring into layers ('strata'). In biomedical statistics, it implies grouping members of the group/cohort into more homogeneous subgroups ('strata') before attempting to calculate the effects in the framework of an association study. These subgroups should be mutually non-overlapping and exhaustive (that is, the entire set of the subgroups includes the entire cohort). During confounder analysis, the sample cohort is divided into subgroups or strata using the list of potential confounders. Then, the effects of interest are calculated for individual subgroups using univariate and multivariate analyses. As patients are grouped into strata according to the potential confounders (which would be age, gender and more specific CVD risks), we can expect that these subgroups are more balanced in terms of the confounders. Formally, since the total number of participants in each subgroup is reduced, stratification worsens statistics of the study. In actual studies, both regression and stratification strategies are used. In order to assess an observational study, it is necessary to note in what manner these strategies were used, how clearly they were described, whether they produced consistent results and what is plausibility of the results in respect to confounders and the effects. We summarize the general criteria we just discussed in the Box 3 and illustrate an application of these criteria on the example of a particular genetic association study (ref. [22]) in the Box 4.

Box 3. General criteria of assessment of observational studies

- Were groups clearly defined?
- How precise were the definitions/selection criteria?
- What were possibilities of selection bias?
- What were potential confounders?
- Were at least some of the potential confounders analyzed?
- What was the stated magnitude of the effects?
- Were the analytical strategies clearly described?
- Did the different analytical strategies produce consistent results?
- Were the results clinically plausible?

STUDY ASSESSMENT: THE STUDY COHORTS

While considering the general criteria for the assessment of observational studies, we might have noticed that an adequate selection of the study cohort is crucial for obtaining statistically significant and scientifically sound results. Indeed, a genetic factor is a life-long risk. From informational point of view, genetics is a "weak signal" in any periods of time shorter than the life span (or at least, several decades). In science, detection of any weak signal requires an excellent signal-to-background ratio or, alternatively, a long period of the registration. Both these approaches aim at lowering the levels of the "noise". In case of a medicinal study, this noise is the action of the factors other than those considered in the study (*i.e.*, so-called "unknown confounders"). The "low noise" setting is often the primary aim in physics and other exact sciences. However, only rarely a "low noise" setting can be achieved in a clinical study, be it randomized or observational. Treatment of the inevitably high levels of the noise in a biomedical study is only partially alleviated by stratification. Thus, a long-term observation (including post-mortem examination) of a cohort of patients is, perhaps, the most reliable way to improve the signal-to-background ratio in an observational study such as a genetic association study.

Box 4. A brief study of a study

Here, we illustrate the application of the general criteria (Box 3) to a particular phenotype-genotype association study. The paper under consideration [22] features a decent combination of post-genomic technologies (DNA biochips, image scanners, multiplex genotype determination through minisequencing) with a reasonably well-designed clinical study. This cross-sectional survey is done on Finnish subjects that represent genetically stable and relatively isolated population. Despite a few minor weaknesses, it is an exemplary genotype-phenotype association study of the kind not so often found in the area.

Were the groups clearly defined? 'FINRISK' participants (http://www.kttl.helsinki.fi/content.asp?route=4.750.4920.5323), an elaborated questionnaire, the study protocol was the same as that established by the 'MONICA' project. The cohort is small-to-medium (152 patients, 152 matched controls).

How precise were the selection criteria? History of MI according to InternationalClassification of Diseases 9 ed. (ICD-9, http://www.who.int/classifications/icd/en/), diagnoses 410 and 412 for cases, absence of the diagnoses for the controls.

What were possibilities of selection bias? The main possibilities include age, gender and geographical area. The study group is strongly skewed towards males (80%). The stratifications were analyzed.

What were the potential confounders? Were at least some of the potential confounders analyzed? Gender, age, region, genetic heterogeneity and the most common CVD risks. Genetic heterogeneity was, in part, taken care of by common CVD-related polymorphisms (8 genes) and by analyses of monogenic disorders common in this population (LDLR variants that account for at least 75% of the familial Finnish hypercholesterolaemia). The selected MI cases and controls were matched for age, gender and region. BMI, diabetes, HDL-cholesterol and triglycerides were other potential confounders analyzed.

What was the magnitude of the effects? Unadjusted OR for the GPIIIa variant was 1.73 (95% CI: 0.96-3.19); for the PAI-1 variant OR=2.1 (95% CI: 1.15-3.84); gene-gene interaction OR = 4.5, P = 0.001.

Were the analytical strategies clearly described? t-test, $\chi 2$ test, Hardy-Weinberg, pooling of the genotype data, parameters of logistic regression described.

Did the different analytical strategies produce consistent results? Were the results clinically plausible? Adjustments for BMI, diabetes, HLD-cholesterol and triglycerides stratifications did not significantly affect the regression. The differences between the cases and controls remained significant after restricting the study subjects to males only. For the APOE gene, a clear trend of MI risk-increasing effect of the APOE*4 allele was observed.

Having said this, we admit that the long-term-observation is not the only way to improve the signal-to-background ratio in a clinical setting. The noise can be reduced by studying relatively isolated populations (as it was the case of the Finnish study, Box 4). This principle can be applied in a number of other ways. For example, one cross-sectional study selects as the study cohort the elderly Caucasians from a population living in a secluded valley [23]. The study analyzes the implications of the C-344T polymorphism in CYP11B2 for the blood pressure. In view of what we mentioned in the Chapter II concerning urbanization and higher incidences of CVD, such a study cohort is, indeed, of great interest.

Carefully crafted questionnaire, analysis of the medical history, anthropometrics, biochemical tests along with analysis of the polymorphic variants clearly suggested that in this highly homogeneous population sample the aldosterone-to-renin ratio was 19% lower among -344CC genotypes than among -344TT genotypes. A decent signal-to-background ratio that characterizes the study also allowed to establish a dose-dependent correlation (Chapter V) between systolic blood pressure and the number of the C-alleles in the genotype: thus, blood pressure was significantly lower in subjects with the CC genotype, higher in the TT and intermediate in the CT subjects [23]. Adjustments for age, gender, smoking and antihypertensive treatment as potential confounders did not affect this association. Another interesting finding was that a significant increase of systolic blood pressure with increasing age and with increasing skinfold thickness was observed in the TT homozygotes but not in the C-carriers. These and many other positive associations reported for this quite peculiar cohort within just one study [23] underline the necessity to subtract, figuratively speaking, the higher levels of the physiological noise that characterizes most of the urban subjects. Unfortunately, accurately done genetic association studies of the kind are extremely rare.

The drugs particular patients take can also be significant confounders of a study leading to null or insignificant associations in the framework of a genotype-phenotype association study. Theoretically, the "noise" produced by the drug treatment can be reduced by analysis of the untreated patient groups. Studies of the kind are also exceptionally rare, although they do exist. Let's take the CYP11B2 polymorphism as an example. In a French study [24], the presence of the C-allele corresponded to significantly higher heart rate and lower stroke index. The results are consistent with a local role of endogenous aldosterone on both heart and vessels in the absence of the effects of the drug treatment [24]. This particular study also illustrates how the various levels of function fit in when an association study features both proper management of information and also reasonably accurate experimental work.

These studies directly aimed at the reduction of the statistical noise in the study cohort. Attempts such as these are very rare. More common are retrospective and long-term prospective studies of a particular cohort of subjects. There are many cohorts of the kind. The Framingham/Framingham Offspring cohort, is, perhaps, one of the best known in the area of the genotype-phenotype associations related to CVD. This cohort(s) includes North American Caucasians only [25] and studies of it allowed confirmation of a number of important genotype-phenotype associations in this Caucasian population for the genes of the coagulation cascade (F7, PAI1, ITGB3), apolipoproteins and lipases (APOA1, APOA5 APOB, APOC3, LIPC), MTHFR *etc.*

In the framework of a prospective cohort study, the reduction of the statistical noise can be achieved by selecting the groups of patients in direct relevance to the outcome of interest. A carefully planned selection and analysis can reliably detect associations with genetic factors even in the absence of other conditions. For example, in a Dutch study relationship between a number of common polymorphisms and the gain weight was investigated [26]. As cases, the patients who gained average of 12.8 kg during a 7-year follow-up were selected as 'cases' while the matched 'controls' of the same cohort were those who did not show a significant weight change during the same period. Albeit analysis of questionnaire data did suggested that weight gainers were largely those who consumed 'more savory snacks' and were 'less active during leisure time', the procedure also allowed detection of two genetic factors of the weight gain: G16R of ADRB2 gene and K109R of LEPR gene.

Despite that prospective studies dealing with the same cohort are more common than the other examples we cited earlier in this section, the cohort studies of the sorts still comprise only a small percentage of all the available studies of genotype-phenotype associations. The most common procedure for making a cohort for an association study is to select a number of consecutive patients characterized by the presence of certain multifactor disease (say, CHD) and then to select the age- and gender-matched controls. Often, the patients of the same clinic who do not have the disease registered in their medicals are selected as controls. Alternatively, a population sample of healthy blood donors is selected as a control group.

The fact that this procedure is the most common does not in any way suggest that it is the most proper or even valid. It's apparent that unless more strict criteria of the cohort selection are enforced, this procedure is bound to result in an exceptionally low signal-to-background ratio. Accordingly, the results obtained on different cohorts selected in such a way will be, for the most part, incompatible and contradictory ("elephant-in-the-dark" syndrome, Introduction)

Indeed, the heterogeneity of the different study cohorts is often put to be one of the main reasons behind the contradictory results obtained in independently made genetic association studies. A meta-analysis technique (Chapter V) can hardly be considered as a substitute for a comprehensive analysis of a prospective patient cohort made during several decades. As these studies are rare, what can be done in the framework of an average association study? *The Case of IHD200,* presented at the end of this Chapter, illustrates some of the answers. Large-scale analyses of the thousands of reliable associations, such as the analyses of the *medSNP* database (Chapter IV) can also provide some general recommendations concerning the optimal planning strategies of the association studies (Volume II).

STUDY ASSESSMENT: STATISTICAL ASPECTS

"Do not worry about your difficulties in mathematics. I can assure you mine are still greater."
A. Einstein

The view that puts statistics to be the only objective tool of scientific investigation is common and has been around since the differentiation of statistics into an individual discipline of numerical mathematics. Statisticians, however, generally tend to avoid discussing the relationships between the statistics and the scientific method. What is "statistical method"? Is it the same as "scientific method"? The reader might refer, for example, to the paper [27] for an excellent discussion concerning the place of statistics within much larger context of scientific research. This paper describes the general structure of a statistics-based scientific experiment on the example of the famous Michelson's study of the light speed.

In brief, statistics is *not* to be identified with the scientific method (Volumes V&VI). Statistics is, rather, a formalized approach to investigate phenomena as they relate to collections of units in order to obtain numerical estimates of the unknown 'true values'. The five stages of the *statistical method* (problem, plan, data, analysis and conclusion) do provide a practical and a uniform framework for managing scientific investigations [27]. Such a framework can ensure that important issues were, at least, considered. However, clearly not all kinds of the scientific study will fit into the structure of the statistical method. For instance, the case reports in biomedicine (or any other science, for that matter) do not study a 'collection of units'. Rather, they study a single 'unit' but study it in greater detail, by means of using substantiated logic and fundamentals of the particular science. Generally, statistical methods are not likely to be successful without careful exploration of the data [27]. Statistics is very common and useful method employed in scientific research but it is only one of the methods.

Having said this, we have to admit that most of the data on the genotype-phenotype associations are available within the framework of the scientific studies that are strongly biased towards using the statistical argumentation. This bias is quite understandable if we take into account the high levels of both the inherent and the apparent complexity in biology (Volumes V&VI of the present book series) as well as the systematic lack of the fundamental theories in contemporary biology. Accordingly, the straight-forward use of logic, justified and productive in the case of "exact sciences" is not always possible in biology and medicine. At the same time, statistics provides a certified instrument for assessment of the significance of associations and magnitude of the effects. In this section, we consider some of the general statistical criteria commonly used in the clinical research within the framework of a population study. Then, we consider some of the problems that purely statistical approach might engender (and usually does).

P-value is one of the oft used measurements of the statistical significance of particular findings. However, P-value *per se* cannot be any kind of the criteria of significance of a particular finding. Sir R. Fisher, the originator of the P-value testing, saw P-value merely as a simplistic index measuring the strength of evidence against the null hypothesis. The smaller is the P-value, the stronger is the evidence against the null hypothesis (*i.e.*, negative finding). As a general rule-of-a-thumb, he proposed $P > 0.05$ to be a standard level for concluding that there is some evidence against the hypothesis tested. Dr. Fisher also strongly argued that interpretation of the actual P-values is entirely at the hands of particular researchers and is in no way any sort of "objective evidence".

Later, his cautions were apparently forgotten and the magic rule "$P<0.05$" became a rigid dogma in accordance, apparently, with the doctrine-of-signatures symptom (Introduction). It is out of question that an arbitrary division of results into "significant" or "non-significant" in accordance with this numerical parameter was the intention of the founders of statistics [28]. Results of medical research should not be reported as "significant" or "non-significant" but should be interpreted in the context of the biomedical type of study without leaving out of view other available evidence (including the analysis of the confounders and the other points we summarized in the Box 3).

Odds ratios are used as a common measure of the effect size and are routinely reported in various biomedical studies. An odds ratio (OR) reflects how many times 'cases' are likely to have certain trait than the 'controls': say, an OR>1 indicates the trait is associated with the presence of the disease, while OR<1 indicates a protective trait. There is recognized problem that odds ratios do not approximate well to the relative risk when the initial risk (that is, the prevalence of the outcome of interest) is high [29]. The *relative risk* of a disease event is the number of those who experience the event divided by the total number of people at risk of having that event. If the odds ratio is interpreted as a relative risk, the effect size will be somewhat overstated. However, serious divergence between the odds ratio and the relative risk occurs only with large effects and in groups at high initial risk. Therefore, qualitative judgments based on interpreting odds ratios as though they were relative risks are unlikely to be seriously biased in most of the cases.

Odds ratios are always supplemented by the **confidence intervals** that independently reflect the statistical quality of the data in the study. In statistics, a confidence interval (CI) at a specified level of confidence characterizes the data dispersion inherent for the study. Usually, 95% confidence intervals are used. The wider is the confidence interval, the greater is the data dispersion of the study and the worse is the statistics of this study. When used along with odds ratios, both the high and the low values of the confidence should be either higher than 1 or less than 1. However, no rigid assessment can be allowed in this area, too. For the example cited in the Box 4, unadjusted OR for the GPIIIa variant was relatively high (OR=1.7) and the confidence interval was relatively narrow (95% CI: 0.96-3.19) though the lower CI border (0.96) was slightly less than 1.0. At the same time, OR for the individuals carrying three or four of the risk-increasing GPIIIa and PAI-1 alleles when compared with those carrying none or one allele was 4.5 and the confidence interval was much wider (95% CI: 1.82-12.72). Nevertheless, the lower border value (1.82) was considerably higher than 1.0 thus reflecting that the gene-gene interaction is likely to have more pronounced effect than the separate effects of the gene variations.

It might also be said that the border values of the confidence interval partially reflect what we might call as "the statistical stability" of the results. With some reservation, border values of the confidence interval can be used for a rough assessment of the relative quality of association. For instance, in the Case Study of the *IHD200* cohort (further in this Chapter), we have obtained associations of PAI1 5g/4g with CHD (P=0.0001; OR: 2.6; CI: 1.7-4.1) and of AGTR1 A1166C with DVT (P=0.0002; OR: 5.0; CI: 2.0-12.0). Despite that the OR values of these associations differ by the factor of 2, both these associations can be said to be of similar "statistical stability" since the lower borders for both of the confidence intervals are close. In the Volume II, we consider the scrambling code and variation techniques for the assessment of the statistical stability of a study and then compare the results with the usage of the conventional confidence intervals.

The statistical parameters we just considered are used for the assessment of the quality of the statistical data in a particular study. Although taking a look at these values in univariate and multivariate analyses, along with the application of the general criteria (Box 3) can often give a fair assessment of a study, there are several problems that purely statistical approach to biomedical research engenders.

First of all, there are no truly independent factors in medicine and the factors assumed to be "independent" are only regarded as such for the sake of simplicity. The statistics *per se* does not allow to find the truly independent factors so biomedical expertise has to be used in planning the experiment and analyzing the statistical data, if we are to obtain any results of clinical significance.

Secondly, statistics, by the very nature of its approach, searches rather for the general trends thus tending to obscure many unique cases as "not fitting the model". Therefore, in order to be complete, any statistical generalization should be complemented by a careful analysis of the identified exceptions and that is rarely done in biomedical studies. Such a complementary analysis can underline (or, on the contrary, considerably downplay) the scientific validity of the entire study. As an example, we used this technique of "statistical complementation" (Volume VI) for discovery of the "exotic" hydrogen bonding in proteins (see Chapter IV of the Volume V and the references therein).

The third and, perhaps, the most important point is that self-consistency of the biomedical data should never be left out of view. We illustrate this principle on the example of the APOC3 association in the study of the IHD200 study at the end of this chapter. Often, however, conclusions based exclusively on statistics are given precedence over the logic, common sense and the currently available biomedical knowledge. A purely statistic association only indicates that in the population the researchers studied, there was a prevalence of a certain genotype in patients with a certain condition. However, genotypes do not act by sheer magic, as some researchers in the area appear to think. Instead, there is a clear stratification of the levels of biological function (Figure 1-5): genotype adjusts the biochemical variables through the functional changes and it is the biochemical variables that, under a complex set of other conditions, adjust predispositions to diseases. The consistency of the clinical observations, lab tests and other data across the levels of function (Chapter V) is essential both for the correct diagnostics and success of association studies. Nevertheless, only rarely the biochemical variables are thoroughly analyzed for the population under study.

STUDY ASSESSMENT: A FORGERY?

Mundus vult decipi...
"World wants to be deceived..."

The history of quacks in science is as long as the history of science itself. In medicine, various nostrums were advertised for millennia and still are widely available on the global market. This aspect of the history of science is by no means exhausted and in today's journals, even highly acclaimed ones, the science papers with the data proven to be forged are not uncommon. The recent scandals with microelectronics papers published in "Nature" and "Science" that were later found to be made on non-existing samples with the data collected in non-existing computer files [30]; almost annual publications of "missing links" invariably followed by the reports that these were forgeries as well as the remarkable examples of the failures to tackle even explicitly identified scientific fraud [31] make us keenly aware to the fact that the history of fraud in science is far from being over. Social studies, both official and unofficial ones, do indicate that over 30% of individuals that call themselves as "scientists" admit (anonymously) that they committed an act of questionable scientific integrity at least once in their carrier. The suicidal drive of publishing-perishing, submission to "hidden curriculums", change of ethical integrity, acceptance of hierarchy, competitive atmosphere [32], unscientific practices in modern science [8] and other factors will almost warranty that a certain percentage of the biomedical studies will be forged. And, the common human weakness to respond to an effective outward presentation instead of looking at the truly objective factors will warranty acceptance and publication of these forged studies.

It must be remembered that so-called "peer review" does not necessarily detect fraud. This statement, rather than being a particular view, was amply illustrated by the Sokal's hoax and other "case studies" of the sort. A similar point was even proven statistically in the case of biomedical funding: the way in which the results of randomized controlled trials and systematic reviews are *presented* strongly influences health policy decisions [33]. It is presentation thus matters, not so much the scientific validity of studies or study proposals for a study to be funded.

No doubt, the criteria of study assessment we've mentioned so far can point out objective weaknesses in an honestly made research report. However, no one warranties that these and other criteria can be used by unscrupulous individuals to forge studies on an entirely new level of forgery. Accordingly, in addition to the criteria we considered so far, additional methods are certainly required to ensure the study reliability and consistency.

Actually, there is no rule-of-a-thumb to detect forged data. There were a few attempts by statisticians to put forward a common technique that would allow detection of statistical abnormalities of the digit data presented in the initial data set. These techniques, although being brilliant academic expositions (especially for students attending a statistics course in a university), are hardly viable as methods for everyday application. For example, in a study of the kind data from two trials were analyzed [34]. The authors of this publication, though, were wary enough to admit that such kind of digit preference in itself is no evidence of misconduct. Nevertheless, the researchers gone as far as analyzing the original trial data provided on handwritten sheets (for one randomized trial) and as a computer file (for another trial). They analyzed the frequencies of the various digits and concluded that the data of a drug trial (presented in a computer file) are 'genuine' whereas the data of a diet trial (presented on handwritten sheets) are 'abnormal' since they indicate some digit preferences.

For a student in a lecture hall, that sounds like a marvelous technique. For a computer scientist with two decades of practical experience, that "technique" would sound, using a parable from English literature, like Insp. Lestrade's attempt to frame Dr. Watson for the Milverton's murder. That is, all of these statistical intricacies will be utterly vain if we just consider the possibility that the data in that computer file might have been forged---using a random digit generator within a 20-line program. At the same time, a simple-minded but, nevertheless, an honest researcher (or a lab assistant or technician, for that matter) might well show some personal preferences for rounding the least significant digits. In other words, it is rather the expert assistance of Sherlock Holmes that would be required than some kind of a statistical study and not merely counting digits.

Since we cannot call on Mr. Holmes for any paper to be read, we'd rather use the logic, the expertise and the general scientific sense. First of all, it is important to recognize that a data forgery might not always be conscious. Common statistical errors, even in the papers published in some "top journals" (http://www.newscientist.com/article.ns?id=dn5051) as well as the distinct possibility that large studies might not be always right [10] hardly follow from conscious forgery. We can ascribe problems of the sort to the great complexity of biological systems combined with inefficient organization and poor informational management of the research. In other words, a "forged" finding in an article might be result of human error or a gross accumulation of the human errors, especially in the case of large-scale projects. Nevertheless, for practical purposes, a study that features a grave systematic error or a crucial error in numerical computations would be as good as an entirely forged study.

Secondly, what might be called as "scientific quality" of a publication isn't only about "statistical significance", as we already stated in the previous section. A well-developed scientific sense of fitness of things allows detection of inconsistencies and also of reasonable patterns in each particular publication. Often, consistency over the levels of information presented in an article can be far more important a criterion of scientific validity than any sorts of statistical criteria (Chapter V). *Ideally*, the researchers in the post-genomic era should never miss the fundamental post-genomic tetrad (Figure 1-3) and structure their research according to the corresponding effects at the levels of function. *Ideally*, again, after an adequately planned and properly performed study a consistent picture of correlations over these levels of function would emerge. Practically, a few examples we present in the Chapter V illustrate that self-consistency of the biomedical and genetic data across the levels of biological function can be a decisive criterion in the assessment of validity of a biomedical study. In the most general manner, this principle was implemented in *The Case of IHD200* cohort which we are about to consider.

Analyzing a Cohort

The last dozen of pages in this book dealt with assessment of the quality of biomedical studies. It's all fine to assess and evaluate. Sometimes, entire societies and even countries specialize exclusively in evaluating, judging and criticizing rather than in doing anything. So, it would be interesting to see how the principles we've formulated and discussed so far can be used to avoid traps in one's own research and not just applied for assessment of other's work. This is done in the framework of *The Case of IHD200*, a Case Study that investigates the potential links between common genetic factors and incidence of thrombosis in IHD/CHD. A detailed report of this research effort is presented in the text of this Case Study; here we will discuss some general points.

In the scope of this Case Study, we intentionally did not elaborated on *all* of the important points we've discussed in the present chapter. This was done in order to draw attention to the most significant issues that need to be remembered, again and again. The most significant feature of the IHD200 study is that it directly illustrates the crucial influence of the accuracy and completeness of the clinical diagnosis on the relative success or failure of a genetic association study. The conclusions of the Case Study illustrate once more the importance of clinical diagnostics which we discussed in the Chapter I. The clinical diagnosis should not only be accurate in respect to some rules (such as those in the Box 5) but also should include data on other disorders and, at the very least, the results of the most common lab tests (Box 2). This study also illustrates that the results of even the most common lab tests (such as CBC, for instance) can be helpful for the analysis of the failures in genetic association studies.

The study also answers the question which we raised in the Introduction and which we also discuss elsewhere (Chapter V, "Para-analysis"): why in the case when many genetic polymorphisms are taken, associations with only a few of them are usually identified? The main answer found in the course of the present Case Study is, more or less, obvious: because of an incomplete clinical characterization of the study cohorts. The incomplete clinical characterization implies neglect of significant factors and such a neglect inevitably leads to the higher levels of the statistic noise thus obscuring the associations of the primary interest.

This Case Study is, of course, not without drawbacks: the list of potential confounders is far from being exhaustive, the physiologically questionable association with APOC3 polymorphism, relatively small cohort size which prevents more detailed stratifications *etc.* Also, a few questions remain. Why no associations were found with the polymorphisms that definitely lead to functionally altered proteins or genes (such as F5 R506Q or LPL S447X)? Thus, even after adjustments for the congenital heart defects, still only 3 out of 11 well-studied polymorphisms were associated with CHD. The answer, apparently, varies from gene to gene and from polymorphism to polymorphism. For instance, the F2 and F5 polymorphisms are characterized by relatively low frequency. Therefore, in order to detect an association, considerably larger cohorts should be studied. To some extent, this consideration is also relevant in the case of LPL S447X and ITGB3 L33P.

However, the population frequency of polymorphisms cannot explain why an important and quite common genetic marker such as FGB -455 G/A was not associated neither with CHD nor even with fibrinogen levels even after various adjustments were tried. Generally, while systematic reviews indicate that increasing blood levels of fibrinogen, hematocrit and/or erythrocyte sedimentation rate may be predictors of CHD risk, their causal relevance is not always clear [35]. Accordingly, interplay of the factors on the top levels of function (Figure 1-5) is likely to mess the association between the FGB polymorphisms, fibrinogen levels and the incidence of CHD. Further stratification of the data might have helped to clarify this and similar issues but because of the relatively small size of the cohort more complex stratifications quickly run out of the statistical significance.

Once again, the present Case Study is just an example of what usually happens when rigid mechanistic criteria (such as statistics *per se*) are applied in the field of biomedical science without any further scientific reasoning. And what might happen, if just a small bit of natural human intelligence is added to the analysis. *Natura non contristur* which means, of course, "the nature is not sentimental". Neither is the scientific research. A scientific attitude requires active, logic and flexible use of the human intelligence rather than adherence to statistical dogma or to emotional prejudices so common among current "experts in the field".

THE CASE OF IHD200

"Your colleague, a clinician, consulted you concerning plausibility of an association study on a cohort of patients with ischemic heart disease. The primary aim is to analyze the correlation between genetic polymorphisms and presence of deep venous thrombosis in CHD patients. However, initial analyses did not seem to show any significant correlations between the genetics and the clinics..."

Introduction

As we have reiterated several times throughout this chapter, the major problem with any genotype-phenotype association study is that the relatively weak signal that comes from the genetic component tends to be obscured by the high levels of noise from the extraneous influences (such as other diseases, lifestyle, diet, drugs *etc*). Consequently, in order for an association study to be both physiologically meaningful and statistically significant, it is necessary to reduce or to compensate for these sources of the informational noise. The operations within the standard statistical model of the case-control study do allow compensation for some of the noise. However, the results of a "case-control" study greatly depend on the selection of the patient and control groups and the statistical theory *per se* can give only very general recommendations in this regard. Therefore, compensation and reduction of the noise can be done only outside the statistical model: in our case, by using our clinical, biochemical and genetic expertise. Accordingly, the causes for failure of an association study to find any correlations with the genetic polymorphisms should also be researched outside the confines of the formal statistics.

An example of analysis of such failure is presented here, in this Case Study. A cohort of 200 CHD patients (Table 3-2), approximately $1/3^{rd}$ of them having DVT, was studied in order to find potential genetic markers predisposing to thrombosis in CHD, as well as to establish prospective genetic markers of CHD in this population. A number of genetic variants predisposing to thrombosis were analyzed in the cohort of the patients but no statistically significant correlations were found. With the exception of the trend association PAI1 4G/5G polymorphism with CHD ($P<0.08$), there wasn't any correlation found in a case-control study that included IHD200 cohort and 200 age- and gender-matched controls.

Table 3-2. General clinical characteristics of the CHD patients and of the control group

Characteristic	CHD patients (N=200)	Controls (N=200)	P-value	OR and 95% CI
Age, yr (range)	60.5 (37-91)	56 (27-81)	P=0.98	-
Gender, %males	65%	56%	P=0.07	-
Smoking	70%	55%	P=0.002	1.9 (1.3-2.9)
Daily alcohol intake	75%	65%	P=0.03	1.6 (1.1-2.5)
Hypertension	50%	35%	P=0.002	1.9 (1.2-2.8)
Anemia (low hemoglobin)	29%	10%	P<0.0001	3.7 (2.1-6.4)

Similar situations, actually, occur quite often while attempting genetic association studies. Usually, stratification of the cohort, according to the known demographic and clinical parameters is used in order to compensate for the lack of correlation. However, this can be done only when most of the significant parameters are accounted for (which was, apparently, not the case with the cohort under consideration) and when the cohort is large enough. The unique feature of the present case study, that's almost never can be found in a standard academic publication, is that it shows how almost a complete failure during an association study can transform into a reasonable success. An essential stage of this transformation is a careful analysis of the entire array of data for each patient as an individual case with the purpose of elucidating the diagnosis.

Materials and Methods

Study cohort. General demographic and clinical characteristics of the patient and the control groups are presented in the Tables 3-2 and 3-3. Briefly, among 200 consecutive Caucasian patients diagnosed with CHD, 67 patients (34%) were also diagnosed with DVT. The CHD and DVT diagnostic criteria are delineated in the Box 5. In addition, each patient was characterized by a panel of routine lab tests (for non-specialists, the short descriptions of these lab tests are presented in the Box 2). Controls were CHD-free Caucasians from general population.

Selection of the polymorphisms. In the initial study, we decided to include a few genetic markers that, nevertheless, would cover several biochemical pathways related to CHD and thrombus formation. The 11 genes were selected as being of mutual interest both to the clinicians and the bioinformatician; these genes were AGTR1 (renin-angiotensin system), highly pleiotropic gene MTHFR, few genes of lipid metabolism (LPL, APOC3), a number of genes involved in hemostasis (F2, F5, F7, FGB, ITGB3 and PAI1) and the gene involved in pharmacogenetics of anticoagulants (CYP2C9). More information on these genes is available in the *CVD GeneScape* diagram in this volume (see Figure 2-5 and Appendix I). The genetic profile of the patients was compiled using the following selection of 11 polymorphisms: AGTR1 A1166C; MTHFR C677T; LPL S447X; APOC3 -482 C/T; F2 20210 G/A; F5 R506Q; PAI1 5g/4g; ITGB3 L33P; FGB -455 G/A; F7 R353Q; CYP2C9*2 R144C.

Data analyses. The $\chi 2$ testing was used to estimate differences in prevalence among the groups and for analysis of the potential confounders. The univariate and multivariate analyses were performed in Excel tables and using statistical software written by the author. The essential feature of this study was the expert analysis of the entire array of data prior to association study itself. This analysis included five main stages: 1. Initial analysis of the data array for each patient in the group (to elucidate the individual diagnosis); 2. Requesting additional clinical data on the patients with incomplete diagnostic data; 3. Including patients into groups for the association study; 4. Checking the patient and control groups for the genetic equilibrium (Hardy-Weinberg testing); 5. Performing the association study itself.

Table 3-3. Clinical characteristics of the initial cohort of CHD patients with and without DVT

Characteristic	Patients with DVT (N=67)	Patients without DVT(N=133)	P-value	OR and 95% CI
Age, yr (range)	65 (54-84)	62 (37-91)	0.6	-
Gender, %males	58%	70%	0.06	-
Hypertension	50%	50%	1.0	-
Recent surgery	10%	3%	0.03	3.7(1-13)
Anticoagulant therapy	33%	33%	0.97	-
Bed rest	<1%	3%	0.5	-
Atrial fibrillation (AF)	77%	58%	0.02	2.2(1.1-4.3)
CHD, class >=2	72%	81%	0.1	-
CHD, class >=3	55%	63%	0.3	-

Results

At the first attempt of association study, a number of conventional CHD risk factors were found to be statistically significant in IHD200 group using gender- and age-matched group of controls. These factors were smoking, daily alcohol intake and hypertension (Table 3-2). Anemia (inferred from the low CBC hemoglobin) was also strongly statistically significant factor. However, none of the 11 genetic markers was found to be involved in a statistically significant association with CHD with the exception of the trend for PAI1 4g/5g ($P<0.08$).

Subdividing the IHD200 group in two subgroups (patients with and without DVT) has shown that only recent surgery and atrial fibrillation had statistically significant associations with DVT in the initial cohort of patients (Table 3-3). Association between CHD and alcohol was statistically significant ($P=0.03$) and, at the same time, there was an inverse correlation between the alcohol intake and DVT ($P<0.0001$). Again, none of the proposed 11 genetic markers was associated with the presence or absence of DVT.

Thus, according to the data present in the Tables 3-2 and 3-3, the initial attempt of the genotype-phenotype association study was a failure. Neither blunt attempt of the association study nor analysis of the apparent subgroups produced any results of statistical significance. Therefore, it is the situation when straight-forward application of the statistical apparatus is not an option and biomedical expertise is essential for the data analysis.

Indeed, even a cursory view at the entire array of data can shows several apparent abnormalities of the initial array of the clinical data. The most remarkable of them was that anticoagulant therapy (warfarin, acenocoumarol and other drugs) was not a statistically significant factor for DVT (Table 3-3)--- highly irregular a finding. There was an observable proportion of patients with early CHD (age <50y) and early CHD in advanced stages (CHD classes 3 or 4). A number of patients appear to have severe hepatic problems, as indicated by high bilirubin levels and by other lab tests.

The subsequent expert analyses of the clinical and biochemical data (to save the printing space, we'll state only the results here) has shown that IHD200 group is not a clinically uniform group but should be split into three subgroups: CHD patients *per se* (N=155), CHD patients with inferred congenital heart disease (N=35) and CHD patients with severe hepatic problems (probably, viral hepatitis, N=10). These groups should be treated separately during the association study. For example, let's look at the results of the comparison of the patients with the inferred congenital heart disease (N=35) and the rest of the CHD patients (N=155) which we summarized in the Table 3-4.

In fact, these two groups strikingly differ in a large number of clinical parameters as well as in the results of the lab tests (Table 3-4). Thus, it is clear that the patients with the inferred congenital heart disease were mostly males, less than 50 year of age and, typically, had severe form of CHD (NYHA class 3 or 4). These patients were also characterized by prevalence of the chronic atrial fibrillation (P=0.02). Hemoglobin, AST&ALT enzymes, bilirubin and two of the coagulation markers were distinctly different from the rest of the patients. In particular, it is low hemoglobin in the absence of higher hematocrit (usually, higher hematocrit occurs in erythremias) and the early age of CHD that suggested the presence of congenital heart disease during the bioinformatic analysis. Actually, the observations summarized in the Table 3-4 even allowed to elaborate a sort of a prediction scheme that would "diagnose" the congenital heart deficiency using the analysis of the above-mentioned parameters. After contacting the clinicians concerning these 35 patients, it turned out that 31 of them were indeed diagnosed with congenital heart disease, though these particular clinicians did not think it to be particularly important fact to consider in the association study (nor, by the way, to inform the bioinformatician about the matter without additional enquires).

As for the patients with a considerable liver damage (as suggested by bilirubin, AST, ALT and fibrinogen tests), the presence of viral hepatitis was induced on the basis of the AST/ALT ratio (which is known to be much lower that 1.33 in patients with viral hepatitis). These patients were also excluded from the association study since they tend to have abnormal fibrinogen and APTT. Inherited thrombophilic gene mutations might increase the thrombotic tendencies in patients with impaired liver function [36].

The exclusion of the patients with congenital heart defects and with hepatitis had several immediate and profound effects on the associations of the primary interest. Thus, the associations of the clinical characteristics of the adjusted cohort of CHD patients with and without DVT, as well as the associations of the genetic polymorphisms both with CHD and DVT became not only meaningful but also strongly statistically significant (Tables 3-5 and 3-6). One of the most peculiar abnormalities of the initial data (that anticoagulant therapy was not a statistically significant factor for DVT, Table 3-3) was resolved: in the adjusted cohort, the inverse association between anticoagulant therapy and DVT was statistically significant (P=0.01). After the adjustment for the heart defects and hepatitis, hypertension was found to be a statistically significant risk factor (OR=1.9, 95% CI 1.0-3.5, P=0.04), which also seems to be physiologically plausible when taking into account the results of the association study with genetic markers (Table 3-5).

Table 3-4. Clinical and biochemical characteristics of the CHD patients with the congenital heart disease *vs* the CHD patients without congenital heart disease (only statistically significant associations are presented)

Characterisitc	CHD patients with congenital heart disease (N=35)	CHD patients, no congenital heart disease (N=155)	P-value	OR and 95% CI
Clinical data				
Age <50y	67%	7%	0.00001	25.0 (10-64)
Gender, %male	82%	65%	0.04	2.6 (1.0-6.6)
CHD class>=3	85%	68%	0.03	2.9 (1.1-7.8)
Chronic atrial fibrillation	51%	31%	0.02	2.4 (1.1-5.0)
Daily alcohol	65%	33%	0.0003	3.9 (1.8-8.5)
Biochemical data **1. Basic lab tests**				
Anemia (CBC hemoglobin <10 g/dl)	49%	11%	0.00001	7.7 (3.3-17.6)
AST>41 U/L	51%	32%	0.03	2.2 (1.1-4.7)
AST/ALT >1.33	68%	47%	0.02	2.5 (1.1-5.3)
Bilirubin >20 μM/L	85%	25%	0.000001	17.8 (6.5-49)
2. Coagulation markers				
APTT>40s	71%	18%	0.00001	11 (4.9-26)
D-dimer>250 ng/ml	51%	25%	0.002	3.1 (1.5-6.7)

The adjustment for the congenital defects and hepatitis lead to detection of several associations with the genetic markers. Thus, out of the 11 genetic polymorphisms, 3 were statistically associated with CHD in a case-control study, as the data in the Table 3-5 suggest. These polymorphisms were MTHFR 677 C/T, PAI1 5G/4G and APOC3 -482 C/T. There was also a statistically significant gene-gene "interaction" (MTHFR-PAI1, P=0.03). These associations remained significant after adjustments for smoking, alcohol, hypertension and anemia (potential confounders in the initial cohort, see Table 3-2).

While the physiological significance of the MTHFR and PAI1 associations with the CHD is, more or less, clear, the physiological significance of the APOC3 association needs to be discussed. The frequency of the APOC3 -482T variant (that is, formally, the variant predisposing to CVD) was approximately 2 times higher among the controls than the patients (Table 3-5). In other words, association with CHD was statistically significant but in the opposite direction. Analysis of the data collected in the *medSNP* database has shown that the published data for APOC3 -482C/T might, indeed, be contradictory. However, as shows ortho-analysis of the levels of function of APOC3 and other kinds of ortho-analyses (Chapter V), this association is not likely to be of biomedical significance despite the indication of the strong statistical significance of this association (P<0.001).

Even more important is that the adjustment for the congenital heart defects and hepatitis has allowed achieving the primary aim of the study: identification of significant associations of the genetic markers with DVT in CHD. These associations remained significant after adjustments for the potential confounders (surgery, atrial fibrillation, anticoagulants, hypertension). According to the data in the Table 3-6, the presence of AGTR1 1166C variant was a statistically strong risk for DVT (OR=5.0, P=0.0002) while the presence of the CYP2C9*2 variant was, apparently, a protective factor. Both results are quite unusual but do not appear to contradict the results of the corresponding ortho-analyses.

Table 3-5. Study of association between CHD diagnosis and the genetic markers; allele frequencies and the statistical parameters are presented

Polymorphism	CHD patients (N=155)	Controls (N=200)	P-value	OR, 95% CI
AGTR1 A1166C	18%	21%	0.5	-
MTHFR C677T	47%	25%	0.0001	2.7 (1.7-4.2)
LPL S447X	7%	7%	0.7	-
APOC3 -482 C/T	11%	25%	0.001	0.4 (0.2-0.7)
F2 20210 G/A	3%	1%	0.5	-
F5 R506Q	3%	2%	0.5	-
PAI1 5g/4g	74%	52%	0.0001	2.6 (1.7-4.1)
ITGB3 L33P	13%	8%	0.1	-
FGB -455 G/A	18%	16%	0.6	-
F7 R353Q	11%	15%	0.2	-
CYP2C9*2 (R144C)	24%	19%	0.25	-

Table 3-6. Study of association between DVT and the genetic markers

Polymorphism	Patients with DVT (N=65)	Patients without DVT(N=90)	P-value	OR and 95% CI
AGTR1 A1166C	31%	8%	0.0002	5.0 (2.0-12.0)
MTHFR C677T	41%	50%	0.2	-
LPL S447X	10%	4%	0.1	-
APOC3 -482 C/T	10%	14%	0.5	-
PAI1 5g/4g	65%	60%	0.1	-
ITGB3 L33P	11%	13%	0.6	-
FGB -455 G/A	14%	21%	0.2	-
F7 R353Q	12%	6%	0.2	-
CYP2C9*2 (R144C)	5%	25%	0.002	0.2 (0.04-0.6)

AGTR1 1166C. No association of DVT with AGTR1 polymorphism was previously reported. The AGTR1 A1166C polymorphism is a well-known hypertension marker. Hypertension itself was a risk factor for DVT (OR=1.9, 95% CI 1.0-3.2, P=0.04) in the adjusted cohort and thus might seem to be a potential confounder. However, multivariate analysis indicates the polymorphism to be an independent risk factor. Indeed, a genetic polymorphism is more fundamental variable than hypertension *per se*: in other words, while polymorphism can influence hypertension, hypertension can't influence polymorphism. Also, AGTR1 polymorphism was associated with hypertension risk in this cohort even after adjustments ($P<0.01$). Thrombotic rather than hemorrhagic events are likely to represent a major complication of hypertension [37] and the pathophysiology of the thrombosis in hypertension involves, in particular, the renin-angiotensin system (of which AGTR1 is the part). Chronic shear stress, especially aggravated by chronic inflammation, may convert normally anticoagulant endothelium into a procoagulant surface. Also, it appears that treatment with ACE and AGTR1 antagonists does not only influence the blood pressure but also reduces risk of thrombosis [37].

CYP2C9*2. The association of the CYP2C9*2 variant with attenuated risk of thrombosis is also an intriguing result. CYP2C9*2 is a common pharmacogenetic marker of the sensitivity to warfarin therapy. Against the background of warfarin treatment, the CYP2C9 variant *2 can be considered as a 'protective factor' (since the variant corresponds to a pronounced bleeding tendency). However, only 4 subjects out of 65 (6%) patients with thrombosis were on warfarin therapy. Therefore, the physiological pathways by which thee *2 variant might, potentially, exert influence on the thrombotic processes, are far from being obvious. These pathways might involve some other drugs taken by the patients. Alternatively, the endothelial cytochrome 2C9 is known to influence the bioavailability of NO, which is a potent vasodilatator. In a very small study (16 CAD patients, [38]), it was found that CYP2C9 inhibition dose-dependently enhances blood flow response to acetylcholine. In this study, a CYP2C9 inhibitor enhanced endothelium-dependent vasodilator responses in patients with manifest coronary artery disease, the effect being potentially related to an increase in the bioavailability of NO [38]. As inhibition of CYP2C9 corresponds to the lowered activity of the *2 variant of CYP2C9, a change in the bioavailability of NO might be the cause of the reduced risk of thrombosis associated with this variant in the present Case Study.

Thus, the adjustments for the congenital heart defects and hepatitis immediately lead to a number of statistically significant and physiologically plausible (for the most part) results within just one association study. It seems fit to consider why these two subgroups should have been excluded *a priori* from the association study we describe here. It is true that all of the 200 patients in the initial cohort were CHD patients. However, the presence of the subgroup with congenital heart defects skewed several otherwise "independent" parameters such as the male/female ratio, the age, CHD class and quite a number of important biochemical variables/markers (Table 3-4) thus acting as an "unknown confounder". None of the patients with the congenital heart disease seemed to have thrombotic complications whatsoever thus additionally obscuring the association between DVT and the genetic factors analyzed. Furthermore, clinical experience indicates that the congenital heart disease is, more often than not, determined rather by adverse exposures during gestation (maternal alcohol use and poor maternal nutrition, before all) than by some vague genetic components. According to PubMed searches, most of the genetic association studies attempting to link congenital heart disease with any genetic polymorphisms were, apparently, negative findings. And, by the way, absence of thrombotic complications in this subgroup of patients might be partially explained by the daily alcohol use (as indicated by questionnaire).

In the case of patients with the inferred hepatitis, most of them were also without DVT but, nevertheless, with coagulation markers observably affected. Apparently, including the patients from both the congenital heart defects and the hepatitis subgroups into one CHD cohort inevitably leads to the unnecessary increase in the statistical noise. Nevertheless, bioinformatic analysis of the data integrity suggested that certain retrospective analyses of these patients can be used to elucidate the medical diagnosis thus leading to a more substantiated association study. Also, constant cooperation between clinicians and bioinformaticians is essential to improve reliability of the data analysis in the framework of a genetic association study.

Box 5. The diagnostic criteria for coronary heart disease (CHD) and deep venous thrombosis (DVT)

Coronary heart disease criteria as defined by the World Health Organization (see Nomenclature and Criteria for Diagnosis of Ischemic Heart Disease Circulation, 1979, Vol. 59, No. 3, 607-9). Generally, CHD (synonymous with "ischemic heart disease") is defined as myocardial impairment due to an imbalance between coronary blood flow and myocardial requirements caused by changes in the coronary circulation. There are five distinct classes of CHD that include primary cardiac arrest, angina pectoris, myocardial infarction, heart failure and arrhythmias. In the Case Study, the notion of CHD included only angina pectoris and arrhythmia (atrial fibrillation, both paroxysmal and chronic).

Angina pectoris included both the stable effort angina (chest pain, precipitated by emotional or physical stress, relieved within 10 minutes by rest or with sublingual nitroglycerin, >1 month in duration) and spontaneous angina (episodes of chest pain that occur without apparent relation to increased oxygen demand of the myocardium, less readily relieved by nitroglycerin, no enzyme changes).

Arrhythmia may be the only symptom of CHD. The most common cardiac arrhythmia is the *atrial fibrillation* (AF). The symptoms of AF included palpitations, irregular heart beat, shortness of breath, chest discomfort, and dizziness and, often, the general feeling of weakness. The rhythm problems of the heart were initially detected by using the stethoscope and confirmed by electrocardiogram.

For each patient, the *functional class* of CHD was assigned in accordance with the "New York Heart Association Functional Classification", which categorizes patients into one of the four classes on the following basis- *Class 1*: no symptoms and no limitation in ordinary physical activity; *Class 2*: mild symptoms and slight limitation during ordinary activity, comfortable at rest; *Class 3*: marked limitation in activity due to symptoms, even during less-than-ordinary activity; *Class 4*: Severe limitations, symptoms even while at rest.

Deep vein thrombosis (DVT) is a condition characterized by the presence of blood clots in the leg veins. These blood clots can dislodge and move throughout the vessels causing damage to vital organs. DVT is found mainly in the veins in the lower leg and the thigh. Deep venous thrombosis is most common in adults over age 60. The major symptoms included changes in skin color of one leg (redness), leg pain, tenderness and swelling, especially asymmetrical (of one leg only). In the Case Study, the diagnosis of the deep venous thrombosis was confirmed by plethysmography of the legs (systolic blood pressure of a leg compared with that of an arm) and d-dimer blood test.

Chapter Summary

The present chapter deals with the search of the **published biomedical evidence** and with the **criteria of assessment** of the biomedical studies. The **trends of the information growth** in physiology and biomedicine suggest that that it is paramount to implement adequate strategies of the literature search and of the study assessment to avoid information overload characterizing modern biomedicine and physiology. Elaboration and implementation of the adequate strategies in search and assessment of the published evidence is a cornerstone of the **purpose-oriented bioinformatics**. The chapter describes general strategies of the efficient literature search and also presents some general points pertaining to the assessment of the quality of biomedical studies: the **use of hierarchies**, **questionnaires**, general **kinds of biomedical studies**, appraisal of **observational studies**, study **cohorts** and **statistical considerations**. The principles formulated in this chapter were then applied to the assessment of an **actual analysis of genetic associations** presented in a Case Study that investigated potential links between commonly studied genetic polymorphisms and incidence of thrombosis in coronary heart disease.

References

[1] Sanders S, Del Mar C. Clever searching for evidence. *BMJ.* 2005;330:1162-3.

[2] Ramos K, Linscheid R, Schafer S. Real-time information-seeking behavior of residency physicians. *Fam Med* 2003;35:257-60

[3] Green M, Ruff T. Why do residents fail to answer their clinical questions? A qualitative study of barriers to practicing evidence-based medicine. *Acad Med* 2005;80:176-82).

[4] Tang H, Ng J. Googling for a diagnosis--use of Google as a diagnostic aid: internet based study. *BMJ.* 2006 Nov 10; PMID: 17098763

[5] Haynes RB, Wilczynski NL. Optimal search strategies for retrieving scientifically strong studies of diagnosis from Medline: analytical survey. *BMJ.* 2004;328:1040.

[6] Haynes RB, McKibbon KA, Wilczynski NL, Walter SD, Werre SR; Hedges Team. Optimal search strategies for retrieving scientifically strong studies of treatment from Medline: analytical survey. *BMJ.* 2005;330:1179.

[7] Mishra GR, Suresh M, Kumaran K, Kannabiran N, Suresh S, Bala P, Shivakumar K, Anuradha N et al. Human protein reference database--2006 update. *Nucleic Acids Res.* 2006;34:D411-4.

[8] Smith R. Unscientific practice flourishes in science. *BMJ.* 1998;316:1036.

[9] Kleiner K. Most scientific papers are probably wrong. New Scientist 30 Aug 2005 (http://www.newscientist.com/article.ns?id=dn7915)

[10] Ioannidis JP. Why most published research findings are false. *PLoS Med.* 2005; 2(8):E124.

[11] Bacchetti P. Peer review of statistics in medical research: the other problem. *BMJ* 2002;324:1271-1273

[12] Torshin I. Shifting accents: from "anonymous" to "peer review". (http://bmj.bmjjournals.com/cgi/eletters/329/7457/113-c#66360)

[13] O'Donnell M. Why doctors don't read research papers: Scientific papers are not written to disseminate information. *BMJ*, 2005; 330(7485):256.

[14] Greenhouse G, Peering into review. *Curr. Biol. 10*, 23:R849-R850.

[15] Glasziou P, Vandenbroucke J, Chalmers I. Assessing the quality of research. *BMJ* 2004;328:39-41.

[16] Boynton P, Greenhalgh T. Selecting, designing, and developing your questionnaire. *BMJ* 2004;328:1312-5.

[17] Ioannidis JP, Haidich AB, Lau J. Any casualties in the clash of randomised and observational evidence? *BMJ* 2001;322:879-80.

[18] Concato J, Shah N, Horwitz R. Randomized, controlled trials,observational studies, and the hierarchy of research designs. *N Engl J Med* 2000;342:1887-92

[19] Gurwitz J, Sykora K, Mamdani M, Streiner D, Garfinkel S, Normand S, Anderson G, Rochon P. Reader's guide to critical appraisal of cohort studies: 1. Role and design. *BMJ* 2005;330:895-897

[20] Mamdani m, Sykora k, Li p, Normand s, Streiner D, Austin P, Rochon P, Anderson G. Reader's guide to critical appraisal of cohort studies: 2. Assessing potential for confounding. *BMJ* 2005;330:960-962

[21] Normand S, Sykora K, Li P, Mamdani M, Rochon P, Anderson G. Readers guide to critical appraisal of cohort studies: 3. Analytical strategies to reduce confounding. *BMJ* 2005;330:1021-1023

[22] Pastinen T, Perola M, Niini P, Terwilliger J, Salomaa V, Vartiainen E, Peltonen L, Syvanen A. Array-based multiplex analysis of candidate genes reveals two independent and additive genetic risk factors for myocardial infarction in the Finnish population. *Hum Mol Genet* 1998;7:1453-62.

[23] Casiglia E, Tikhonoff V, Mazza A, Rynkiewicz A, Limon J, Caffi S, Guglielmi F, Martini B, Basso G, Winnicki M, Pessina AC, Somers VK. C-344T polymorphism of the aldosterone synthase gene and blood pressure in the elderly: a population-based study. *J Hypertens*. 2005;23:1991-6.

[24] Safar ME, Cattan V, Lacolley P, Nzietchueng R, Labat C, Lajemi M, de Luca N, Benetos A. Aldosterone synthase gene polymorphism, stroke volume and age-related changes in aortic pulse wave velocity in subjects with hypertension. *J Hypertens* 2005;23:1159-66.

[25] More data on the Framingham Heart Study/Framingham Offspring cohorts are available at http://www.nhlbi.nih.gov/resources/deca/descriptions/framoff.htm

[26] van Rossum CT, Hoebee B, Seidell JC, Bouchard C, van Baak MA, de Groot CP, Chagnon M, de Graaf C, Saris WH. Genetic factors as predictors of weight gain in young adult Dutch men and women. *Int J Obes Relat Metab Disord* 2002;26:517-28.

[27] Mckay R, Oldford R. Scientific Method, Statistical Method, and the Speed of Light. Working Paper 2000-02, Dept of Statistics and Actuarial Science, University of Waterloo. (http://www.stats.uwaterloo.ca/~rwoldfor/papers/sci-method/paperrev.pdf)

[28] Sterne JA, Davey Smith G. Sifting the evidence-what's wrong with significance tests? *BMJ*. 2001;322:226-31.

[29] Davies HT, Crombie IK, Tavakoli M. When can odds ratios mislead? *BMJ* 1998;316:989-91.

[30] So-called "Schon scandal", see the entire story at http://en.wikipedia.org/wiki/Jan_Hendrik_Sch%C3%B6n

[31] E. Young. Faking it. New Scientist, 12 Dec 2000 (http://www.newscientist.com/article/dn253.html)

[32] Abbasi K. The four pillars of global academic medicine. *BMJ* 2004 Oct 2; 329 (http://www.bmj.com/cgi/content/full/329/7469/0-g?ijkey=5091cdf94d088fcd0e6131204ad3993e536c032b&keytype2=tf_ipsecsha)

[33] Fahey T, Griffiths S, Peters TJ. Evidence based purchasing: understanding results of clinical trials and systematic reviews. *BMJ.* 1995;311:1056-9.

[34] Al-Marzouki S, Evans S, Marshall T, Roberts I. Are these data real? Statistical methods for the detection of data fabrication in clinical trials. *BMJ* 2005;331:267-70.

[35] Lowe GD, Rumley A, Whincup PH, Danesh J. Hemostatic and rheological variables and risk of cardiovascular disease. *Semin Vasc Med* 2002;2:429-39.

[36] Erkan O, Bozdayi AM, Disibeyaz S, Oguz D, Ozcan M, Bahar K, Karayalcin S, Ozden A, Bozkaya H, Yurdaydin C, Uzunalimoglu O. Thrombophilic gene mutations in cirrhotic patients with portal vein thrombosis. *Eur J Gastroenterol Hepatol* 2005;17:339-43.

[37] Dielis AW, Smid M, Spronk HM, Hamulyak K, Kroon AA, ten Cate H, de Leeuw PW. The prothrombotic paradox of hypertension: role of the renin-angiotensin and kallikrein-kinin systems. *Hypertension* 2005;46:1236-42.

[38] Fichtlscherer S, Dimmeler S, Breuer S, Busse R, Zeiher A, Fleming I. Inhibition of cytochrome P450 2C9 improves endothelium-dependent, nitric oxide-mediated vasodilatation in patients with coronary artery disease. *Circulation* 2004;109:178-83.

Chapter IV

MEDSNP: A MIDDLE GROUND BETWEEN PHYSIOLOGY, GENETICS AND MEDICINE

TESTING GENETIC RISKS

Previously, we have considered the basics of physiology (Chapter I) and, in particular, of the cardiovascular physiology and genetics (Chapter II). We also considered the general principles of the retrieval and assessment of biomedical studies (Chapter III). These discussions were intended, in particular, to provide background for an adequate and an efficient retrieval/analysis of the phenotype-genotype association studies. In this and the following chapters as well as in the Volume II, we attempt to illustrate that the large-scale analyses of the published data on the genetic association studies can be very helpful for the modern research in clinical genetics. These analyses considerably assist planning, performing and data analysis of the association studies in the post-genomic era. The next chapter, in particular, describes the most important modes of comparative and integrative analyses of the genetic association studies. The *medSNP* database, described in the present chapter, serves the essential material basis for these analyses.

It must be noticed that the simplistic approach, assumed in a common phenotype-genotype association study, is not enough for the purposes of the practical medicine. Firstly, a common association study attempts to dissect the effects of only a few polymorphisms in a few genes and in a group of patients. At the same time, in the case of a *particular patient* in a clinic, there are many polymorphic variants to be analyzed and in many different genes but in just one patient. Secondly, there are numerous non-genetic factors crucial for the disease etiology and our summary of CVD physiology can make this contention clear (Chapter II). Thirdly, in the common medical practice there are clinical observations (such as the patient's symptoms, patient's physical constitution *etc*) and there are lab tests. Both clinical observations and lab tests can only rarely be directly linked to genetic data, if at all. At the same time, this linkage of the data from the various area of biomedicine is essential for an adequate interpretation of the results of the gene tests that tend to become more and more common though their medical significance is not entirely clear. Consequently, in post-genomic perspective, it is essential to adopt a proper strategy of information management in order to link what the doctor knows about the patient (clinical observations, lab tests, questionnaire data, diet *etc*) with the results of the genetic analyses. Elaboration of such strategies for the data linkage is the prerogative of bioinformatics.

Before approaching this task of the data linkage, it is important, first of all, to keep in mind the relative role of genetics in the etiology of a disease. Dieting factors, lifestyle habits, personality type and amount of the emotional stress: all these and many other factors are known to be essential health factors. The factors of risk for particular diseases are known not only from the individual clinical expertise of particular doctors but are also confirmed by many randomized trials, observational studies and meta-analyses. Therefore, it would be naïve to expect that association studies that report effects of only a few polymorphisms from the entire genome and the subsequent single-polymorphism genetic testing will give any definitive answers concerning predisposition to *common* multifactor diseases such as CVD. In this sense, it does not matter much, for an average practical clinician, whether human genome was sequenced or not, whether it is "pre-genomic" or "post-genomic" era. Since we are, technically speaking, in the *post-genomic* era of the human genome research, it does behoove, then, to take into account the genetic data *along with* other clinical data.

Any attempt of such data linkage should take into account that the sciences of physiology, genetics and medicine vary in their scope (Figure 1-2) and operate within the three sets of data which distinctly differ from each other (Figure 1-3). Although the level of genetic information is, undoubtedly, the most fundamental level (from which, using formal logic, all the upper levels can be considered as "derivatives"), it is also the level that is farthest from the level of the clinical manifestations (Figure 1-5). It is, therefore, no surprise that tracing the effects of the common genetic polymorphisms to particular diseases is far from being a straight-forward task. There are always several other levels of biomedical function between the genetic makeup and the clinical manifestations, so these intermediary levels greatly attenuate the effect of the patient's genetic makeup on the state of health. Accordingly, attempts to find associations that involve only the genetic effects ignore the actual context of the clinical science.

An important peculiarity of the genetic factors as "risk factors" for the multifactor diseases is that they represent latent and life-long risks. Both points indicate a special place of the genetic factors among the other factors of risk. Yes, a particular genetic defect (say, our common example of the F5 "Leiden" variant) might be found to be a risk factor for a particular disease in one association study and, perhaps, not in some other association study. The *latency of the risk* implies that, given some combination of the external circumstances of the patient, the risks associated with the genetic defect become several times higher. A remarkable example of this effect is the association between F5 "Leiden" and venous thromboembolism as well as related disorders. In the case of the sinuous thrombosis, the presence of F5 variant "Leiden", by itself, increases the risk 2- or 4-fold. When a female patient also takes oral contraceptives, the risk grows almost an order of magnitude thus making an incident of the sinuous thrombosis almost inevitable [1]. We consider more examples of the kind throughout this Chapter [2-7].

Genetic factors also represent *life-long risks.* A patient, resolved to deal with his or her health problems, might take some drugs or abolish taking of other drugs, change the diet and even reconsider the disastrous effects of his/her precious lifestyle on the state of health. All of the external factors such as diet and lifestyle can be properly adjusted to minimize the health risks. The same, however, can never be said about the genetic makeup (apart from the concept of "gene therapy" which, at present, is rather an area for science fiction novels and imaginative movie scenarios).

Many *pros* and *contras* can be produced for the case of the genetic testing as a common clinical practice (see, for instance, [8] and the materials available at http://www.genetests.org/). Indeed, genetic tests are already advertised on the Internet for a number of cardiovascular phenotypes such as thrombosis, hyperlipidemia or atherosclerosis. The validity of the tests is defended by references to a couple of particular association studies and we already considered the problems which plague most of the currently available association studies (Chapter III). Accordingly, the value of particular gene tests should be carefully assessed. Nevertheless, since the genetic factors are both latent and life-long, a particular patient might indeed be at a great risk, down to the premature lethal outcome, if the routine clinical practice ignores the results of the genetic testing.

The issue is far from being just another debate in academic medicine or another advertisement of pharmaceutical industry. Sometimes, it does make bigger headlines. For instance, families of British holidaymakers who suffered deep vein thrombosis on the British Airways flights have taken a legal claim for compensation to the highest court in the UK. The hearing took place at the House of Lords and it was argued that the genetic testing could have been used to avoid the lethality knowing that the airplane travel is a risk factor (*Daily Telegraph*, 29.10.2005, see this news item at http://portal.telegraph.co.uk/global/main.jhtml?xml=/global/2005/10/29/etnewsdvt29.xml).

Given that the clinical importance of the genetic testing is accepted, there are two principal tasks that have to be solved prior to making the genetic testing a valuable clinical tool and not just another media hype. The first task the practical genomic medicine faces is the selection of the genes and of the polymorphisms to be studied in order to arrive to medicinally plausible and clinically useful conclusions. The second task is how to provide a substantiated and reliable biomedical interpretation of the results of genetic testing of a particular patient.

Bioinformatical techniques can be very useful for both of the tasks. The present chapter concentrates on the informational methods that are used to solve both the tasks. We'll start with the compilation of the *medSNP*™ database that is directly relevant for the post-genomic applications in clinical diagnostics. In order to do so, we'll need to know for whatever purpose(s) this database is made, what kinds of data are required, from what sources the data should be obtained and how to locate and to sieve those data out from tens of thousands of publications. In the next chapter, we discuss the methods that allow verification of the biomedical consistency of the data from independently made association studies by taking into account the biochemical and molecular-biological studies of the corresponding genes and their products. The task of verification is accomplished by using comparative and integrative methods of analyses of the database and of the particular sets of studies. In the Volume II of the present book series, we present various clinical applications of the *medSNP*, among them: elaboration of the genetic profiles for analysis of predisposition to certain diseases, interactions between genetic polymorphisms and non-genetic risk factors and an expert system for the analysis of the genetic profiles of the patients.

medSNP™: A Database for Clinical Applications in Genomic Medicine

The most important question that has to be answered prior to making any database is that of **the purpose.** To some, collection and analysis of all imaginable SNPs (or any kinds of NPs) for all possible diseases might seem an attractive purpose and the most fundamental task. Such a task, however it would simulate imagination of those talented individuals, is nothing short of the "panacea" from the archaic past. Moreover, such an endeavor is highly impractical: it is too general and will take, at least, several decades, (apart, of course, from the matter of the quality and reliability of the data that will be collected into those huge databases). Therefore, we'll start with a particular group of multifactor diseases, which, in the case of the present volume, are cardiovascular diseases (CVD).The primary purpose of the *medSNP*™ database can be generally formulated as making an inventory of the most reliable medical data from the association studies linking genetic polymorphisms with various forms of CVDs.

Having set the primary target, we need to select the genes in which the polymorphisms should be studied. Again, we are not interested in all of the genes that potentially can, in the long run, contribute to CVD. As we have briefly mentioned in Introduction, the huge post-genomic studies that set this kind of the aim tend to be huge and spectacular failures. At its best, such a study would produce a very few associations. These associations, although statistically justified, are often incompatible with previous research and, not less often, are questionable from physiological point of view. What is even more disconcerting that such implausible associations are found only for polymorphisms in only very few genes out of thousands of the genes analyzed! Therefore, we are interested in collecting and thoroughly analyzing the data already available from the previous decades of the research in the genetic associations: namely, the published data on the polymorphism-CVD associations. The compelling informational overload characterizing the modern human physiology (Figure 3-1) suggests that this endeavor might be not a simple task. Then, after the CVD-associated genes are established (Chapter II), there comes the stage of the selection of the studies using certain selection criteria. The Chapter III provided a general background for performing this task.

The next question concerns the **kinds of data required** for the inclusion into the database. Since the database is oriented towards actual clinical applications in genomic medicine, its data pool should contain all of the information required for elaboration, verification and interpretation of the genetic tests. Apparently, the backbone of the database is a list of the CVD-associated genes. We have compiled this list in the Case Study of the Chapter II. Generally, there are several common polymorphisms for each of these genes. For each of the polymorphisms, there are many associations with the specific CVDs and other diseases. Each polymorphism needs also to be described with the exact nucleotide position: either by a dbSNP identifier (www.ncbi.nih.gov/SNP) or, if dbSNP data are not available (which is very often the case), by an exact numerical position in a certain sequence of the gene available in GeneBank (www.ncbi.nih.gov/Genbank/). Then, the frequencies of the polymorphism in certain populations, the trivial identifiers of the polymorphism, cross-links to other databases such as OMIM *etc* are required. The forgoing set of data might be called as a "technical description" of the polymorphism.

Apart from this technical description, also a "biomedical description" for each polymorphism in the database needs to be provided. The medical information is taken from the selected association studies and includes CVD phenotype, population, cohort parameters, risk factors analyzed, OR, P-value, CI *etc*. This medical information is to be complemented by the gene-gene or polymorphism-polymorphism interactions detected during the association studies as well as associations of the polymorphisms with non-genetic factors (such as "gene-nutrient" associations). These and other kinds of data (illustrated in the Figure 4-1), are far from being some information of general interest. Each of these items of information, as we'll see further in this Chapter, in the Chapter V and in the Volume II, will be required either for the elaboration of the genetic tests or for the verification and analysis of the information included into the database and/or for interpretation of the results of the genetic tests.

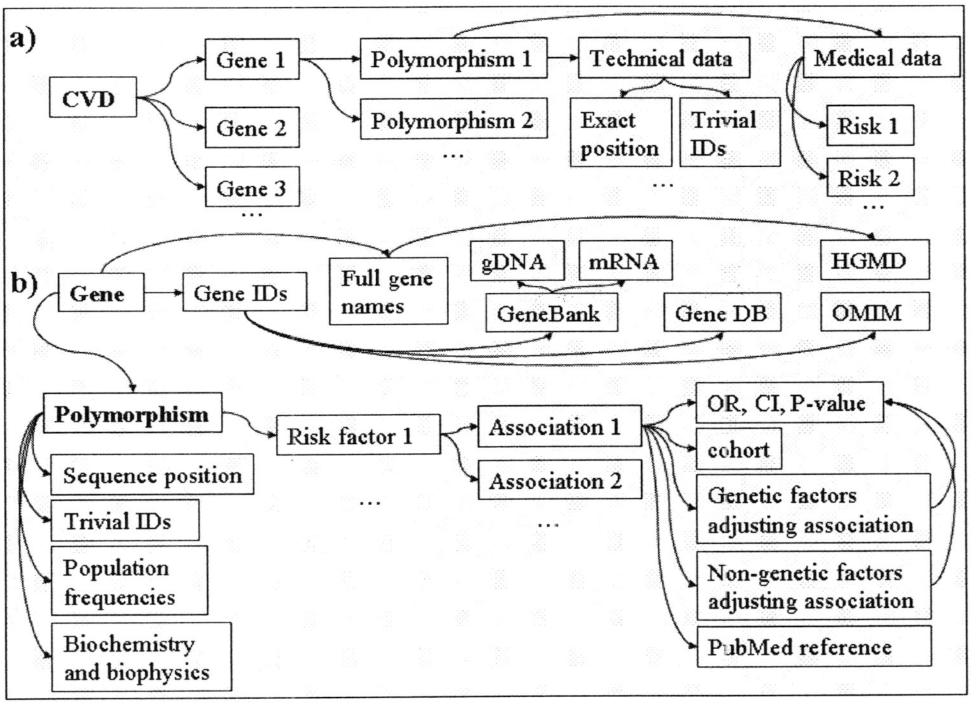

Figure 4-1. Data structures in the *medSNP* database. a) General data structure; b) detailed data structure for a particular gene/polymorphism.

Having decided on what kinds of data are to be included, we need to know **from what sources the data should be obtained.** Bioinformatics deals with storage, retrieval and analysis of biological and biomedical information. Therefore, it is important to use the most efficient methods of the information retrieval each particular case of research. We are aware, reading this book so far, of the enormous amount of biomedical information, represented in terms of the PubMed abstracts alone. In relation to the genetic determinants of diseases, we can conceive several rational ways to retrieve those arrays of information (Chapter II): a direct search of a database of annotated human genes; a comprehensive analysis of a database of monogenic hereditary defects that are known to lead to severe CVD problems; a direct literature search for the association studies using an abstract/full text database of publications. Apparently, each of these bioinformatic approaches has a number of advantages and disadvantages. It should be noted, before all, that not all of the well-known huge biological databases, usually associated with bioinformatics, are exceptionally regularly updated. Moreover, these databases are almost never checked for errors and inconsistencies of the data crammed into them. Therefore, the Pubmed/MEDLINE database of publications remains the primary source of most of the published biomedical information. What is the strongest side of using MEDLINE is that this database is, at least, exceptionally regularly updated. The technical information, such as nucleotide sequences, can be procured from specialized databases such as GenBank and others.

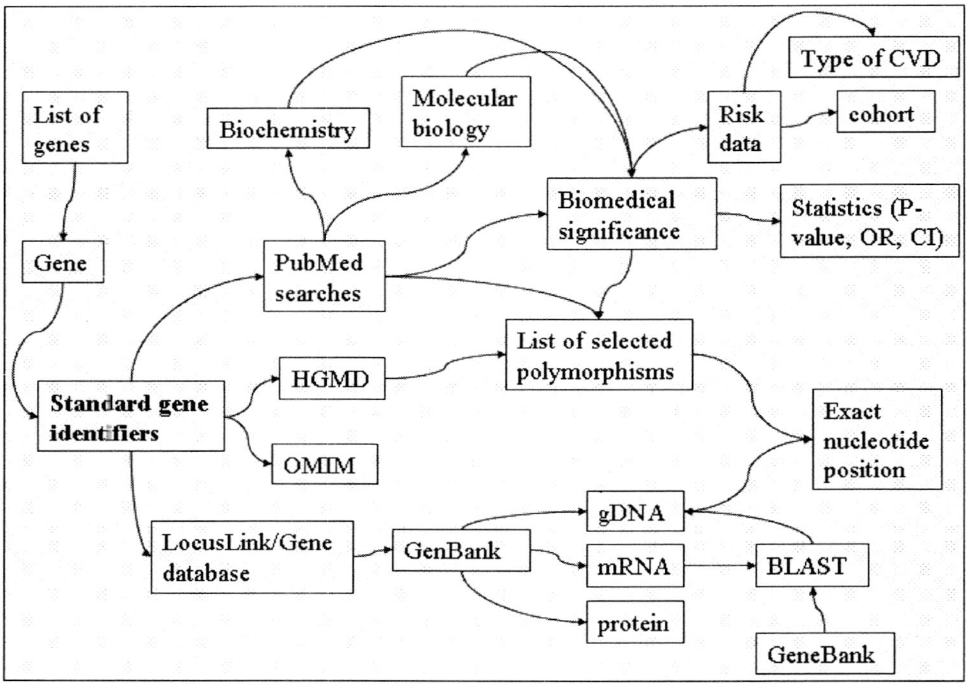

Figure 4-2. The data retrieval scheme for the compilation of *medSNP* database. OMIM, Online Mendelian Inheritance in Man, a database of human genes and genetic disorders (www.ncbi.nlm.nih.gov/omim/); HGMD, Human Gene Mutation Database is also a database of the human gene mutation data that features references to few publications and locus specific databases (www.hgmd.cf.ac.uk/); LocusLink/Gene is a searchable database of the annotated genes (www.ncbi.nlm.nih.gov/entrez/query.fcgi?db=gene); GenBank (www.ncbi.nlm.nih.gov/Genbank/) is a database of nucleotide sequences, that contains the data on the genomic DNA (gDNA), matrix RNA (mRNA) and coding DNA (cDNA); PubMed is the search service that provides access to over 11 million citations in MEDLINE and other related databases (www.ncbi.nlm.nih.gov/entrez/); BLAST (The Basic Local Alignment Search Tool) is a common tool for the sequence similarity searches (www.ncbi.nlm.nih.gov/blast). The program finds regions of local similarity between sequences and sometimes is necessary to select the most accurate sequence for the gene among those available in the GenBank.

Having defined the purpose, contents and the sources of information, the last issue to be decided upon is **how to locate and to extract the required data** among many thousands of publications. We have covered the issue, in part, while considering the "studies of the studies" in the previous Chapter. Given the data structure of the database we planned (Figure 4-1), it's apparent that to obtain both technical and biomedical pieces of the required information we might need to use a number of auxiliary databases. A general structure of the data retrieval for the database compilation is presented in the Figure 4-2. This scheme of the data retrieval process, though it does contain many of the items of the data structure, is distinctly different from the data structure itself. The structure of the data retrieval process illustrates the important items to be found and the order of their extraction.

One of such important items is the list of identifiers. It's no secret that every researcher would give the name of his or her own choosing to any gene or polymorphism. Consequently, there is no gene or polymorphism that would have only one unique and commonly acceptable identifier. For example, the gene known as "tissue factor pathway inhibitor" has "standard" identifiers "TFPI" or "TFPI1". Some other researchers might use other synonyms in their publications. In the case of TFPI the most common synonyms are "extrinsic pathway inhibitor" ("EPI") and "lipoprotein-associated coagulation inhibitor" ("LACI"). TFPI inhibits the tissue factor coagulation pathway. The gene known as "tissue factor" (as well as "thromboplastin" and "coagulation factor III") has a number of synonymous identifiers: TF, F3, TFA and CD142. This and other similar rules can be helpful also for the retrieval the results of molecular biological and biochemical research which are required for further analyses of the validity and relevance of the biomedical data (Chapter V). Facing the actual abundance and variety of the published data, it is clear that no computer algorithm or AI system will be as efficient and selective as the expert judgment aided by the skill of the rapid and efficient reading through thousands of the publications.

MAKING THE DATABASE

The database outline, presented in the previous section, is a brief summary of what needs to be decided *prior* to making a database. Many issues mentioned there might seem as "banal" or "self-evident". Hopefully, all the readers who abundantly use these and similar words did already drop out after reading our Introduction and the preceding Chapters... Briefly, there are no "banal" issues when it comes to actual database compilation. Here, we'll discuss some of the important practical points that became clear during the actual making of the author's *medSNP*™ database.

The primary purpose of the database is to make an inventory of the reliable medical data from the genotype-phenotype association studies dealing with CVD. Although the published data on the polymorphism-CVD associations are quite extensive, bioinformatic techniques can significantly alleviate the extraction of the data using the methods described in the previous chapter. In *The Case of A GeneScape for CVD* we had, actually, obtained a selection of a hundred of the genes for which association studies were made. These data were compiled into the *CVD GeneScape* diagram, which features the genes involved into various aspects of the cardiovascular physiology (Figure 2-5). The genes include those involved in vasoconstriction and vasodilatation, renin-angiotensin system, maintenance of the electrolyte balance, vascular constitution/remodeling, formation of the atherosclerotic plaques (lipoproteins and inflammation-related genetic determinants of atherosclerosis), adipose metabolism and, of course, the genes of the blood coagulation cascade (Appendix I). For most of these genes, many association studies of acceptable quality are available. The *CVD GeneScape* diagram, actually, serves as one of the main entry points into the *medSNP™* database.

For each of these genes, a certain data structure with the data fields in accordance with the fields in the database outline (Figure 4-1) has to be filled in. The Figure 4-2 presents the general schema of how this is to be done. For each gene in the list, there is a single entry comprised of a number of different records (Figure 4-3). As there are two kinds of the information (the technical and the biomedical), there are two distinct types of the records: those containing the technical data (such as the records GeneID, GBgDNA, GBmRNA, OMIM *etc*) as well as those containing the biomedical data (FRMUT, RISKF) including biochemical, biophysical and physiological data (BIOCHEM, BIOPHYS, PHYSIOL *etc*). As it can be seen from the Figure, all of the data are organized in the tagged text format, much like the original tagged text format of the SwissProt and many other databases (http://www.expasy.ch/sprot/). It is known for decades that this kind of the database format significantly alleviates both expert and computer analysis. The biomedical data need to be collected and analyzed manually, from the abstracts and the full texts of each publication, taking into account the points we discussed in the previous Chapters. At the same time, the analysis of some of the technical data is alleviated by means of the computer and a detailed procedure for this analysis is presented in the Figure 4-4.

```
>load F5.msnp
>gene F5
F5>dump ent
GeneID   F5
Synonym  "coagulation factor V"
Synonym  "protein C cofactor"
Synonym  "proaccelerin"
GBgDNA   AY364535
GBmRNA   NM_000130
OMIM     227400
MEDSNP   "R506Q","Leiden";
NPPOS    gdna:38592;na:G>A;cdn:506;aa:Arg>Gln;
BIOCHEM  alters anticoagulant response to APC (APC resistance)
FRMUT    f1:92;f2:8;pm:8566967;co:Germany;pt:180;pp:Caucasian;
FRMUT    f1:96.5;f2:3.5;pm:7803250;co:UK;pp:Anglo-Saxon;pt:144;
RISKF    rfid:"genetic";OR:3.7;ph:"thromboembolism, venous";co:USA;pm:11929758;pt:335+688;
RISKF    rfid:"smoking";OR:1.7;ph:"thrombosis";gender:M;age:40-84;pm:7586244;co:USA;pt:77;
RISKF    rfid:"genetic";OR:2.8;ph:"myocardial infarction, without significant coronary ste
RISKF    rfid:"F2 20210G>A";ph:"venous thrombosis";OR:2.6;pm:10477778;co:Italy;pt:624;
RISKF    rfid:"oral contraceptive";OR:30;ph:"thrombosis, cerebral sinus";gender:F;age:18-5
.
.
.
```

Figure 4-3. A database entry (on the example of the F5 gene). This is a fragment of a screenshot taken in the OS *medSNP*.

Since the database is oriented towards clinical applications in genomic medicine, even seemingly "minor" technical details will be very important for the application to be successful. As the technical details are exceedingly important, some of the algorithm features given in the Figure 4-4 ensures the double-checking of the consistency of the technical data. We'll illustrate the point by using some of the examples of the data inconsistency that were found while making the database when discussing the ortho-analyses of the data in the Chapter V.

Thus, after collection of the data from publications and the analyses of the consistencies of the technical data (Figure 4-4), all of the gene-related and polymorphism-related the data are represented in the tagged-text format (Figure 4-3). What comes next is the stage of the selection of the polymorphisms and of the reliable publications representing the biomedical data. This is done during the preliminary database analysis.

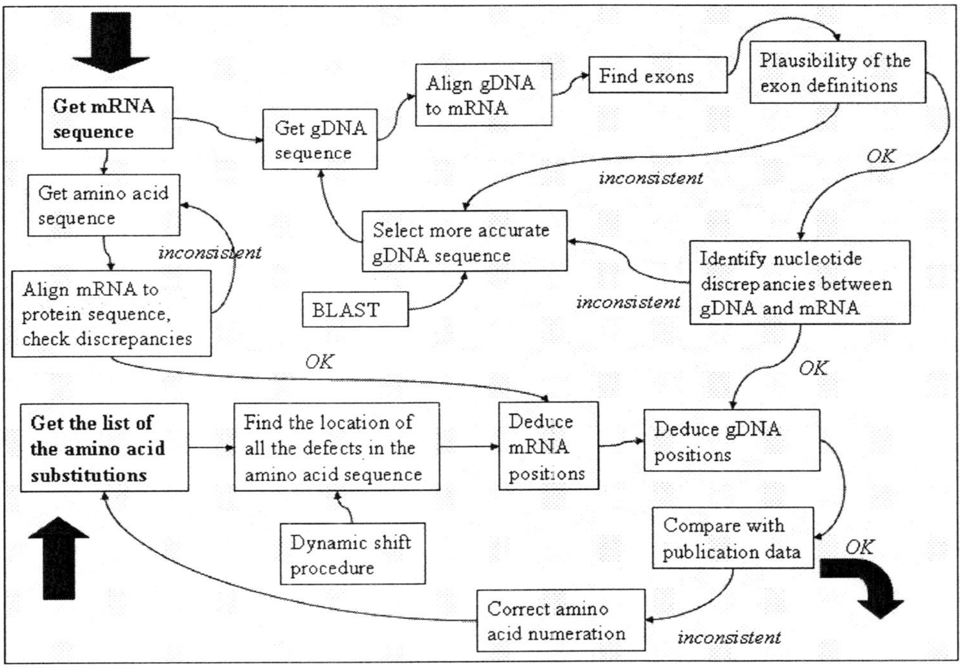

Figure 4-4. The core procedure for the analysis of the consistency of the technical data in OS *medSNP*. This fragment of the OS *medSNP* represents, mainly, the analysis of the polymorphisms corresponding to the amino acid substitutions. The data on the promoter (regulatory) nucleotide substitutions, deletions and insertions can only be extracted directly from the publications.

Preliminary Database Analysis: Selecting the Studies and the Polymorphisms

Given that all of the described procedures were carefully performed for each of the genes in the list of the CVD genes (Figure 2-5), we will be left with thousands of the RISKF and other records in the database file (Figure 4-3). Even if the spurious studies were largely excluded through expert analysis assisted by the criteria we've discussed in the preceding Chapters, still there will be quite an amount of information encoded in the "technical" and "biomedical" records of the *medSNP*™ database. This information is available for a number of polymorphisms known for the gene under consideration. Prior to conducting the comparative and the integrative analyses of these data, which we consider in the next chapter, there are several preliminary analyses to be done. At this stage, the polymorphisms are sorted and classified according to a certain set of criteria that can allow selection of the most important genetic markers for the clinical applications of interest.

Two points in relation to this task are clear. First is that the selection of the polymorphisms is dependent on the selection of the studies. Second is that a set of criteria to select the most important genetic markers is largely empirical. That is, each of these criteria was identified during the actual information digging as being a helpful aid. Some of these criteria deal with the "technical" data (such as polymorphism type and population frequency), while most of them address the sort, the quality and the availability of the biomedical information (Box 6). The most reliable, from the biomedical point of view, will be those data in *medSNP* that comply with all the criteria.

Box 6. The criteria of the polymorphism selection

The summary of the empirical criteria which we discuss in the text

Criterion 1: Preferred polymorphism type: SNP. The polymorphism frequency in the targeted population should be not less than 0.2..1%. Preferable populations: Caucasian.
Criterion 2: Preferred type of biomedical evidence: data from case/control studies.
Criterion 3: Scientific significance of the findings of each particular association study; statistics and common sense.
Criterion 4: Biomedical significance of the findings; whether the study takes into account other factors.
Criterion 5: Reproducibility of the association in independent studies of the similar population (say, Caucasians).
Criterion 6: Preference of the functionally important polymorphisms (in distinction from mere genetic markers).

The *first criterion* concerns the **type of polymorphism**, the population frequency of polymorphism and the targeted population. Although "SNP" tends to become more and more of a media word than a term of biomedical significance, most of the polymorphisms in *medSNP* are, indeed, SNPs (*single* nucleotide polymorphisms). However, it does not mean that SNP is the only type of medically significant genetic polymorphisms. There are a number of important common polymorphisms of insertion/deletion type, such as PAI-1 5G>4G; ACE I/D, APOB SPins/del and others. Nucleotide repeats (such as penta-nucleotide repeats in lipoprotein (a)) are also known to be important biomedical markers. Generally, the type of polymorphism cannot be a criterion for selecting polymorphisms for their biomedical value.

Apparently, polymorphisms occur with different frequencies in various human populations. For instance, polymorphism E487K of the gene ALDH2 (acetaldehyde dehydrogenase) occurs with frequencies of 16%-25% among various Asian populations. The K487 variant (or "ALDH2*2") was associated with pronounced alcoholic dependency in numerous independent studies. At the same time, this polymorphism was never found to occur in Caucasians at frequencies higher than 0.1%. Contrariwise, the F5 polymorphism R506Q "Leiden", a common example mentioned throughout this book series, occurs at frequencies 2-5% among various Caucasian populations but it was never found to be a common polymorphism among Asians or Native Americans.

The cutoff value for the *populational frequency* of the polymorphisms was deliberately made somewhat vague (Box 6). By itself, any kind of such cutoff is an arbitrary value of no biomedical significance whatsoever. For example, in a US study the variant L28P in the apolipoprotein-E was found to be a strong predictor of the Alzheimer disease (OR >13), especially in combination with APOE*4 (OR=29.1). The population frequency of L28P was <0.2% [2]. In an independent Italian study [3] the L28P was also found to be a significant risk factor of the disease (OR>10) with the population frequency <0.5%. Clearly, using "1%" as a stringent cut-off would not be an adequate practice of polymorphism selection. The population frequencies of polymorphism for various populations are important for the comparative analyses of the results of the association studies (Chapter V).

The preferable *population* in *medSNP* database is Caucasian. Nevertheless, the data are not restricted to Caucasians. Generally, the population/ethnicity is an important biomedical variable since the biomedical effects of polymorphism deduced from association studies tend to differ even among different cohorts, not mentioning the inter-population differences (Chapter V). The data collected in *medSNP* suggest a number of examples of the kind. One of them is the HindIII (H+/H-) polymorphism of the LPL gene (lipoprotein lipase). The HindIII H+, which is a G-to-T SNP, occurs at frequencies of 50-65% among European populations. It was found to be a reliable genetic marker for atherosclerosis-related myocardial infarction in Caucasian populations (Russian, French, Italian, white Anglo-Saxons and others) and Asians (Japanese). Despite the similar frequency of occurrence (50%) is distinctly non-significant as MI marker in, for example, Saudi Arabs. Though dietary and life-style differences among these populations are likely to be the major cause for the discrepancy, LPL H+ appears to be a population-specific CVD marker.

The *second criterion* establishes the preferred type of the **biomedical evidence**: the data abstracted from case/control association studies. As we remember (Chapter III), the observational study design is the most appropriate for an epidemiological study. In an *observational study*, the variables of interest are observed and not directly manipulated. The most common feature of an observational study, independently of its scale, is that it usually examines past events (retrospective). The scales of an observational study may vary greatly: it can be limited to a single patient ('case study'), may involve a family, a number of families or a large cohort of patients/healthy controls observed for a certain period of time. The types of the observational phenotype-genotype studies are summarized in the Figure 4-5. Each of the three main types of the phenotype-genotype studies (case study, family study and association study) has its own function in characterization of effects of a polymorphism. A future genetic marker is often first discovered in a single patient, then analyzed in a family study and then is studied in the larger framework of an association study.

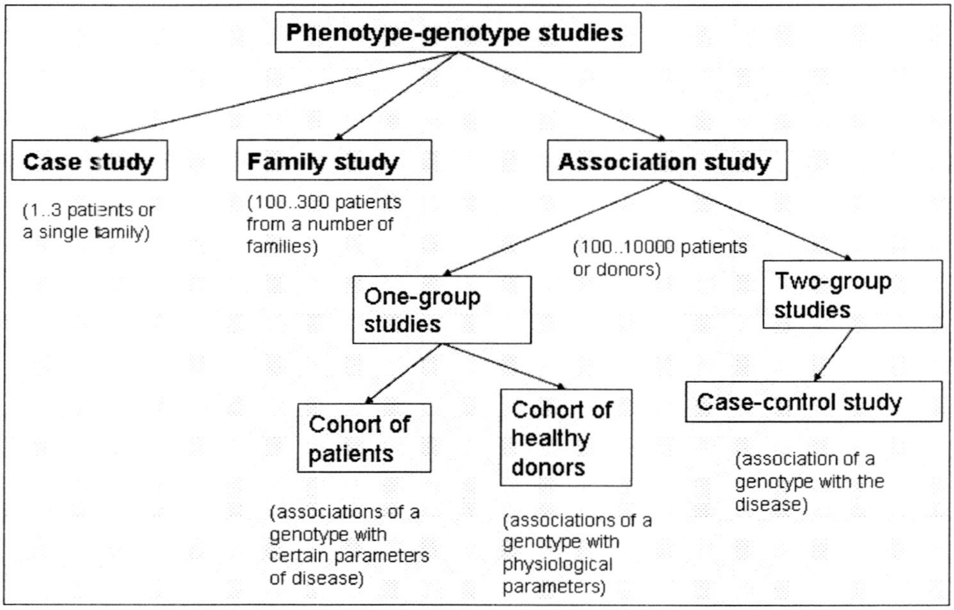

Figure 4-5. The general types of the observational phenotype-genotype studies. The diagram includes the experimental studies and does not include theoretical analyses (such as meta-analysis). See details in the text.

The three kinds of observational studies produce results which are complementary. For example, if a polymorphism occurs at very low population rate, then the case studies of the polymorphisms represent, often, the only available evidence of clinical significance. If a polymorphism is of low occurrence in the general population but occurs at considerable rates in groups of patients, then a family study might become possible. At higher population frequencies, both family study and association study become possible. A family study and association study might produce contradictory results (see the alpha-2-macroglobulin example, "Para-analysis", Chapter V). Generally, the case studies and the family studies are helpful to indicate potential usefulness of a polymorphism as a genetic marker but the availability of the reliable results of the published association studies is the central point of this criterion.

As the Figure 4-5 indicates, there are three major kinds of the phenotype-genotype association studies: the two-group case-control studies, and the one-group studies that can include a single group of patients or a single group of healthy donors. All of these types of the association studies produce complementary data that are valuable for the integrative analyses and for the elaboration of subsequent clinical applications. Our Case Study of the *IHD200* cohort provided an illustration for both the two-group and one-group studies: the two-group study was the case-control study of the genetic predisposition to CHD while the one-group study was the analysis of the group of the CHD patients concerning the genetic predisposition to DVT. Clearly, both association studies produced different but complementary data.

As an example of a one-group study dealing with healthy donors we can use the analysis of the effects of the APOC3 promoter polymorphisms on the plasma levels of non-ester fatty acids and insulin response [4]. In this prospective cohort study (223 male, 279 female) the effects of the -482C>T, -455T>C variants of APOC3 gene were studied in relation to the lipid phenotypes. It was found that levels of the non-ester fatty acids were higher in the carriers of the homozygotes of the rare variants and in men only. A strong correlation between the polymorphisms and smoking affecting fasting triglyceride was identified both in men and women. In particular, the data suggest that men who carry the rare APOC3 promoter variants tend to have a disturbed glucose homeostasis and less favorable lipid phenotype [4]. These data well complement the data from other association studies of the APOC3 variants in *medSNP* and are used in the subsequent integrative analyses of the data for this particular polymorphism.

The *third criterion* concerns the **scientific significance of the findings** reported in a particular association study. The studies of genetic association are based mostly on statistical analyses and any results thus obtained characterize only probabilities of interaction between the factors involved. Accordingly, these associations *per se* do not imply any sort of direct causation. Statistically, the validity of a particular association is judged by the $\chi 2$-test and is expressed in terms of P-values, ORs and CIs. Previously, we noticed that one cannot judge a value of the study merely on the base of the P-value alone, using Neanderthal thinking of the kind "all studies with $P<0.05$ are scientifically valid". Apparently, scientific validity of a study includes statistical significance as only one of the criteria. Integrative analyses of the data from various levels of function and of the data from various studies can well illustrate the point. Indeed, what's the point of talking about the "statistical significance of the association of F2 20210 G>A with thrombosis" when, in a particular study, the researchers, actually, studied entirely different polymorphism? (see Chapter V for additional details). Nevertheless, only in some cases scientific significance or lack thereof can be unequivocally judged by using the cross-level comparisons such as this and, generally, this is the matter of the expert judgment which cannot be described just in few words.

The rest of the criteria in the list (Box 6) also deal with the problem of the scientific significance. The scientific significance is closely related to the general **biomedical significance of the findings**, which is the *fourth criterion* for the selection. Supposing that in a publication there are no obvious contradictions and the reported statistics is acceptable (at least, according to the reported P-values). Does this mean that a particular study has a distinct *biomedical* significance? Not necessarily. The analysis of the P-values can, indeed, exclude clearly "negative findings". However, as we've already seen in the *Case of the IHD200*, a negative finding can be the direct result of the lack of biomedical characterization of the patients (or, at least, lack of that information during bioinformatic analyses). Although we still consider publications reporting the negative findings at the stage of "para-analysis" (Chapter V), a negative finding by itself is of no biomedical value.

On the contrary, even if a particular P-value is well below 0.05 thus indicating a statistically significant correlation, P-value by itself it is not an indicator of the extent of the biomedical significance of the association. The OR-values *cum* CI provide more information concerning the quality of the statistics of the particular study. We already discussed the interpretation of OR-values in the Chapter III. Here, let's consider a very crude formula which, nevertheless, can give a general idea of the bioinformatic significance of the OR-values. This formula defines a parameter which we call here "prediction accuracy":

Prediction accuracy=100-50/**OR**=50(2-1/**OR**) (*for OR>1*)
Prediction accuracy=100-50***OR**=50(2-**OR**) (*for OR<1*)

Ideally, this "prediction accuracy" represents the percentage of the correctly made diagnoses using the conditions stated for a particular association. For OR=2, the "prediction accuracy" parameter is estimated to be 75%. Let's elaborate on the following two examples from *medSNP* (both dealing with F5 Q506 "Leiden"): a venous thromboembolism (VTE) study [5] and previously mentioned study of cerebral sinus thrombosis [1]. The first is a US-based multicenter study of 335 patients who developed VTE during 8 years of follow-up and of 688 controls. The odds ratio of VTE was 3.7 in participants carrying factor V Leiden compared with non-carriers. According to the above formula, the "prediction accuracy" of VTE using the results of the genetic testing on F5 would be at 86%. As most of the bioinformatic methods, in any area from protein structure prediction to clinical genetics, rarely produce prediction accuracy better than 70-75%, 86% is a high figure indeed. Of course, this is (1) an *average* "prediction accuracy" of particular finding; (2) it is derived from a *particular study* and (3) it is calculated on the base of OR value alone disregarding CI.

Another example is the association between the oral contraceptives, the F5 Q506 polymorphism and incidences of the cerebral sinus thrombosis found in a Dutch study we already mentioned [1]. According to the results, the OR for cerebral sinus thrombosis in the case of females taking oral contraceptives was 13 (age-adjusted, 95% CI 5-37). In the case of females taking oral contraceptives *and* having a prothrombotic defect (such as F5 "Leiden"), the odds ratio for cerebral sinus thrombosis were about 30.0 [1]. Therefore, this particular combination of the conditions of association (gender: female, drug: oral contraceptive, SNP: F5 "Leiden", age-adjusted), in this particular study, gives a prediction of accuracy of 98.3%, which is an exceedingly high value for predictions in medicine and biology. Although this kind of the crude interpretation of OR-values is a subject to a number of plausible criticisms, these estimates can illustrate the bioinformatic significance the OR-values and provide an *upper* limit for the actual prediction accuracy of a particular study.

Further analysis of the data collected in *medSNP* database suggests that most of the reliable studies report OR-values greater than 2.0 or less than 0.5 which corresponds to the "prediction accuracy" better than 75%. The Figure 4-6 details these data. Thus, OR-values >2.0 and <0.5 comprise 77% (925 out of 1200) of the associations registered in the database while the OR-values in the 0.5<OR<2.0 interval (the least significant findings) comprise about 20% of the raw array of the *medSNP* data. These figures imply that at least 77% of the reliable association studies *would* allow to achieve at least 75% "prediction accuracy" on a single genetic marker. Figures like 75% and 77%, from bioinformatical point of view, suggest a good prospect for elaboration of the computational diagnostic techniques. Having said this, we ought to remark that OR-value alone is a poor descriptor of the statistical quality of the reported association. CI-values provide more complete a description of the statistics.

According to the results of the *medSNP* analyses we present in the Volume II, the statistics presented in the Figure 4-6 somewhat overestimates the percentage of the reliable association studies with strong statistical justifications.

Another aspect of the biomedical significance of the findings is *whether the study accounts for other factors* (in particular, polymorphisms in other genes and external factors). In other words, the notion of the biomedical significance of a reported genetic association includes the data on the "interaction" of the polymorphism of interest with those other factors. Although, generally, the studies reporting significant interactions of polymorphisms are not very abundant, the reliable studies dealing with "gene-nutrient" or "gene-environment" types of interactions are considerably more often. Statistics on these kinds of data collected into *medSNP* is presented in the Table 4-1. Despite the common concerns addressing the "study power" which usually arise in the course of the stratifications, associations reporting other conditions (such as involvement of specific external factors) are more valuable for the clinical applications than mere "genetics-only" associations (Volume II).

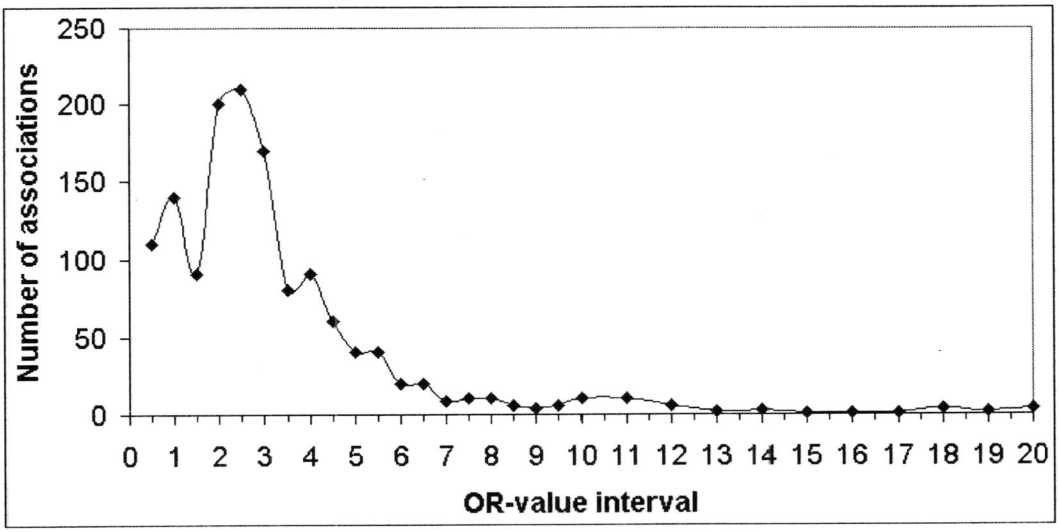

Figure 4-6. Statistics of the OR-values for the associations registered in the *medSNP* database. The data were collected from over 1200 of the RISKF records reporting results of reliable association studies. Each dot represents the number of studies for the fixed interval of OR-values. For example, "1.0" in the diagram stands for OR values in the interval $0.5 \leq ORs < 1.0$ registered in the database. Studies with OR=1.0, corresponding to "negative findings", are not registered in *medSNP*. Since there were only 6 studies reporting OR>20.0, the diagram represents OR-values less than or equal to 20.0.

Table 4-1. Phenotype-genotype associations: "genetics-only" vs "interactions" registered in *medSNP*. "Gene-environment" includes interactions with smoking, alcohol intake and other environmental hazards. "Other" includes associations with physiological parameters analyzed in clinical lab tests (such as lipid and coagulation profiles)

Type of association		No. of associations in *medSNP*
Genetics-only		**797**
Gene-environment	Smoking	95
	Alcohol	65
	environmental hazards	20
	other diseases	45
	Totally	**205**
Gene-gene		**115**
Gene-drug (pharmacogenetics)		**40**
Gene-nutrient		**38**
Other		**73**

The data in the Table 4-1 indicate an apparent prevalence of the "genetics-only" genotype-phenotype associations. Albeit this result might seem to be an argument in the favor of the genetic association studies, from the point of view of the levels of function (Figure 1-5), the "genetics-only" associations are of extremely limited value for biomedicine. A "genetics-only" association implies an association of one polymorphism with only one phenotype and without any adjustments for any additional factors except age, gender and population. Such an association, apparently, does not explicitly describe the influence of all the other important factors such as diet and life habits. Age, population, genetic composition and gender are non-modifiable risks of CVD (Chapter II). At the same time, according to WHO reports, it is the management of the modifiable risks (dietary measures, physical activity, smoking abstinence) that considerably reduces CVD incidence thus being of the primary interest of the clinical medicine. Accordingly, the "genetics-only" associations are very likely to downgrade the importance of the modifiable risk by hyping contribution of the genetic factors to the CVD etiology.

Nevertheless, although the over-simplified associations of the "genetics-only" kind comprise the bulk of the reliable associations, the associations involving additional factors are far from being infrequent (Table 4-1). From a medical point of view, the associations involving other factors are considerably more valuable for diagnostics and other practical purposes than genetics-only associations of the kind "polymorphism A1 in gene AAA increases the risk of MI (OR=1.5, 95% CI=1.1-2.9)" *etc.* Indeed, even if we remain merely at the dogmatic stage of thinking (which states that *both* genetic makeup and environment influence predisposition to disease), it is still clear that the value of the genetic-only associations is extremely limited since they are not able to provide any patient-specific context except age, gender and population. In the case of the gene-environment, gene-gene *etc* associations, the clinical context is somewhat more elaborated so the value of such associations for biomedical interpretations grows (albeit, surely, the statistical power of the association decreases). Accordingly, it is the polymorphisms with this kind of association data available that are likely to be more important for elaboration of the disease-specific genetic profiles for the clinical gene testing (such as the ones we present in the Volume II, the volume which, as the readers might have guessed, deals much with the practical applications of the *medSNP*).

The fifth criterion of the polymorphism selection deals with reproducibility of the associations found in the course of **independent studies** performed on similar population cohorts (say, Caucasians). First of all, the statistics of a particular study is relevant only to that particular cohort in that study. P-values characterize the probability that associations reported in the study are random findings, while OR-values and confidence intervals further describe the statistical quality of the found associations. Nevertheless, the cohort-related statistics of a particular association study does not provide any other means to verify, shall we say, the "statistical stability" of the results. In other words, what will happen with the "positive" association, if a few specific patients are removed, without affecting much the group averages? How significant will the association continue to be if another random group of controls is chosen for the very same study? Usually, association studies do not involve similar analyses so the "perturbation" and "scrambling code" techniques we consider in the Volume II of the present book series are quite novel in the area. Hence, as a particular published study cannot always be characterized in terms of what we called here as the "statistical stability", there are inevitable discrepancies and sometimes controversial conclusions among the results of the independently made association studies. The controversies often occur even if the independent studies were made on the different cohorts of the same population. Accordingly, the availability of the concurring studies of the polymorphism that were made by independent groups of researchers is an indication of the potential value of the corresponding phenotype-genotype association. There is a variety of polymorphisms in the *medSNP*; some of these polymorphisms were studied only for several years and are characterized by but a few association studies. However, for the bulk of the polymorphisms in the database (>85%) several independently made association studies are available.

Finally, the *sixth criterion* concerns the **preference of the functionally important polymorphisms** to mere genetic markers. This criterion requires a short excursion into genetics, molecular biology and biochemistry. A "*genetic marker*" is any polymorphism that has been associated with certain medical condition. A genetic marker itself may or may not affect any physiological changes. Given a certain population (say, Caucasian), a polymorphism can be associated with the condition of interest simply because of the phenomenon of linkage disequilibrium. Thus, a marker might not have any direct physiological significance and only reflects a pattern of hereditary transmission of the genetic material describing the population under study.

In the case of a *functionally significant polymorphism*, the nucleotide change directly affects either the function of the gene or of the gene product(s). Such a nucleotide change, when placed in the promoter, can affect a several-fold change in the gene transcription. Some other nucleotide change, located at an intron-exon border, might influence RNA splicing. A polymorphism in the DNA area coding for the protein might lead to an amino-acid substitution, truncation or even a deletion of the polypeptide sequence which, quite often, observably affects the protein function. To elucidate the cellular or biochemical functions of a particular polymorphism, a number of biochemical and molecular biological studies are to be performed.

The following examples detail the point. For instance, variant -455C of the gene APOC3 was found to be defective in its response to insulin treatment, the loss of insulin regulation was mapped to changes in an insulin response element [6]. Our common example of the F5 Q506 variant "Leiden" is known to affect the protein-protein interaction between F5 and protein C. A C-to-G transversion in LPL gene directly affects the amino acid sequence resulting in a premature termination codon (Ser447X) and thus is a functionally important polymorphism. At the same time, the latter polymorphism is in strong linkage disequilibrium with Hind-III RFLP (restriction fragment length polymorphism) in intron 8 which, at best, can serve only as a genetic marker for a particular Caucasian population [7].

In sum, the criteria of the polymorphism selection and analysis we just considered can serve as preliminary means for selecting polymorphisms that are likely to be the most reliable for applications of genetic testing in clinical setting. Further indications of the biomedical significance of a polymorphism in a gene and the association reported for it follow from the comparative and the integrative analyses we consider in the following chapter. Before that, let's make a brief overview of the *medSNP* database.

EXPLORING *MEDSNP*: A BRIEF LOOK

The *medSNP* database contains all of the genes mentioned in the *CVD GeneScape* diagram (Figure 2-5). In addition, the database features data on the polymorphisms in genes of alcohol metabolism (ADH1B, ALDH2); cytochromes and other genes involved in detoxification (CYP1A1, CYP1A2, CYP1B1, CYP2C19, CYP2C8, CYP2C9, CYP2D6, CYP2E1, CYP3A4, CYP3A5, GSTM1, GSTP1, GSTT1, EPHX1, NAT2); interleukins and chemokine receptors (IL1A, IL1B, IL2, IL6, IL10, IL13, IL15, IL18, IL1RN, IL18BP, CX3CR1, CCL2, CCR2, CCR5) and a number of other genes (see, for instance, Chapter VI). In the current version, there are entries for over 370 polymorphisms in 145 genes. The data on these polymorphisms were collected from about 4000 publications which, in turn, were selected from tens of thousands of publications. Although "medSNP" is, apparently, acronym for the "Medicinally-important Single Nucleotide Polymorphisms", other types of polymorphisms are also registered, such as repeats, insertion/deletion polymorphisms *etc*.

There are three different entry points to the database: the GeneScape diagram, the polymorphism selection menu and the command-line interface. An updated GeneScape diagram (similar to the one shown in the Figure 2-5) allows browsing the polymorphism data in accordance with the general physiological mechanism of interest. The selection of the general mechanism (for instance, blood coagulation cascade, Figure 2-5b) is followed by the selection of the gene (say, F5). Each gene is associated with a list of polymorphisms which are represented in the "gene data" screen (Figure 4-7).

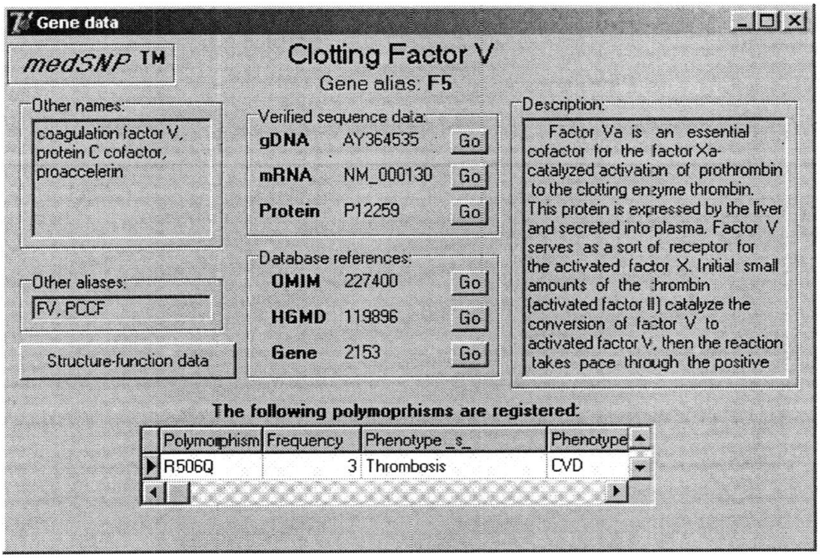

Figure 4-7. "Gene data" screen for F5 in medSNP database.

Then, selecting a polymorphism in the table leads to the "polymorphism data" screen which represents the array of the technical and biomedical data for a particular polymorphism (Figure 4-8). The data in the screen are those described in the database structure outline (Figure 4-1) and also include short summaries of the known biochemistry and biophysics of the effects associated with particular polymorphism.

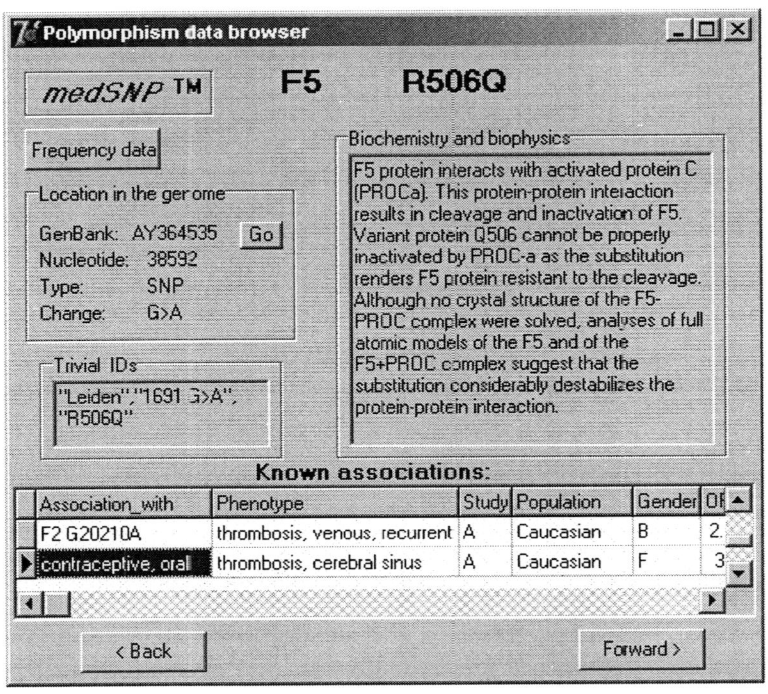

Figure 4-8. The "polymorphism data" screen for the F5 R506Q polymorphism.

The *second entry point* to *medSNP* is the polymorphism selection menu (Figure 4-9). This entry features the criteria we have considered in the previous section and allows efficient selection of the polymorphisms in accordance with the availability of the technical and biomedical data of interest (such as population, population frequency, molecular function of polymorphism, availability of association with certain factors, relative quality of those of associations *etc*).

The *third entry point* is the primary database interface: a command-line database engine that includes the expert system of the OS *medSNP*. Command-line interface was designed in order to achieve greater flexibility in the database operation, scripting as well as to allow more efficient re-design of the database structure. As the menu-driven interface, the command line interface allows polymorphism selection and browsing. In addition, it allows scripting, modifications of the filters, direct corrections of the data in the tagged-test format and a number of options important for biomedical interpretations of the genetic profiles.

In sum, the *medSNP* contains a large volume of the systematically analyzed and critically assessed information on the clinically-relevant association studies of the genetic polymorphisms. This information is complemented by the biochemical, cellular and biophysical data on the functional effects of genes and polymorphisms. Despite that the database is, primarily, CVD-oriented, information on all phenotypes for each gene and polymorphism was analyzed. Thus, a number of non-CVD phenotypes are available which is important both for comparative and integrative analyses of the data consistency which we are about to consider in the chapter to come.

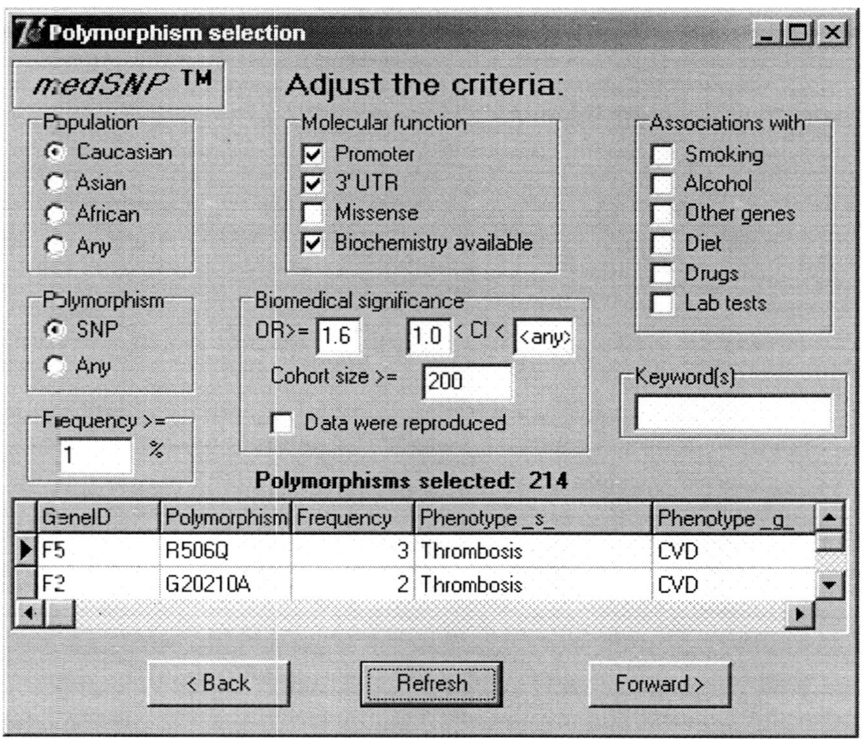

Figure 4-9. The "polymorphism selection" screen.

Chapter Summary

The simplistic approach that most of the published **studies of the genetic associations** feature is hardly sufficient to allow elaboration of the practical biomedical applications. Most association studies report effects of only few polymorphisms and tend to neglect the **major modifiable risks** of the corresponding phenotypes. Thus, unlike other biomedical studies, results of particular association studies rarely bear any direct implications for the clinical practice. The major problem that plagues practical application the **gene testing** is absence of any substantiated means to provide the **data linkage** between the **clinical observations** on a particular patient and the **genetic profile** of the patient. The ***medSNP* database/expert system**, described in the present chapter, is based on the results of thousands of the published association studies and serves as a bioinformatic base for linking the clinical and the genetic data in order to **interpret the results** of the gene tests. As such, the database has numerous **clinical applications** in post-genomic biomedicine.

References

[1] de Bruijn SF, Stam J, Koopman MM, Vandenbroucke JP. Case-control study of risk of cerebral sinus thrombosis in oral contraceptive users and in [correction of who are] carriers of hereditary prothrombotic conditions. The Cerebral Venous Sinus Thrombosis Study Group. *BMJ* 1998;316:589-92.

[2] Kamboh MI, Aston CE, Perez-Tur J, Kokmen E, Ferrell RE, Hardy J, DeKosky ST. A novel mutation in the apolipoprotein E gene (APOE*4 Pittsburgh) is associated with the risk of late-onset Alzheimer's disease. *Neurosci Lett*. 1999;263:129-32.

[3] Scacchi R, Gambina G, Ferrari G, Corbo RM. Screening of two mutations at exon 3 of the apolipoprotein E gene (sites 28 and 42) in a sample of patients with sporadic late-onset Alzheimer's disease. *Neurobiol Aging* 2003;24:339-43.

[4] Waterworth DM, Talmud PJ, Luan J, Flavell DM, Byrne CD, Humphries SE, Wareham NJ. Variants in the APOC3 promoter insulin responsive element modulate insulin secretion and lipids in middle-aged men. *Biochim Biophys Acta* 2003;1637:200-6.

[5] Folsom AR, Cushman M, Tsai MY, Aleksic N, Heckbert SR, Boland LL, Tsai AW, Yanez ND, Rosamond WD. A prospective study of venous thromboembolism in relation to factor V Leiden and related factors. *Blood* 2002;99:2720-5.

[6] Li WW, Dammerman MM, Smith JD, Metzger S, Breslow JL, Leff T. Common genetic variation in the promoter of the human apo CIII gene abolishes regulation by insulin and may contribute to hypertriglyceridemia. *J Clin Invest* 1995;96:2601-5.

[7] Stocks J, Thorn JA, Galton DJ. Lipoprotein lipase genotypes for a common premature termination codon mutation detected by PCR-mediated site-directed mutagenesis and restriction digestion. *J Lipid Res* 1992;33:853-7.

[8] Humphries SE, Ridker PM, Talmud PJ. Genetic testing for cardiovascular disease susceptibility: a useful clinical management tool or possible misinformation? *Arterioscler Thromb Vasc Biol* 2004;24:628-36.

Chapter V

GENETIC ASSOCIATIONS: ANALYSES OF THE DATA CONSISTENCY

INTRODUCTION

As we mentioned in the Chapter III, scientific quality of a particular biomedical finding is never simply the matter of the statistical significance of the corresponding association. Apart from the point that it's quite easy to forge statistical significance (by using selective evidence or otherwise), a statistical description by itself is hardly an adequate substitute for logical thinking and unprejudiced scientific argumentation. We also mentioned in passing that detecting consistency over the levels of function might be an important criterion of scientific validity. Also, we maintained that adoption of the post-genomic perspective can allow detection of the data self-consistency or, *vice versa*, of numerous inconsistencies. In this chapter, we discuss certain methods of scientific analysis that allow verification of the biomedical consistency of the data from independently made association studies as well as of the data from the studies of the corresponding genes and gene products. These methods of analyses are routinely applied to the data collected in the *medSNP* database that was compiled on the base of thousands of actually made association studies and which also incorporates a certain amount of data from studies in biochemistry and in molecular cell biology (Chapter IV). Verification of the data consistency is accomplished by using comparative and integrative analyses of particular association studies or of particular selections of association studies. While the comparative analysis (meta-analysis) always requires a set of highly similar (or "homogeneous") studies, the integrative analyses (ortho-analyses and para-analyses) can be applied not only to a group of heterogeneous studies but also to any particular study.

META-ANALYSIS

The prefix "meta" appears in words from Greek or in words composed after the Greek model of word formation. The prefix has a broad range of meanings including "after", "among", "along with" *etc*. Thus, the term "meta-analysis" implies a comparative analysis. In statistics, meta-analysis combines results of several similar studies that address closely related research hypotheses. Traditionally, the statistical meta-analysis is employed to overcome the problem of reduced statistical power in studies featuring small sample sizes. In statistics, there are several alternative methods for estimation of the pooled odds ratios (inverse variance, Mantel-Haenszel *etc*). In the evidence-based medicine the procedures of statistical meta-analysis are common techniques. Indeed, meta-analysis is the most obvious way to bring together the results of many independently made studies.

The statistical method of meta-analysis is based on operating the data exclusively in terms of statistical categories. Accordingly, it has at least three major peculiarities which, if neglected, become weaknesses of the method. Firstly, for a meta-analysis to produce adequate results, the studies pooled together should be quite similar or "homogeneous" in a number of aspects. Secondly, the definitions of the terms "small" and "large" in respect to the cohort sizes are, more than often, exceedingly arbitrary. Accordingly, substantiated definitions of these terms are essential. Thirdly, sources of bias cannot be controlled by the statistical method. Therefore, only methodologically sound studies should be included in a meta-analysis. This latter principle is known as the "*best evidence meta-analysis*".

These three points often pose significant problems for the meta-analyses performed within the framework of the evidence-based medicine. The problems might not always be widely discussed but they are there, as we'll see from the examples of meta-analyses we've used to illustrate this section of our book. The major difficulty with many of the currently published meta-analyses, especially those dealing with genotype-phenotype associations, is that there are no reliable and, at the same time, commonly accepted criteria to assess heterogeneity across the studies, to distinguish as objectively as possible between the "small" and "large" studies and to select the "best evidence". According to the task in hand and to the extent of the personal capacity, each biomedical researcher often elaborates his or her own criteria to deal with these three points.

When inconsistency or **heterogeneity among studies** in a meta-analysis is high, it lowers the confidence of the results of meta-analysis. Inconsistency is usually assessed with a $\chi 2$ test for heterogeneity. This kind of test is susceptible to the number of studies included in the meta-analysis. A recently formulated measure of the study heterogeneity (I^2–value, described in [1]) was proposed to measure the degree of inconsistency across studies. This statistical parameter is said to allow comparisons across meta-analyses of different sizes, of different types of study, and using different types of outcome data. Despite that this recently suggested measure of the study heterogeneity is claimed to be a reliable numerical estimator of the study "heterogeneity", the examples we analyze below clearly suggest that I^2-value analysis alone is not likely to be a sufficient criterion for identification of less "heterogeneous" or more "homogeneous" studies.

The elusive terminology of **"small" and "large" samples** is also a considerable problem in association studies and meta-analyses. Researchers or the reviewers often complain that "sample is too small" for the statistics to be "truly significant" or tend to produce other pseudoscientific statements of the sorts (strangely enough, nobody ever seemed to complain that "the sample is too large". But, then, everything is possible in post-genomic era)... From statistical point of view, the "power of study" grows with the cohort size so the "larger" samples would be preferable, from this formal point of view.

Apart from the point that the matter of the "sample size" is not so simplistic and a "large" cohort might be even more disadvantageous than a "small" one (because of the substantial increase in the unknown confounders, see "Planning of association studies" in the Volume II), there is never a non-arbitrary definition that would allow to distinguish between "small", "large" and other cohort sizes while comparing various studies. The study power and sample size calculations from the statistics provide one flexible way of dealing with the problem. Alternatively, we can analyze thousands of actually available studies and to look at the actual cohort sizes and other characteristics. The four-partite classification of the studies into "very small", "small", "medium" and "large", which we present in the Volume II, was based on such an analysis of the data in *medSNP*.

As for **selection of the best evidence,** the check for the methodological soundness is, apparently, outside the scope of meta-analysis as a statistical technique. In this book, we call this type of checks as an "ortho-analysis". There are at least two types of the integrative or "ortho-analyses" which we consider in the following sections. One of them checks for consistency of the physiological manifestations among various association studies related to the gene (not necessarily restricted by a particular polymorphism or a particular disease) and also requires a large number of independent association studies. The other can be done within a singular association study by bringing together the available information about the different levels of biological functions and, again, checking the consistency of the available information across these levels of function. These three and other moments pertaining to validity of the results of a meta-analysis are discussed further in some detail (see Box 7 for a summary).

Box 7. Common errors in meta-analyses of genotype-phenotype associations

- Inadequate measures of heterogeneity among studies
- Neglect of substantial heterogeneity in the clinical details
- Neglect of dependence of associations on additional variables
- Inadequate definitions of "small" and "large" studies
- Inadequate criteria for selection of the "best evidence"
- Neglect of the basics of population genetics including:
    - Lack of analysis of population strata
    - Neglect of the differences in the male/female ratios
    - Not considering the numerical effects of the polymorphism frequency
    - Omission of either allele or homozygote effect calculations

In the case of genetic association studies, meta-analyses are made for particular polymorphism in a particular gene and all of the studies to be included should report the effects of that particular polymorphism in relation to the same or similar outcome. As examples for our present discussion, we used several published meta-analyses of the genotype-phenotype association studies. Discussion of these examples illustrates how an overview of the potential problems inherent for meta-analyses can allow us to avoid common errors leading to "negative findings" and other kinds of unsubstantiated conclusions. In these examples, we remade the meta-analyses of interest rather than simply criticized the published data. The points we raise are the actual "study limitations" though these limitations are far from the commonly mentioned ones such as exquisitely vague notions of "publication bias", "small sample size" and others.

Firstly, let's consider an example of meta-analysis that occurs most frequently. In this meta-analysis [2], relationships between the F2 20210 G/A polymorphism and different forms of ischaemic heart disease was analyzed. The meta-analysis included 19 studies, totally 4944 patients and 7090 controls (n=12034). Sample size, inclusion criteria, geographical location, clinical presentation, age, cardiovascular risk factors, and angiography-determined extent of disease were extracted from each study and are presented. The $\chi 2$-test was used to assess heterogeneity among studies; the data from the studies were pooled according to Mantel-Haenszel. In the case of the 20210A variant, the OR for unspecified IHD was 1.21 (95% CI 0.99-1.59); for unstable angina and myocardial infarction OR was 1.24 (95% CI 0.98-1.63). The analysis also indicated higher risk of myocardial infarction for younger groups of patients (<55 years OR=1.8, CI 1.2-3.4; <45 years OR=2.3, CI 1.3-4.6) that have the variant.

This is a reasonably well written meta-analytic study. The results also do sound to be of clinical significance. What interests us here is to what extent these results are, actually, substantiated in the case of this particular polymorphism in this particular meta-analysis. Before considering any other argument, we ought to recall that the polymorphism 20210 G/A in F2 is characterized by relatively low population frequencies (2..5% in various Caucasian populations). Accordingly, statistically significant odds ratios even in a cohort this large (n=12034) can be produced even if the differences between the groups do not exceed a dozen of individuals. For example, in many association studies the authors included into their meta-analysis, all the differences between the cases and controls were only 3..5 patients (for instance, 20210A found in 5 patients and 2 controls, total cohort of 200 individuals). This kind of evidence is, clearly, limited.

Let's illustrate the foregoing by further exploration of the data. Totally, the data include 160 patients and 194 controls with 20210A allele and Mantel-Haenszel odds ratio was OR=1.21. However, if there would be not 160 but just 155 cases with 20210A, the results would become statistically insignificant (P>0.1; OR=1.18, 95% CI 0.95-1.45). This is an imaginary and unrealistic scenario since we just can't knock even five patients off the study.

However, the results also become statistically non-significant if we exclude just one study and analyze the data for the remaining 18 studies. For example, excluding a study featuring 16 cases and 7 controls with 20210A ([3], Italy) leads to statistically insignificant results of the entire meta-analysis (P>0.2, OR=1.15, 95% CI 0.92-1.41). Clearly, the significance of the entire meta-analysis of 19 studies crucially depends on the results of only one association study.

The above discussion was not intended to be a criticism of a reasonably well-written paper. Rather, it serves here as an illustration of one extreme in performance of meta-analysis: *the selective inclusion of studies*. Selective inclusion of studies is, indeed, required in accordance with the paradigm of the best-evidence meta-analysis. This selective inclusion is supposed to be done on the base of clearly defined and scientifically objective criteria (see, for instance, the examples we cite in the following section that deals with quantitative analyses). However, arbitrary decisions in the study selection will result in selective picking of evidence and analysis of selective evidence is nothing short of extremism.

Let's consider another meta-analysis dealing with genetic polymorphisms and predisposition to CVD. It might be instructive to see how different meta-analyses, analyzing the similar problems, using largely the same set of studies still produce different and even contradictory results. This meta-analytic study [4] attempted to analyze the correlations between various forms of CVD and several genetic polymorphisms such as F2 G20210A, F5 R506Q, F7 R353Q, ITGB3 L33P and MTHFR C677T. No correlations between the F2 or F5 polymorphisms and the coronary artery disease (CAD) were found in a meta-analysis including 5431 patients and 5607 controls; correlations of polymorphisms from other genes (ITGB3, MTHFR, F7) with CAD were found to be statistically significant.

The authors claim that the main limitation of their study was that some of these control subjects may have had asymptomatic CVD. Despite the possibility that those would comprise a very small percentage (less than 3%, most likely), that may well be true. However, what interests us more is that the meta-analysis of the relationship between the CAD and the F2 polymorphism was made on a partial set of publications that were also used in the meta-analysis we discussed previously (IHD meta-analysis [2]). It's remarkable that a meta-analysis done on a partial set of publications (i.e., [4]) leads to a negative finding while a meta-analysis made on a larger selection of studies (i.e., [2]) indicates a positive association. When we consider yet another meta-analysis that features the same polymorphisms [5], we might wonder why there is little consensus between the results of these meta-analyses involving the same polymorphisms. A number of explanations can be proposed but the most obvious one is that in the case of [4] paper, the evidence used was even more selective than in the case of [2]. The causes for this selectivity might be numerous and one of them might be the neglect of the basic piece of biomedical information: that the terms IHD, CHD and CAD are synonymous. As we already discussed, the "trivial" details of the sorts can be exceedingly important at the stage of the retrieval of the studies (Chapter III).

These and other difficulties with the attempts to meta-analyze selective evidence can lead to another kind of extremism: *inclusion of "all the available" studies* without regard to basic principles of population genetics and of biomedical research. We illustrate this other extreme on the example of another meta-analysis. In this study [5], the authors analyzed the correlations between polymorphisms of the genes of the coagulation cascade with CAD/IHD/CHD. The authors report that only associations involving F2 and F5 polymorphisms were found to be statistically significant and the correlations for the rest of the polymorphisms (PAI1 4g/5g F7 R353Q, GPIa 807T, GPIb –5C, and GPIIIa 1565T) showed no significant overall associations with coronary disease. For all of these genes, over

190 association studies were analyzed and for each of the polymorphisms the available evidence on each comprised at least 5000 cases and at least 5000 controls. That gives an impression of a comprehensive and an accurate study. Also, the authors of this work were very kind to provide the entire table they used in their meta-analysis which includes dozens of various parameters extracted from the published association studies. This table allows us to take an alternative look at the alleged results.

The major though not the only problem with this attempt of meta-analysis is that the authors used a purely statistical criterion (the I^2-value we mentioned previously) without any apparent regard to the basics of the population genetics. I-square is quite a generic criterion [1] for assessment of the heterogeneity. Indeed, the authors state that "there was little heterogeneity" among studies for F2. A closer inspection of the table the authors kindly provided has shown that these "relatively homogeneous" studies included, actually, data for such various populations as Caucasians, Chinese, Japanese, Indians, Amazon Indians, Turks and others thus covering the entire globe. Considering the data for such various populations as having "little heterogeneity" and, then, pooling them into one cohort to make meta-analysis would, speaking figuratively, surely make the heart of a population geneticist fail. Indeed, on several occasions I did mention this kind of approach to meta-analysis, no geneticist believed me that this kind of "analysis" can ever be made and even published: until, of course, they were shown a printed copy of the paper we discuss here. After showing the copy, I heard phrases similar to "How is it possible to pool the data from so different populations the way they did?" made usually after a prolonged pause…

Nevertheless, the paper [5] is a piece of work and is, definitely, an outstanding effort, especially in respect to the search and retrieval of the studies. Also, the authors sought confirmation of genotype frequencies and genotyping procedures by correspondence with investigators, which is also an important issue rarely taken care of during meta-analyses. Let's remade this entire meta-analysis for each of the polymorphisms using the data (http://www.thelancet.com/journals/lancet/article/PIIS0140673606682639/abstract, "extra" tables) the authors generously provided. A remake of this meta-analysis, of which a brief account is presented below, distinctly shows correlations for most of the polymorphisms. We remade the meta-analysis using Mantel-Haenszel formula.

The authors reported association between CAD and F2 20210 G/A polymorphism. Over 40 studies for the F2 were retrieved, twice the number of studies used in all of the meta-analyses we previously analyzed. These studies included, totally, 11416 cases and 14245 controls. Our calculations on the entire selection produced OR 1.31 (CI 1.16-1.59) for the association between 20210A and CAD ($P<0.01$). These results were close to those reported in the paper (OR=1.31, CI 1.12-1.52). However, only 27 studies from over 40 were explicitly specified as dealing with Caucasian populations while the rest of the studies either included Asians and other highly heterogenic populations. Therefore, we can make reasonable assumption that, if the results of meta-analysis were statistically significant on the entire pool, they undoubtedly should shown even better statistics if we use stratification analysis. Indeed, restricting the analysis to Caucasians only (7835 cases, 10196 controls) produced higher OR (OR=1.39, CI 1.17-1.65) and much lower P-value ($P<0.0002$). This association was confirmed by the age and gender adjustments which were possible to be made according to the data presented in the supplementary tables. The Caucasian subgroup was the largest and analysis of the rest of the population subgroups (African Americans, Asians, Indians and others) did not appear to produce statistically significant results. Logically, these observations suggest that the results of the entire meta-analysis of this polymorphism reported in [5] were strongly biased by the data for Caucasians so the results on the entire selection were also statistically significant, regardless of the population factor.

Largely the same was the situation with F5 polymorphism, for which 60 studies were available, with only a half being the studies on Caucasians (and yet showing "little heterogeneity" according to the I^2 criterion). Thus, our reader's appraisal confirmed the two positive findings (for the F2 and F5 polymorphisms). Now, we might want to reconsider the negative findings for the rest of the polymorphisms in the genes of blood coagulation cascade.

One of the most disturbing things about this particular study [5] was that authors took quite a strange and awkward criterion to define a "small" association study: the small were labeled the studies that dealt with less than 500 cases. It's surely an arbitrary number not based on any kind of evidence nor experience. Our analysis of the *medSNP™* database that includes data abstracted from thousands of association studies clearly indicated the definitions of the cohort sizes based on the statistics of the *actually made* association studies. According to this statistics (see details in the Volume II), the cohorts of the genotype-phenotype associations studies can be classified into one of the four following categories: "very small" (<20 cases), "small" (<50 cases, total cohort<200), "medium" (<400 cases, total cohort <1000) and "large" (>500 cases, total cohort>1000). Therefore, against the background of the thousands of the association studies, the cut-off "<500 cases" will, in fact, select "medium-to-large" studies so the funny rule akin to "small studies are those with <500 cases" can never be used to account for the bias from the "small studies". Taking into account the statistical definitions of the study size we presented above, let's reanalyze the bias from "small studies" in the associations considered in [5].

For example, in the case of ITGB3 L33P ("GPIIIa C1565T" or PLA1/2 polymoprhism), there were 43 studies (including two unpublished studies). For the entire set of studies, there was no association either in the entire sample (P=0.8) or when we select Caucasians-only group (P>0.2). However, after adjustments for the potential bias of the small studies (*i.e.*, those with cases<50, total cohort <200) and using only case-control studies, the results for the Caucasians became strongly statistically significant. Thus, meta-analysis of the 33 case-control studies totally including 13269 cases and 11087 controls produced statistically significant results (P=0.002) albeit indicating small average effects (the OR for the 33P variant was OR=1.1, CI 1.0-1.2). By the way, similar adjustments for the small study bias further improved the statistics of the F2 G20210A association with CHD (OR 1.36, CI 1.14-1.63 P<0.0001, 7418 cases and 9825 controls, Caucasians).

Neglect of the population strata and of the small study bias are two of the most common drawbacks in meta-analyses of the genotype-phenotype associations. Being the most common, these two are, of course, not the only flaws in the data exploration. Let's take, for instance, another meta-analysis of 15 studies of the GPIa C807T (ITGA2 F224F) polymorphism reported in the same publication [5]. For the entire set of studies, there was indeed, no correlation (P=0.4). Nor was there a correlation in Caucasians (P=0.4) and neither the analysis of the bias of the small studies was of any help. However, hastily piling up data from various studies, seeing no results and then rushing to produce an infantile cry "Daddy, there is no association!!" is hardly an excuse for a scientist. A little more research is required.

The polymorphism GPIa C807T is the case opposite to the F2 and F5 polymorphisms. That is, while both F2 and F5 polymorphisms are characterized by relatively low population frequencies of the rare allele (<5%, Caucasians), the frequency of the rare variant of C807T polymorphism is relatively high (30...40% in Caucasian controls). Accordingly, a momentary glance on the well-known formulas for calculation of $\chi 2$ and of the odds ratio will immediately suggest that the high allele frequency is likely to obscure any associations if they are calculated using the alleles. However, we can accommodate for the high allele frequency by checking the associations using homozygotes and not individual alleles (since the allele counting will inevitably result in high levels of the statistical noise). The homozygotes of C807T polymorphisms were less frequent in the populations studied and the results of the homozygote analysis were quite clear-cut: meta-analysis of 9 studies (all Caucasian), totally including 3833 cases and 5851 controls produced OR of 1.4 (CI 1.2-1.6) and a low P-value (P<0.0001) for the homozygotes. Thus, from epidemiological point of view, it is homozygotes of this polymorphism that represent CAD risk.

In the case of the -5 T/C polymorphism of GP1Ba, mere pooling of the 14 studies did not lead to any associations [5]. The polymorphism frequency is 10..16% in Caucasian controls, so it would be safe to analyze allele frequencies. Indeed, restricting the studies to Caucasian-only population (8 studies, 5809 cases, 4398 controls) we obtain a statistically significant correlation (P<0.01) indicating small average effects (OR=1.2, CI 1.1-1.3). The data for this polymorphism presentedin [5] might also be used to illustrate low sensitivity of the I^2 as a criterion for study heterogeneity. The authors write that there was a "marginal evidence of some heterogeneity among the 14 available studies". This statement looks quite appalling when we look at the data and see that population frequency of this polymorphism in Japanese population was twice higher (31%) than the Caucasian population frequencies. And yet, clearly incompatible studies (made on such distinctly different populations as Asians and Caucasians) were labeled as being only "marginally heterogeneous". Looking at the exceedingly generic formula for the I^2 [1], we might understand why there was only a "marginal evidence" in the case of strongly heterogeneous studies.

These our remakes of the meta-analyses attempted in [5] indicated that out of the seven polymorphisms analyzed only one (F7 G10976A R353Q) has not shown any association with the CAD/CHD. Indeed, population stratifications, adjustments for the small study bias and analysis of the alleles/homozygotes of the 24 association studies have not shown any statistically significant correlations albeit P-value dropped observably when the studies were restricted to Caucasian data only and then adjusted for the small study bias (from P=0.9 to P=0.09). Generally, this polymorphism is held to be a minor protective factor. We considered the aspects of biophysical function of this polymorphism in the Volume V of the books series. In brief, biophysical analyses of the F3-F7 complex do suggest a functional role for this polymorphism and biochemical data on the protein levels/activity confirm this. However, the known biochemical effects are, indeed, moderate: there is only a 25% drop in the levels of the 353Q variant in comparison to the native homozygotes. Given functional effect this small, the diagram of the levels of function (Figure 1-5) does suggest that the clinical manifestations of the effect of this polymorphism will be greatly obscured by the other factors. And, given that there were very few studies of this polymorphism that would include an analysis of these additional risk factors (such as smoking *etc*), it is at best doubtful that genetics-only analysis would produce any significant associations. Therefore, further studies of this polymorphism should be complemented by the analyses of the other factors of CVD risk.

Thus, the examples we presented above illustrate that by using simple logic and by not forgetting about the basics of population genetics, we can find statistically significant associations during meta-analyses of the published data. Our analyses also illustrated a number of common flaws in conduction of meta-analyses. One of the major flaws is, apparently, pooling data without regard to the fundamentals of biomedical sciences. Although one might agree with this kind of "generic" approach as a medical doctor (who treats patients without race or gender prejudice), one will never agree to such an arbitrary approach as a geneticist (who studies specific populations without race or gender prejudice). Also, pooling the data from different populations into one meta-analysis and not even attempting to analyze the population strata is hardly a strategy appropriate for scientific research.

Neglect of the biasing effect of the very small studies as well as discarding the distinction between case control studies, prospective and other types of biomedical studies was one of the most likely reasons for another example of negative report (ITGB3 L33P association with MI [6]). In addition, this attempt of meta-analysis [6] also appears to neglect the difference in the male/female ratios of the studies. Analysis of the 25 studies cited in [6] also shows that there was considerable heterogeneity in the clinical details of the studies pooled up with no regard to the clinical data. For instance, the studies differed in the use of angiography to confirm the MI. Some of the included studies make a difference between fatal and non-fatal MI while some others do not. The same concerns the difference between low-risk and high-risk patients. Not all studies included MI as the primary outcome. In some cases, the associations depended on additional variables such as BMI, fibrinogen levels and other non-genetic factors.

The neglect of these and other principles of meta-analyses of genetic associations (Box 7) is not typical only for these very few papers on meta-analysis we used as examples. These relatively simple principles appear to be neglected again and again thus leading to numerous "negative reports" and various sorts of scientifically irrelevant concerns such as "publication bias". As the few analyses we presented above might suggest, a "negative finding" might indicate anything, from sloppiness in the conduct of the scientific research to a profound lack of logical capcacity but hardly a "lack of association". Since author of this book is far from being the only one who understands the basic points we've just considered, the "publication bias" is quite understandable a phenomenon, since the journal editors are supposed to know their area well and are quite aware of the inherent pitfalls in it. In the opinion of the author, any sort of a "negative finding", especially in the case of a meta-analysis, should not be allowed for publication unless the authors disclose the full data table (as, for example, the authors of the papers [2] and [5] did) in supplementary materials or elsewhere. In this case, an objective base for the "negative findings" can be verified by all the sides interested, including journal readers dutifully paying their subscriptions.

In conclusion, the examples of meta-analyses we considered in this section illustrate that the meta-analysis greatly depends on how the data are pooled and with what assumptions in mind. We also noticed that pooling data without regard to the fact that the studies were made on different populations, without making distinction between the small, medium and large studies and without regard to the population frequencies is likely to result in reporting a "negative association". Such results, in a way, describe more the attitudes of the researchers reporting them rather than any sort of objective assessment. Apart from this, the researchers who conduct meta-analyses rarely check self-consistence (data integrity) for each study they include in their meta-analysis, especially self-consistency in respect to the levels of function. The method of meta-analysis *per se* does not even imply possibility of such checks and an entirely different type of the data analysis is required for the task. In the following sections, we consider the integrative analyses. An integrative analysis can assess the data integrity of a particular study or the data integrity across a group of independently made studies. The first kind of the integrative analysis can and, perhaps, should be done prior any attempts to make a meta-analysis.

ORTHO-ANALYSES?

The statistician said to the logician: "I have statistically determined that all geniuses are totally vain, even when adjusted for the lack of oversimplification and their not talking much." The logician replied: "Nonsense! Geniuses vain and terse? What about me?"

Albeit presented in the form of a gag, this text does mention the two mental skills essential for a sound scientific analysis: the capacity of meticulous observation ("the statistician") and the capacity of the inductive/deductive reasoning ("the logician"). In a normal scientist, both the capacities should, apparently, be in balance and lead to fruitful scientific discoveries. Contrariwise, if these capacities are in conflict or if one of the capacities is largely neglected while the other is overhyped, then no sane conclusions are likely to come forth (as in the story above).

Meta-analysis, which we looked at in the previous section, is a technique that uses statistical methods to pool together the results of different studies. The accent here, of course, is on the word *statistical*. Meta-analysic technique, though being quite common in evidence-based medicine, is by no means restricted to the medical science. It is a standard statistical method for performing a comparative analysis of some "homogeneous" independent studies and, as such, it is applicable in any branch of science. We also mentioned that the "best evidence meta-analysis" is the best research strategy in performing this kind of study.

The prefix "meta-" in words based on the Greek template, implies some sort of a median or an average. As the statistical technique, meta-analysis operates in the *plane* of relatively homogeneous studies, *among* ("meta") these studies, and does not have anything to do with the checking of the validity of each particular study. The studies should be as "homogenous" as possible (have the same outcome measures, similar baseline factors, similar settings) otherwise meta-analysis will, most likely, produce only negative findings. Therefore, the method of meta-analysis *per se* cannot recommend the techniques to check validity of each study to be included in a meta-analysis. Neither it's possible to use the meta-analysis to pool together clearly heterogeneous studies (say, those including distinctly different outcome measures).

As it happens, a meta-analysis of association studies is, indeed, "planar": since it operates with uniform studies dealing with a definite polymorphism and its effect on the definite disease. Our analyses of the *medSNP* database strongly suggest that it is very common for one polymorphism to adjust risks for more than one disease and even for very different groups of diseases (such as CVD, cancers, neurodegenerative diseases *etc*). As a particular polymorphism can adjust risks of various diseases, there are as many planes of meta-analysis as there are diseases associated with that polymorphism. We also recall that a genetic polymorphism can influence the biological function of the gene at various levels of biological function: from affecting gene transcription to altering some key amino-acids in the protein (Figure 1-5). These levels of biological function add to the stack of the planes associated with the polymorphism of interest.

Using this geometrical analogy of the stack of planes (illustrated in Figure 5-1), it is clear that in order to take into account the highly heterogeneous data from the different planes, it is necessary to draw a line perpendicular (or *ortho*gonal) to at least one of the planes. Then we will be able to look at the various planes of function using this particular line as a kind of a reference or a pointer. Using, again, the Greek model of word formation, we call this kind of an outlook by the generic term **ortho-analysis**.

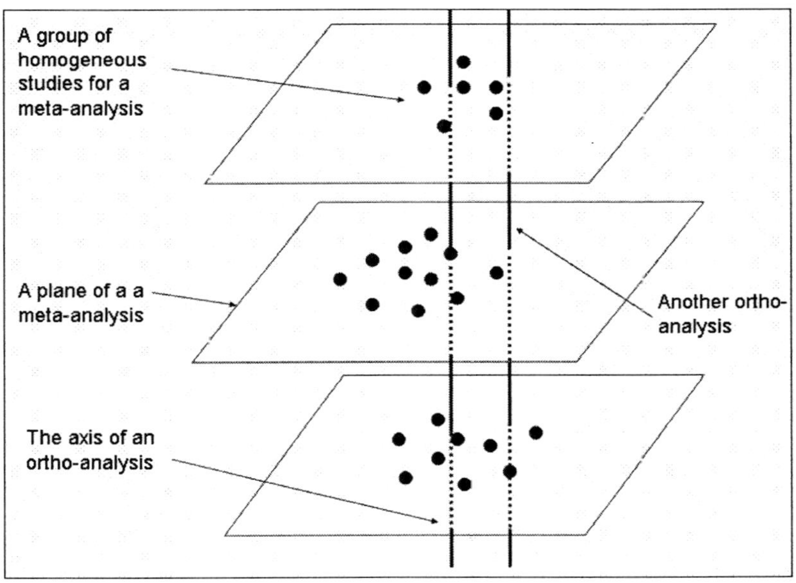

Figure 5-1. A graphical analogy illustrating the difference between "meta-" and "ortho-" analyses. Depending on the purpose of analysis, the planes might represent various levels of biological function and/or levels of clinical observation.

An ortho-analysis implies an integrative form of analysis which goes "beyond" or, rather, *through* the planar simplifications. In meta-analysis, the homogeneous data from multiple independent studies are piled up and analyzed from a singular point of view in a particular plane. In an ortho-analysis, the data from studies of different aspects of one phenomenon are put together and, then, can be analyzed from an integrative point of view.

Clearly, this quite general a definition covers numerous eventualities. For instance, selection of the "best evidence" often requires going through the stack of the planes to acquire an objective view of the problem thus selection of the best evidence is an ortho-analysis. Then, what is known in medical science as a "systematic review" might also fit the definition. Systematic reviews are regarded as the highest level of medical evidence: a systematic review does not only summarize the medical literature on a particular point of interest but also includes a thorough critical appraisal of the individual studies to identify the valid and the most relevant evidence. Although the data synthesis of most of the systematic reviews is, again, based on a meta-analysis of available data, there are also more qualitative reviews dealing with the analysis of the methods used to gather, analyze and report an evidence. An example of a systematic review and meta-analysis was mentioned previously (Chapter II, when we mentioned usage of the d-dimer testing for DVT diagnostics). The Cochrane Database of Systematic Reviews (www.cochrane.org) includes thousands of reviews on various topics: say, from using herbal medicines in CVD to psychological interventions for CHD.

In this book, however, we are interested not so much in performing systematic reviews concerning various aspects of the genetic association studies. Rather, our interest lies in getting a wider overview of the planes of the biomedical significance in relation to the levels of biological function. The three following ortho-analytical aspects will be of especial interest: 1. analyses of the data consistency through the levels of function in an association study; 2. quantitative analyses of physiological variables and 3. qualitative analyses of highly heterogeneous clinical data. These kinds of analyses are sometimes performed in the course of some systematic reviews and a few meta-analyses. On the whole, however, the analyses are performed quite infrequently and are usually not differentiated as stand-alone informational or theoretical techniques. In the following sections, we also will be interested in the implications of the results of the ortho-analyses for the post-genomic medicine.

ORTHO-ANALYSIS: CONSISTENCY ACROSS THE LEVELS OF FUNCTION

A best evidence meta-analysis presupposes actual availability of the best evidence. We have considered some of the criteria that can be used to assess the quality of the data presented in a particular study (Chapter III). However, unlike the methods to perform meta-analysis on a given set of studies (which are pretty standard and include, say, calculating study heterogeneity by $\chi 2$, calculating Mantel-Haentzel ORs, funnel plots and, perhaps, stratification/regression), there is much less uniformity in the methods that deal with the selection of the studies to be included in the meta-analysis.

The few examples from the previous section might well illustrate that the two flawed strategies of either 1. selecting a limited set of studies according to some subjective criteria or, on the contrary, 2. including "all available" studies even without regard to the basic parameters such as population, are hardly efficient strategies. We also have seen that the situation gets aggravated when the basics of population genetics and medicine are left out of view.

Objectively, the quality of the evidence is largely determined by the extent of the functional effects produced by the gene/polymorphism. If, however, an association study appears to neglect the interrelations between the levels of functions (Figure 1-5) or looses the context of the allied branches of biomedical science (Figure 1-2), we might expect that this study will not be likely to report a strongly significant result and thus will not represent a piece of "the best evidence".

Indeed, a recent review attempting to analyze the features of the "good association studies" [7] reiterates that the major source of frustration and confusion in this area is the familiar publication pattern: when initial positive findings for a gene or a polymorphism do not get confirmed or even get disproved in subsequent studies. Further in this chapter, we consider this sociological pattern on the example of the studies associating A2M 5bp deletion with Alzheimer disease. The authors of the review [7] were also right in noticing that the study design, implementation and interpretation are important determinants of the significance of the results of an association study and that selection of the best evidence strongly requires the analysis of a "study quality" (whatever the term of theirs may mean) rather than mere statistical significance.

At present, we won't go into the depth of the features that a valid genetic association study should or shouldn't have. This matter is analyzed and discussed in greater detail in the Volume II of the book series. Here, we'll just summarize some of the ideas in this regard (Box 8). It's worth mentioning that the two most important questions are that of the expectations from an association study and that of the degree to which these expectations can be substantiated *a priori.*

The primary expectation from an association study is that there should be a distinct difference in the allele frequency in the framework of a case-control study. The fulfillment of this expectation, however, greatly depends on how well substantiated is the selection of the "candidate genes", the "cases" and the "controls". If, for instance, the ethnic background of the participants is neglected, then expecting a positive association is unrealistic. If such an association is found, is it very likely to be a statistical fluke. If the "cases" are poorly characterized in clinical terms and the "controls" were not checked for the absence of the phenotype in question, then such an attempt of association study is about to leave the borders of science and is better described under the half-humorous rubric the British Medical Journal uses: "The Limits of Medicine". The last statement is far from being an inappropriate joke: quite a number of published reports do feature poor clinical characterization of the "cases", which is reflected, more than often, in uncannily vague clinical descriptions and in an ultra-concise analysis of the clinical data.

Previously, we mentioned the general criteria of the study assessment (Chapter III). Although these criteria do help to sieve out much of the spurious data, they are not sufficient to ensure the data consistency of each particular study. Here, we consider some of the methods to check the consistency between the levels of function in an association study. These methods are based on the fundamentals of molecular cell biology, biochemistry and biophysics and also fit well with the post-genomic perspective (Figure 1-3). We illustrate ortho-analysis of the levels of function with two examples: APOC3 and MTHFR polymoprhisms.

Box 8. General points for assessment of an association study

A review paper [7] cites a number of features that a properly done association study should possibly have. Unfortunately, none of the existing studies appear to be as good as desired and some of the points we cite below might be not always helpful for selecting the studies representing the "best evidence".

Selection of the "candidate" genes: genes vary in their effect and relative roles in the molecular mechanisms underlying multi-factorial diseases. The impact of the gene can be assessed using available information on the gene function, data on the monogenic diseases, results of positional genome-wide scans and, definitely, the data from the preivous studies of genetic associations with the phenotype of interest.

Selection of the polymorphisms: to guarantee detection of all possible disease-associated variants at a given gene it is necessary to make full sequencing of the entire gene. As this scenario is not realistic, it's necessary to select an adequate set of polymorphisms that can be considered as common genetic risks. See our polymorphism selection/rating criteria in the Chapter IV.

Selection criteria of "cases" and "controls": as we have mentioned more than once in this Volume (Chapter III), this is issue of primary importance for the results of an association study to be significant. Any attempt of selecting a study cohort without adequate expertise in biomedicine and clinical science has a strong chance to result in a negative finding.

Size of the cohort: see the statistical definitions of the cohort sizes in the Volume II. Generally, there are no commonly accepted criteria in this regard.

Genotyping procedures: ideally, the genotype determination should be confirmed by independent experimentation.

Analysis and interpretation: here, the criteria of assessment of observational studies might come in handy (Chapter III). Also, there are several points specifically important for the studies of genetic association (for instance, how adequate the haplotype analyses were performed).

In the Case Study of the IHD200 cohort (Chapter III), we have reported an association of the **-482 C/T polymorphism in APOC3** with CHD. The association was statistically significant (P=0.001) and, statistically, the -482T variant was a protective factor (OR=0.4; 95% CI 0.2..0.7). However, an ortho-analysis of the levels of functions suggests that this statistically strong association is not likely to have a clear biomedical significance. Thus, the -482T is located in an insulin response element of the APOC3 promoter and is associated with a loss of this insulin regulation site [8]. In animals and in cultured cells the expression of the APOC3 gene is downregulated by insulin. Under certain additional conditions, loss of insulin regulation will result in over-expression of the gene leading to hypertriglyceridemia, one of the CVD risk factors. Therefore, there is an obvious contradiction between the fact that the -482T variant disrupts the insulin regulation leading to hypertriglyceridemia (CHD risk) and an association study reporting this variant to be a CHD protective factor. In order to overcome this contradiction, a more thorough analysis of other risks (smoking, fat contents of the diet *etc*) needs to be performed. In particular, the unfavorable effects of the -482T variant on the triglycerides were, indeed, associated with smoking in several studies [9,10].

In the post-genomic perspective, there should be no contradictions between the data across the levels of biological function. Clearly, if a polymorphic variant was reported to be "positively associated" but does not appear to affect gene expression, splicing or protein function, this variant is, at best, only a genetic marker that arisen because of the linkage disequilibrium phenomenon. As we discussed previously, one of the criteria of the polymorphism selection is that it should have strong implications either for gene, gene splicing or protein function (Chapter IV). The available biochemical and other data on the gene and polymorphism should form a self-consistent picture.

For example, the **677 C/T polymorphism in MTHFR** gene leads to a non-synonymous amino acid substitution. According to biochemical data, this substitution considerably affects the protein stability of individual subunits. Therefore, the biophysical analysis of the protein structure should indicate a clear structural basis for this biochemical phenomenon. The cellular studies should show some abnormalities either in the protein processing or in secretion resulting in shortened half-life and lowered protein levels. The alteration of the protein levels/activity should cause a number of clinical manifestations. A consistent picture, corroborated by independent studies made in the allied areas of bioscience (Figure 1-3) gives more credence to a "statistically significant" association. See, for instance, an ortho-analysis of the levels of function of the MTHFR polymorphism (presented in the Box 9 [11-15]). Contrariwise, if studies at these levels of function produce discrepant results (say, there are no distinct changes in the atomic structure of the protein, there is no change of the levels of the protein *etc*), then we can expect *a priori* that the polymorphism might give unequivocal or weak results in an association study.

The ortho-analysis of the levels of function (Box 9) and of the clinical data from the MTHFR polymorphism (Box 10) present the data that fit together and form a consistent picture of the MTHFR function on different levels. Having a consistent picture of the kind built for a particular gene/polymorphism allows to judge with precision how relevant are the results of a particular association study and whether this particular study can be a piece of the "best evidence".

Generally, application of the diagram of the levels of function to the interpretation of the results should not be limited in scope only to the literal aspects. Physiological systems are extremely complex and interpretations of the kind "one cause leads to one effect" are rarely applicable. Thus, the effects of the polymorphism on the level of particular biochemical variable can be counterbalanced by a host of non-genetic factors so the baseline level of this particular variable measured in the lab testing might be in the "normal" range. Moreover, a polymorphism might influence not necessarily baseline levels but the tissue-local levels. Alternatively, the effect of the polymorphism might lie not in the change of the "levels" (baseline concentrations) of the corresponding biochemical variables but in other factors related to the biochemical variables (such as, for instance, particle size in the case of apolipoproteins or the dynamics of the variable change). Accordingly, a polymorphism is still associated with the condition but not through a direct adjustment of the baseline "levels". In this case, additional data on the biochemical variables are required (in the apolipoprotein example, those data will be the measurements of the particle size using sedimentation or other techniques). These are only a few sophistications that can occur during an ortho-analysis of a particular finding.

Box 9. An ortho-analysis of the levels of function of the MTHFR 677 C>T polymorphism

Genetics. A nucleotide polymorphism 677 C>T occurs in many ethnical populations with varying frequency (25%-45% in Caucasians) and leads to a non-synonymous amino acid substitution of Ala222 to Val222.

Biophysical level of function. Structure and function of MTHFR, including the model of the protein structure and the protein mutation map are analyzed in Case Studies of the Volumes II&IV. The Ala222, corresponding to the non-synonymous alanine-to-valine substitution, is located in the TIM barrel domain (residues 47-333 of the MTHFR protein sequence at the beta-turn 220-223. This residue is tightly packed into a hydrophobic cluster formed by I192, V194, V218, S219, G221, I225 and C250. Substitution of the Ala for Val leads to disruption in the protein packing thus lowering stability of MTHFR protein. According to the estimates produced by the method of differential energetic profiles (for description of the method see Chapter II in the Volume V), the stability of V222 mutant will decrease by at least 1.2 kkal/mol.

Biochemical level of function. The A222V mutant is known as the "thermolabile MTHFR". The A222 and V222 proteins show consistent difference in heat stability. For example, the mean residual activities after heat treatment (46 °C, 5 min) were 37% for the A222 and 15% for the V222 enzyme [11]. The MTHFR mutation was associated with somewhat reduced enzymatic activity and increased thermolability in a study of 152 patients [12].

Metabolic/cellular level of function. Metabolic pathway(s) and the metabolic context of MTHFR are discussed in the Chapter VII. The changes in the MTHFR enzymatic activity or amount of the enzyme will lead to a shift of the equilibrium of the corresponding metabolic pathways, in accordance with the diagram of the metabolic context. In particular, lower MTHFR activity will lead to lower levels of folates and higher levels of homocysteine due to dissipation of the folate pool.

Physiology. In MTHFR-deficient mice, plasma total folate was 25% of the wild-type [13]. In humans, the V222 variant was associated with subnormal levels of the serum folate and up to 10-fold increase in plasma homocysteine. In heterozygous individuals, the residual MTHFR activity is below 40% [11,14]. Higher level of homocysteine, known as hyperhomocysteinemia, is a risk factor for vascular pathology [15].

Medicine. Lower levels of folate (either baseline, dynamic or both) and high homocysteine levels affect a wide range of physiological systems. The MTHFR defect was associated with several entirely different groups of diseases: mental and physical retardation, perinatal death, vascular and neurogenerative diseases, cancer *etc*. Smoking tends to aggravate the effects of the polymorphism. An ortho-analysis of the clinical data from *medSNP* is presented in the Box 10.

It should also be noticed that a self-consistent picture of the data from various levels of function (or, at least, an absence of apparent contradictions) cannot always be verified for any particular gene/protein. The main reason for this is that some of the data (say, cellular or biochemical data) may be missing. In fact, data for some or other levels of function would be missing for most of the known polymorphisms in the human genome. However, relying on the principle of the data integrity across the levels of function, one can definitively verify whether the data in a particular paper or a series of papers are consistent or not. Then, it will be possible to come to substantiated conclusions as for the validity of a particular finding.

The principle of the data integrity and data self-consistency can be practically applied even on a very low level: such as checking the consistencies between the amino acid sequence, mRNA, genetic DNA sequences and the genotyping data presented in a particular paper. These checks, routinely done within the *medSNP* system (Figure 4-4), sometimes do detect significant problems with the entire datasets in particular association reports. The problems with the actual genotyping procedures used in a study are the most frequent. Analyses of the data collected in the *medSNP* suggest that not less than 5% of the association studies, initially selected for inclusion into *medSNP*, display severe problems of this sort. To detect this kind of the problem requires careful analysis of the technical data for each study taken into consideration. The following examples of the studies which were "bad" in this respect are given anonymously thus no references to the actual publications are provided.

Our first example concerns the **20210 G>A** polymorphism in the F2 gene. This polymorphism is located in the 3'UTR (an untranslated DNA region close to the 3' terminus). The functional significance of the polymorphism is not clear though polymorphism was found to be associated with higher prothrombin levels. Here, we are interested in the exact location of the polymorphism in a nucleotide sequence of the human gene. It should be noted that the "20210" does not represent some standard position but is an arbitrary numeration used by the authors of the original study of this polymorphism [16]. If we are to find the actual location of the polymorphism in a particular nucleotide sequence that is available in GenBank (for example, M17262), the label "20210" is entirely meaningless. Rather, we need to refer to the experimental data provided by the authors of the original study: the PCR primers (forward 5'-TCTAGAAACAGTTGCCTGGC-3' and "mismatch" reverse primer 5'-ATAGCACTGGGAGCATTGAAGC-3') and that the polymorphism was detected by digestion with Hindlll restrictase (that cleaves sequence patterns 5'-A^AGCTT-3', "^" denotes the cleavage site) that cleaves the 345bp amplicon into 322bp and 23bp fragments [16]. Translating all of this info into simpler form, the native gene sequence ("20210G") would contain the AAGCTC motif while the variant gene sequence ("20210A") would have the AAGCTT motif uniquely located between the two primers. Using any bioinformatic program for sequence pattern search, we locate the first primer to be at positions 26463-26482 in M17262 and the 2nd primer, after correcting for the mismatches, to be at 26793-26814. Then, after some more exertions, we define the actual position of the "20210G/A" polymorphism as being at the nucleotide 26784 in the GenBank entry M17262. This is the actual location of polymorphism, according to the data of one particular publication.

All of this sounds complicated, quite involved and, of course, "trivial". What is more interesting is that even such "banal" issue as the exact nucleotide position of polymorphism may differ from one publication to another! Thus, these technical data on genotyping techniques from most of the genotype-phenotype association studies dealing with this polymorphism consistently produce the position calculated above (26784 G/A in M17262). However, in several phenotype-genotype studies, the similar calculations produced a different result: 26784 G/A instead of 26788 G/A. This number is not a result of a simple typo since the primers, restrictase sites/probe sequences *etc* were analyzed by taking the data from the text of those publications. Apparently, the authors of those studies simply analyzed another polymorphism in relation to the association with the thrombosis though they claimed it to be the "20210" one. Any publication that claims results not for what was actually studied hardly represents a self-consistent piece of evidence.

The second example is the **R219K** polymorphism in the ABCA1 gene. The previous example suggests that the exact sequence locations of the polymorphisms actually studied by the authors of a particular publication need to be determined directly from the full text of the publication to ensure that there were no errors in the basics of the genotyping procedures. Sometimes, particular egg-heads from bioinformatics think that this reasonable statement should be challenged. Their thinking goes like this: for instance, if the polymorphism of interest represents an amino acid substitution, then, given that there are a number of several other amino acid defects in the same protein, their nucleotide positions can then be automatically deduced by using amino acid sequence-mRNA and mRNA-gDNA alignments. Indeed, this procedure does work and *medSNP* OS does contain the procedure for this kind of analysis Figure 4-4. However, there are several *caveat*s which an excited *emptor* of the idea should beware. For instance, amino acid defects in literature are not always numbered uniformly. Some researchers might use numeration according to the complete sequence, while some others might prefer to numerate the defects using amino acid sequence with the signal peptide removed. Consequently, if the list of amino acid defects contains both of these numerations, the above-mentioned procedure will not work or will produce incorrect results.

This is the exactly the case of the R219K polymorphism in the ABCA1 gene. The *medSNP* OS gave error while attempting to locate the polymorphism position from the amino acid substitution data. The putative result (the one obtained despite the error warning), after comparison with the primer/restrictase data from most of the papers, turned to be incorrect. Further investigation had shown that the R219K is a misleading label, which, nevertheless, is persistently used from publication to publication. Using the list of almost 50 amino acid substitutions extracted from HGMD and OMIM, it turned out that the proper label for the amino acid substitution should be, actually, R159K.

Finally, there are studies that contain serious contradictions among the levels of function the origin of which is not entirely clear (at least, according to the data presented). For example, in the case of the association studies dealing with **APOC3 -482** polymoprhism, the analysis of the technical procedures from about three dozen of the selected papers suggested at least three locations for this polymorphism separated by approximately 50 nucleotides. What is even more amusing (or disturbing, depending on the point of view) is that for each of these "alternative" positions there were several independent studies "confirming" it.

The example of the **APOC3 SstI** polymorphism illustrates once more that in order to decide how significant a particular study is the technical analysis of the data is required. Again, we will cite this particular study anonymously, without giving a reference to this paper (published in a high-profile journal, by the way). This study included analysis of a well-known cohort of the North American Caucasians in respect to the effects of the SstI polymorphism in the APOC3 on the lipid profile variables (triglycerides, HDL, LDL, ApoB *etc*). The findings presented appear to be statistically sound and seem to report significant correlations of the presence of polymorphism with the levels of the HDL/LDL cholesterol and triglycerides.

The *medSNP* entry for this polymorphism contains information confirmed by data from about 10 studies. All of the studies registered in database produce the same nucleotide position for the SstI (3238 C/G) polymorphism: namely, pos. 121935 C>G in the GenBank entry AP006216. All, except the study we obliquely refer to here. This particular study presented a position that is several hundred nucleotides off the correct position. What might be the source of this error? Is this a human error (a series of typos in paper, a confusion of the technical procedures *etc*)? Or is this a fabrication of data? It's hard to say given only the text of the paper. It is, however, beyond any doubt that the incorrect polymorphism position totally confounds the results of otherwise good-looking and accurately written study.

It is even more instructive to notice that the fundamental flaw of this and other similar studies was neither the analysis of the known confounders (Chapter III) nor the common problems of association studies (Box 8) but a relatively "trivial" issue: the technical data on the genotyping. Thus, apart from the analysis of the consistency of the biological data (Box 9), an ortho-analysis of the levels of function even at the level of technical details can be a decisive criterion for the quality and reliability of a particular association study.

Ortho-Analysis: Quantitative Analyses of Biochemical Variables

The ortho-analyses of the levels of function we just considered can be used used to verify consistency of a wide range of biological data for particular genes and polymoprhisms. It can be said that ortho-analysis of the levels of function produces a qualitative model for the gene function or of the polymoprhism effects. This model can be used for further investigation: namely, for building corresponding quantitative models. Although studies published under the rubric of "meta-analyses" do sometimes include qualitative and quantitative models of the sorts, they can hardly be called meta-analyses. Sometimes, they are called "systematic reviews" or simply "reviews". Here, we call this kind of a study as a "quantitative ortho-analysis of biochemical variables".

We'll consider the peculiarities of the quantitative ortho-analyses of biochemical variables on the example of the 677 C>T polymorphism of the MTHFR gene. Ortho-analysis of this polymorphism across the levels of function, presented in the previous section, indicated a consistent picture of the effects of this polymorphism. A large amount of the data from association studies is available for the 677 C>T polymorphism. Accordingly, this SNP can be investigated at the next, quantitative level. The studies we cite here analyze, in particular, the relationships between biochemical variables of homocysteine and folate, MTHFR genotypes and the clinical phenotypes. A short summary of these studies is given in the Table 5-1 [17-20]. In this section, we consider only the general features of these quantitative analyses and for any additional details we would like to refer reader to these excellent publications. The features we are about to discuss include adequate selection of the best evidence, analysis of the data both from prospective and restrospective studies and selection of an adequate quantification strategy.

A quantitative ortho-analysis implies a greater responsibility than mere pooling of data in the framework of a meta-analysis does. Therefore, the requirement of the **adequate selection of the best evidence** (Box 7), very important for meta-analysis, becomes paramount in the case of a quantititative analysis. Accordingly, it might be expected that a quantitative study that presents non-trivial associations would feature considerably more consistent set of criteria for the selection of the best evidence than the wacky inclusion criteria in an average meta-analysis.

Indeed, the papers we cited in the Table 5-1 feature elaborated sets of the study inclusion criteria as well as outstanding retrieval and analysis of the published associations. For instance, 443 studies were retrieved in [20] and 250 studies in [19]. The most important point, however, was not the amount of studies to be slammed together but rather how substantiated was the subsequent selection and analyses of these studies. Thus, out of 443 retrieved studies only 70 qualified for the inclusion into the analysis in [20] and only 39 out of 250 in [19]. These numbers are truly remarkable. While less thoughtful researchers either select evidence that emotionally appeals to them or merely slam together all the data without regard to their biomedical relevance, quantitative analyses, *a priori,* cannot tolerate lackadaisical attitudes of the sorts. Thus, substantiated and explicitly defined criteria are required for the reduction of the *clinical* heterogeneity of the studies which otherwise will render a quantitative analysis meaningless. Accordingly, study selection criteria have to be carefully elaborated and implemented. These criteria range from definitions of the outcomes and recognition of the differences in the study design [17], clinical relevance of the data [20] to elaborated sets of criteria.

Table 5-1. Quantitative analyses of the effects of homocysteine and MTHFR 677 C/T polymorphism: a few examples

Phenotype	Main variables analyzed	$N_{studies}$	N_{cases}	OR (95% CI)	Reference
CHD	homocysteine, folate, cholesterol	40	11162	1.21 (1.06-1.38)	[17]
CHD, DVT	homocysteine	12	3210	Dose-response, proportional risks	[18]
Stroke	homocysteine	14	1769	1.8 (1.6-2.0)	[19]
DVT	homocysteine	53	8364	1.2 (1.08-1.3)	[20]

For example, in the systematic review of the relationships between the stroke, homocysteine and MTHFR 677 C/T, three inclusion and six exclusion criteria were specified [19]. The inclusion criteria were: retrospective case-control or prospective study design; clear definitions of homocysteine-related "exposure" and clear definition of the outcome (neurologic syndrome consistent with ischemic stroke or transient ischemic attack, with or without confirmation by neuroimaging). The exclusion criteria were as follows: 1) other study designs (case studies, cross-sectional studies *etc*); 2) studies of stroke in children or adolescents (<18y); 3) studies of stroke in sub-populations not representative of the general population; 4) other outcome definitions; 5) absence of the total Hcy measurements; 6) lack of report of arithmetic mean Hcy, SD, OR, HR, RR or insufficient data to calculate these. This set of strictly-defined and distinctly non-arbitrary criteria allowed to select the 39 relevant studies out of the 250 publications and to obtain consistent results in most of the subsequent qualitative and quantitative analyses.

By the way, non-arbitrary study selection criteria can also benefit the studies that follow the line of more traditional "genetics-only" meta-analyses. Apart from reporting justified and verifiable associations, a meta-analysis operating non-arbitrary selection criteria can also illustrate efficient extraction of the most relevant information. For example, in a meta-analysis exploring relationship between the MTHFR polymorphism and stroke 787 publications were retrieved [19]. After careful analysis using about 10 explicitly defined criteria, only 31 of the retrieved studies satisfied both the inclusion and exclusion criteria. Careful selection of the relevant data, as opposed to citing selective evidence, does have a role in producing consistent and clinically plausible associations. Thus, among 14,870 subjects included in the meta-analysis, the pooled estimated risk of stroke/TIA associated with the 677T allele increased in a dose-dependent manner: T-allele pooled OR was 1.17 (1.09-1.26) and TT homozygote pooled OR was 1.37 (1.15-1.64) [19].

The results of prospective and restrospective (case-control) studies cannot be merged within the standard framework of meta-analysis since these two kinds of studies produce complementary data. At the same time, restricting analyses to only one kind of study, either prospective or retrospective, corresponds to a neglect of important evidence. Accordingly, **analysis of the data both from prospective and restrospective studies** is important to obtain substantiated results, especially when dealing with quantitative correlations. For example, the dose-response plots of the association between serum homocysteine and IHD events [18] were built using meta-analyses of the three study types: prospective studies; retrospective studies and association studies of MTHFR genetic associations. Sometimes, after the separate analysis of the data for prospective and retrospective studies, reliable results might be obtained only for one kind of study. Apparently, retrospective studies are restricted to survivors, whereas prospective studies can include fatal and nonfatal outcomes. Hence, if individuals with MTHFR TT homozygote would have a higher case fatality rate, then the association in retrospective studies would be attenuated. The study [17] illustrated that the TT genotype was associated with increased CHD risk in retrospective but not in prospective studies. At the same time, this conclusion is still subject of a number of criticisms and the main one is the geographical skewing of the data [17].

In order to detect a quantitative association, it's necessary to select **a quantification strategy** adequate enough for the task in hand. A quantitative analysis is not necessarily a logistic regression operating continuous variables. A number of quantification strategies can be used. One of the most common strategies is stratification of the pooled cohort according to the values of the continous variables into quartiles or other percentiles. For example, to explore interaction between the MTHFR genotype and folate status, several subgroups were compiled in [17] whereby folate status was defined as below or above the median serum/plasma folate level. Then, the effect of confounding was explored in a subgroup of studies with available data on age, gender, smoking, hypertension and hypercholesterolemia. Adjustment for these confounding factors indicated that confounding does not appear to affect the overall results. A similar approach was used in [20] where percentiles were used to calculate "absolute odds ratio" for a unit change in homocysteine.

The usage of percentiles in the framework of the standard $\chi2$-testing is the simplest way of analysis of the quantitative dependencies. Further sophistications include compilation of the dose-response plots and analyses of proportional risks. In the study of quantitative relationships between CHD and homocysteine [18], the dose-response plots of the associations between serum homocysteine and ischaemic heart disease indicated that there were linear relationships between serum homocysteine and probability of disease events in the three types of meta-analyses; a given decrease in homocysteine would produce a similar proportional risk reduction from any pre-treatment level. The constant proportional relationships established in this study suggest that the absolute reduction in risk would be expected to be greatest in individuals at highest absolute risk of a cardiovascular event and not necessarily in those with the highest homocysteine. The analyses suggest that, for instance, a 25% reduction in serum homocysteine (which can be achieved using folic acid supplements) may be more important in a patient with average homocysteine levels but at a higher overall CVD risk than in a patient with a high serum homocysteine but at a lower CVD risk (patients at high risk are those, for instance, who had a previous myocardial infarction *etc*).

Other features of the sound quantitative studies include presenting non-trivial study limitations and their careful analysis. Indeed, "small sample size", "unknown confounders", "hidden cases" *etc* are stated to be the "main study limitations" in most of the meta-analyses and association reports. When a reader does not read biomedical publications mechanically, these phrases really dry mouth while setting the reader's teeth on the edge. Indeed, why regurgitate what is established to be the "main study limitations" from the beginning of the statistics as a branch of the numerical math? A sign of a reliable study, made not for the sake of a formal report but in the true spirit of scientific investigation, is citing the *actual* limitations of the particular study. For example, in the analysis of the relationship between the homocysteine levels and the stroke [19], the authors duly cited an actually observed weakness of their analysis: the clinical homogeneity between the subjects in the included studies even despite the thoroughly elaborated study selection criteria. The authors also analyzed another, more specific, potential weakness of their analysis: that an elevated plasma homocysteine may be an epiphenomenon related to the burden of systemic atherosclerotic disease [19].

The few examples of the quantitative analyses we have considered in this section illustrate what care should be taken to include in this kind of analyses the association studies that are both clinically and genetically homogeneous. Accordingly, a sheer statistical evaluation of the study heterogeneity using only a formal criterion like $\chi2$ or I^2 is not likely to lead to any meaningful results. We also might surmise that a careful meta-analysis that takes into account the realities of the population genetics, biochemistry and non-genetic factors will almost never be just another "negative findings" report.

ORTHO-ANALYSIS: DOSE-DEPENDENT EFFECTS

The consistency across the levels of function allows qualitative assessment of the data on physiological mechanisms. The quantitative analyses of biochemical variables can give further elucidations of the particular aspects of the qualitative ortho-analytic model. Both of these types of ortho-analysis serve as the basis of the qualitative analyses of heterogeneous clinical studies which we consider further in this chapter. Before doing so, let's consider another two ortho-analytic techniques that are, from time to time, used in the genetic assocation studies: the analysis of the dose-dependent effects and the comparison of the results of the multivariate/univariate analyses. The dose-dependent effects and multivariate/univariate analysis are closely related and, ideally, should be the integral part of *any* genetic association study or meta-analysis.

We already mentioned dose-dependent effects in the previous section. In relation to the genetic association studies, the "dose-dependent" effects of the primary interest are those of the gene polymorphisms. As we have seen during our discussion of meta-analysis, calculations of the effects or risks can be based either on comparison of the allele frequencies or of the genotype frequencies. In other words, native homozygous, variant heterozygous and variant homozygous genotypes can differ in the extent of effects in respect to the outcome of interest. In the case when these relative risks form a statistically significant trend, it can be said that the genetic polymorphism has a "dose-dependent" ("gene dosage") effect on the outcome of interest. Of course, these effects amount to only three points in the regression line thus being quite feeble a criterion. At the same time, the dose-dependent effects might be without statistical significance in case when the homozygote frequency is relatively low or when the study cohort is not large enough. Nevertheless, presence of a consistent dose-dependent effect in respect to the number of the alleles (0, 1 or 2) strengthens the conclusions of an association study or a meta-analysis.

Alas, in the case of our IHD200 study (Chapter III) only PAI1 4G/5G polymorphism had statistically significant dose-dependent effect ($P<0.05$). For the rest of polymorphisms, the trends were not statistically significant. Our review of the genetic association studies we registered in *medSNP* database (Chapter IV) shows that reports of dose-dependent effects in genetic association studies aren't very numerous. Thus, out of almost 1500 associations in the current version of the database, only 20 (that is, just 1%) report statistically significant ($P<0.05$) and physiologically plausible "gene-dosage" effects. We illustrate some of these "gene dosage" associations in the Figure 5-2. Moreover, only in 119 associations out of 1500 (8%) the dose-dependent effects were *attempted* to be analyzed and in the remaining 92% the gene dosage wasn't even considered by the authors of those publications.

The gene dosage effects can be calculated in a number of ways: in relation to the outcomes (disease risk), in relation to continuous variables (say, PAI1 plasma levels), using alleles, genotypes or haplotypes and can also involve analyses of the gene variants in several different genes. These sorts of calculations are often made during the multivariate data analyses which we consider in the next section.

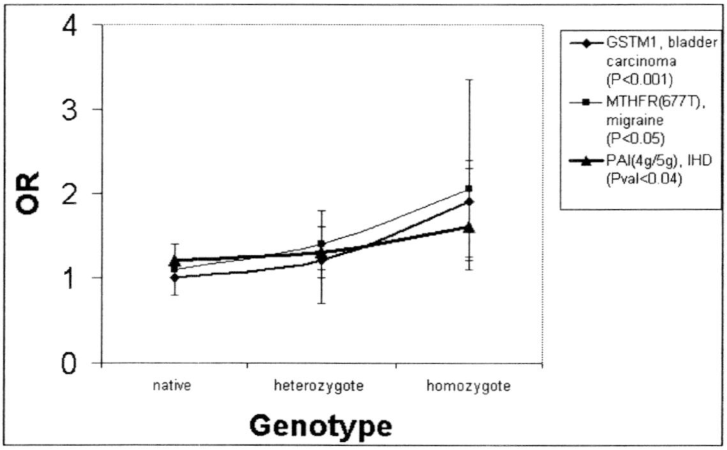

Figure 5-2. "Gene dosage" effects, a few examples from *medSNP*. Examples of discontinuous variables (outcomes) are given. Each association is characterized by OR (ordinate) and CI (error margin). Apparently, each example illustrates increasing risk with the increase of the number of alleles (0,1,2). Albeit the error margins appear to be great, the correlations were found to be statistically significant ($P<0.05$).

Ortho-Analysis: Multivariate vs. Univariate

If we lock again at the diagram of the levels of function in the Chapter I, we'll recall that each level of function is under the influence of numerous variables belonging to various other levels. Say, the level of a certain biochemical variable can depend on age, gender, diet, lifestyle, the medical conditions present in the patient *etc*. A clinical outcome of interest (say, a certain kind of CVD) will depend on these and many other factors. This complex relationship between the levels of function is reflected in the structure of the data analysis in the biomedical studies. Thus, it is not enough to establish a correlation between a particular variable and the outcome of interest (that is, to perform so-called "univariate data analysis"). It is also necessary to consider to what extent the variable of interest and all the other available variables contribute to the outcome in terms of confounding (Chapter III). This task is performed at the stage of the "multivariate data analysis".

The term *multivariate analysis* is a little bit too general. In statistics, there are at least a dozen of various methods to perform this sort of analysis which include *regression* (and its extension, artificial neural networks), *principal components* analysis, *canonical variate* analysis and other computational techniques. In biomedical studies, the term "multivariate analysis" largely means the *regression analysis* which either attempts to describe the phenomenon of interest with a linear formula (continuous response variable) or through logistic regression (binary response variable). First, correlations between individual variables are detected during univariate data analysis. Then, multivariate regression is performed in order to detail the relative significance of the univariate associations. In terms of the cohort studies, the results of the logistic regression determine the extent to which the potential confounders can be used to predict whether an individual is in one group or the other.

Logistic regression also provides an indication of the degree to which the groups are unbalanced in respect to the confounders (Chapter III).

Here, we are interested not so much in the techniques or software that is routinely used for this kind of analysis in biomedical literature and which can be found in books on statistics and in software manuals [21]. Rather, we are interested in using both multivariate and univariate analyses as ortho-analytic techniques. Given that the analyses themselves were performed correctly, what kind of biomedical information can these analyses provide to support the results reported in a particular study?

For example, the R506Q variant of Factor-V was associated with the risk of recurrent VTE [22]. The univariate OR was 4.1 while multivariate analysis shown OR equal to 4.7 (adjustments for age and smoking). In this case, univariate analysis appears to slightly underestimate the effects of polymorphism. Another example: univariate correlation between the PON1 192R and the coronary spasm (OR=1.7, P=0.01) also becomes more pronounced after multivariate analysis (OR=2.5, P<0.002, adjustments for age, gender, BMI, HDL, LDL, smoking, hypertension *etc*) [23]. In these two examples, the multivariate analysis has allowed to reduce the statistical noise and to clarify the association.

There are several points concerning the multivariate analysis that need to be kept in mind while analyzing biomedical literature. First of all, the multivariate analysis isn't always extensive. That is, even if data for dozens of variables were collected for each patient in a particular study, only a few of them tend to be analyzed (or "adjusted for") at the stage of multivariate analysis (like we did in our IHD200 study in Chapter III). Hence, there is the problem of the adequate selection of the set of the adjustment variables. Ideally, those variables should be *all* of the known confounders significant for the cohort under study. Secondly, a multivariate analysis can both underscore the correlation of interest or to show it to be less significant. Sometimes, the multivariate analysis does show greater effects and/or better statistics when compared with univariate analysis (as it was in the two examples above). Sometimes, the multivariate analysis indicates that univariate results were overestimated or even arisen entirely because of the confounders. Thirdly, while statistically significant correlation found during a multivariate analysis supports the conclusions, an absence of such correlation does not automatically disprove the univariate results. Fourthly, multivariate analyses are quite rarely done (or, at least, adequately reported) in the genetic association studies.

Ideally, the adjustments for other factors during multivariate analysis shouldn't change the significance of the univariate results. More often, however, is the case when multivariate analysis over a certain set of adjustment variables leads to a negative association. This brings the matter of numerous problems with multivariate analysis in biomedical literature [21,24]. Apart from the four points we've mentioned above, a brief review of biomedical studies that feature multivariate analysis published a decade ago [21] indicated that assessment of accuracy of regression models in describing the data appears to be neglected in biomedical research. Thus, only 7 out of 111 papers analyzed reported a valid assessment of the adequacy of their multivariate regression model ("goodness of fit"). A more recent review of the kind (published in 2001) still strongly reiterates that authors, reviewers, and editors should pay greater attention to the established guidelines concerning use and reporting of the logistic regression models [24].

Our experience during compilation of the *medSNP* suggests that genetic association studies are not much of exception in respect to the quality of the multivariate analysis and that the recommendations like "to pay greater attention" and others [21,24] are, still, very much actual. First of all, less than 10% of the studies registered in the current version of *medSNP* (103 out of 1500) presented at least some details concerning the effects of the multivariate adjustments. Secondly, most of these studies operate with adjustments for some 2..3 variables (such as age and gender) and <20 out of the 103 studies operate with truly multivariate analysis (that includes analysis of the data on *all* known confounders of the study). These two observations suggest that both gene dosage effects and multivariate analysis, although they could be valuable tools for assessment of the significance of particular genetic associations (and, therefore, of validity of particular studies) are far from being commonly and uniformly applied during the genetic association studies. Accordingly, to get at least a general overview of the validity of a particular study or a group of studies, more general ortho-analytic methods are required. We discussed one of these methods, the analysis of the data consistency across the levels of function, earlier in this chapter. Qualitative analyses of heterogeneous clinical studies, discussed in the next section, represent further elaboration of the ortho-analysis of the levels of function.

Ortho-Analysis: Qualitative Analyses of Heterogeneous Clinical Studies

Apart from quantitative analyses of the biochemical variables of interest, a qualitative model obtained as a result of an ortho-analysis of the levels of function can be used for integration of the data from highy heterogeneous clinical studies. In distinction from the standard technique of meta-analysis, this kind of analysis aims at establishing a maximal set of non-contradictory results of highly heterogeneous association studies rather than at pooling the data from a limited number of quite similar studies. Since each ortho-analysis of the kind considerably depends on the ingenuity and the scope of the particular researcher, it cannot be a technique as "standard" as the meta-analysis. Nevertheless, when properly performed even in a most general manner, the ortho-analysis of the heterogeneous clinical data can provide a number of valuable inferences concerning the genetic association studies of particular gene/polymorphism, their reliability and potential applications of their results. We provide a general illustration of this ortho-analytic approach by describing the clinical effects of the MTHFR polymorphism (Box 10).

Box 10. An ortho-analysis of the clinical data for the MTHFR 677 C>T polymorphism

Associations are illustrated with OR(CI) from the studies registered in *medSNP*. The diagram represents only a fraction of the data on MTHFR in the database. The attribution of the disease either to folate deficiency or homocysteinemia is quite arbitrary. See details in the text.

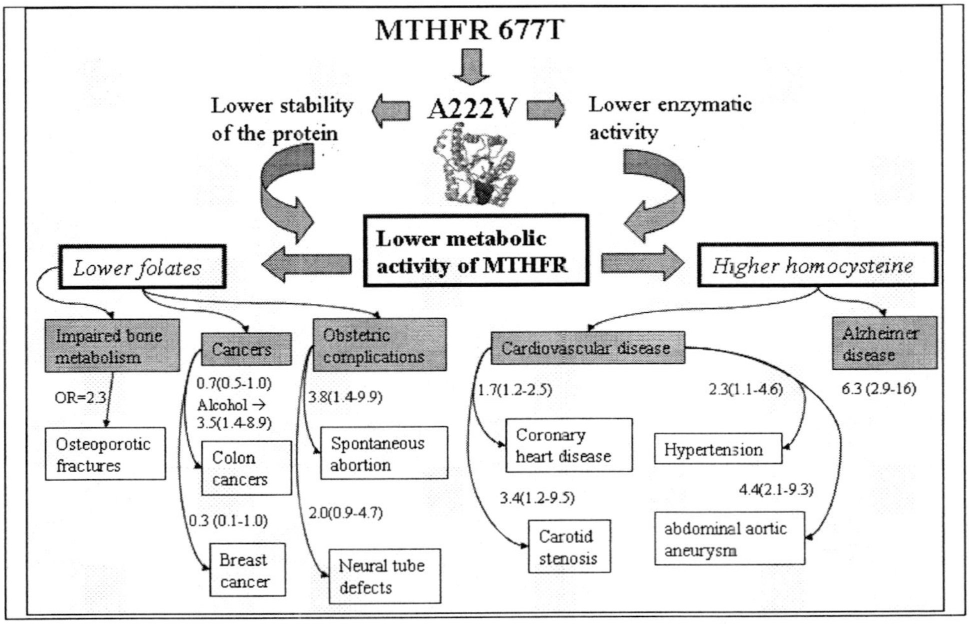

According to the metabolic context of MTHFR (see Chapter VII, Figure 7-3), the lower activity of the MTHFR disrupts conversion of 5-methyltetrahydrofolate to 5,10-methylenetetrahydrofolate. This reaction is the central point in the folate metabolism and its disruption throws the finely tuned folate metabolism off the natural balance resulting in an inefficient usage of all the various forms of folate. This leads to a gradual dissipation and depletion of the entire folate pool, especially when aggravated by a nutritional folate deficiency. As the 5-methyltetrahydrofolate is, in particular, the major substrate for methylation of homocysteine to methionine, the lower levels of 5-methyltetrahydrofolate will lead to higher homocysteine levels (hyperhomocysteinemia) since homocysteine is not consumed in the corresponding reaction. Both factors (low folate and high homocysteine) will influence predisposition to a number of disorders that have very different clinical manifestations through different physiological mechanisms. Most of these disorders are mentioned in the Box 10.

First thing to be noticed is the range of the clinical manifestations of the polymorphism. The folate and homocysteine levels affect the bone metabolism and adjust risks of cancer, obstetrical complications, CVD and Alzheimer's. Albeit epidemiological studies suggest these associations, the present-day understanding of the physiological mechanisms by which the low folate/high homocysteine adjust the risks for these diseases is, often, fairly limited. Ortho-analysis of the heterogeneous clinical data can be helful for the deductions of those physiological mechhanisms. We consider some of the likely mechanisms in the Chapter VII when considering folate physiology in general and the metabolic context of MTHFR in particular.

Secondly, we might notice that for different kinds of diseases the risks associated with the same polymoprhism can be adjusted in opposite directions. Thus, lower folate impairs bone metabolism and so the 677T variant was associated with higher incidence of osteoporotic fractures in elderly [25]. Folate deficiency can harm embryo growth and the

677T variant was also associated with the higher incidence of neural tube defects [26]. Higher homocysteine is a risk factor for CVD, so 677T/677TT genotypes were associated with CVDs in quite a number of studies and meta-analyses made for a number of Caucasian and Asian populations (for example, [27-30]). High homocysteine and the 677T variant were also associated with Alzheimer's disease [31]. Both high homocysteine and low folate are likely to be involved in spontaneous abortions and the 677TT was associated with the condition [32]. These few associations we mention illustrate the deleterious effects of the low folate/high homocysteine genotype/phenotype. However, the 677T genotype also appears to correlate with a lowered cancer risk. Although there is no clear physiological mechanism for this effect and the direction of the association might be arguable, in a number of epidemiological studies it was shown that 677T might be a cancer-protective factor (for example, [33,34]).

Thirdly, the extent and the direction of the effect of polymorphism depend on certain external factors. Despite the apparent prevalence of the "genetics-only" associations (see Table 4-1 in the Chapter IV), there are many genetic associations that include interactions with external factors. One of them is the association between the risk of the colorectal adenoma and alcohol intake [35]. It appears that 677T MTHFR genotypes lower the risk of colon cancer [33]. In the study [35], a putative interaction (P~0.09) was observed between the MTHFR polymorphism and alcohol intake. Thus, the men with 677TT homozygotes who consumed 30+ g/day of alcohol (in ethanol equivalent) had OR=3.5 (95% CI 1.41-8.78) for the risk of adenoma. This particular finding was, statistically speaking, rather putative (P>0.05). Although an increased occurrence of colorectal cancer/adenoma precursor in patients with regular alcohol intake does come to the attention of general practitioners, only further studies may illustrate the MTHFR genotype to be an independent factor of risk.

From what we have seen while discussing the above example of the MTHFR polymorphism, a number of general conclusions can be made. These conclusions are by no means limited to the MTHFR polymorphism. Ortho-analyses of the clinical data for the gene polymorphisms in the *medSNP* do suggest that the following three conclusions are valid almost for 90% of the polymorphisms registered in the database. Firstly, a common genetic polymorphism is not narrowly associated with a particular phenotype or even a disease, as it is the case with monogenic disorders. Instead, a genetic polymorphism appears to influence predisposition to many diseases. Secondly, one and the same polymorphic variant can influence predisposition in different directions: to heighten risk of one disease and to lower risk of another. Thirdly, by no means is the genetic polymorphism the only factor in etiology of a disease (Chapter I) and a considerable number of the genetic association studies do confirm that the genetic associations can be modified by the external factors in various ways and directions. These general conclusions are important for the elaboration of the practical applications of *medSNP* which we consider in the Volume II.

Thus, ortho-analyses of the very heterogeneous association studies (which, often, represent largely epidemiological data) do indicate variety of the effects a genetic polymorphism might have on the human health. An adequate alignment of the epidemiological data with the results of the analysis of the levels of function can serve a general basis for consideration of the physiological mechanisms of the corresponding diseases. A clearer picture of the related physiological mechanisms also allows us to make more specific conclusions from an ortho-analysis of the clinical data for one polymorphism. And, *vice versa*, the results of the ortho-analysis of the epidemiological studies can be used for elucidation and verification of the putative physiological mechanisms of a disease. We return to the subject of the ortho-analyses when discussing the clinical applications of the genotyping and consider an ortho-analysis of ADRB2 in greater detail (Volume II).

PARA-ANALYSIS: TAKING A LOOK AT FAILURES

The prefix "para-" in the words formed after the Greek model implies "at or to one side of". This prefix is used to designate objects or activities auxiliary to or derivative of that denoted by the base word ("parody"), and, sometimes, abnormal or defective ("paranoia"). Here, we use the term "para-analysis" to designate an analysis specifically oriented towards the identification of the potential causes of the data inconsistencies (abnormalities). In view of a self-consistent paradigm formulated as a result of an ortho-analysis, a "negative finding" report is considered as one of such inconsistencies.

Apparently, watching for inconsistencies is an integral part of the meta- and ortho-analyses. Both meta-analysis and the ortho-analysis require attention to the consistency of detail, be it study selection criteria in meta-analysis or the consistency across the levels of functions. Once a non-contradictory and self-consistent picture of a biomedical phenomenon is obtained, there always be a number of exceptions that do not apparently fit into the paradigm arising as a result of meta- or ortho-analysis. It is para-analysis that deals with these exceptions to elucidate the reasons that might stand behind them.

If we were attentive enough while reading the preceding sections, we might have noticed that a negative finding can often be an immediate result of ignoring some or other basic of genetics, biochemistry or clinical science. Alternatively, it may result from human errors or a lack of accuracy when conducting biomedical research. Accordingly, the first stage of the para-analysis is to exclude these common sources of the errors that routinely lead to "negative reports" and contradictions in data. In order to do so, it is necessary to consider a particular negative finding in view of all the criteria of the research assessment we have discussed so far, from the study assessment criteria (Chapter III) and polymorphism selection (Chapter IV) to the basic issues of the common sense we consider in the present chapter. As show para-analyses of the data in *medSNP*, most of the "negative findings" arise because of these common errors. Since most of these errors were already discussed, we reiterate here only the most typical and the most serious. Box 11 presents a summary of the typical features of the "negative reports" dealing with genetic associations.

In addition to these typical features, there are several additional points that have to be mentioned. Firstly, as actual practice shows, errors in genotype determination are not uncommon. Although researchers will never admit that in their papers, the actual errors (which, sometimes, are found years after the genotypes were determined, the papers published and the grant reports accepted) might vary from inadequate selection of the PCR/detection primers and technically inadequate genotyping procedures to such issues as errors in entering the data into a computer file, misnumeration of samples, misnumeration of the samples on the DNA microchip, lack of strict control of the reproducibility of samples *etc*. As this sort of errors can be identified only when an alternative genotype determination is used (or at least, after repeating the procedure several times on the entire set of samples), it is quite possible that some of these errors might substantially "contribute" to the studies reporting negative results. The main problem with this issue that it's virtually impossible to detect these errors having only text of a published report (even though sometimes a published report can provide ample evidence, as illustrated by F2 and APOC3 examples in the "Ortho-analysis" section). The significance of these errors grows with the scale of the study and the large-scale studies of the post-genomic era are remarkably susceptible to this sort of the problem.

Box 11. The most common features of the negative association reports

These conclusions follow from the meta-, ortho- and para- analyses of the data in *medSNP*.

1. Contrary to common beliefs, cohort size does not matter: a full-blown negative finding can be in a large cohort (>1000 patients) as well as in a much smaller cohort (<100 patients).

2. Most of the reports featuring negative findings are characterized by poor (or, at least, by an incomplete and imprecise) clinical characterization. Some of these reports are apparently written by individuals who lack any clinical expertise in the subject matter.

3. Negative finding reports tend to omit or to suppress the results of stratification analysis. In many cases, the stratification is limited only to age and gender.

4. Negative finding reports often demonstrate a considerable lack of understanding the hierarchy of the levels of function.

5. Many of the negative finding reports become available when, paradoxically, there are already quite many independent reports that confirm positive association.

Secondly, the very form and language style in which the negative findings are often reported do suggest certain biases that the researchers involved seem to have. It's quite possible to detect those biases while reading thoroughly the text of a particular publication. Although human nature is notoriously difficult to judge, especially on the base of a few written sentences, some of the most typical features of the negative association reports look quite appalling. For instance, what assumption could be behind the reluctance of the investigators to mention (let alone conduct) stratification analysis of a relatively large cohort for which they report a negative finding? Is it basic laziness? Is it due to the time constraints while writing the report? Or is it the operation of just another infantile assumption which can be formulated as "I want to receive the results in the form I want and I want them to be totally consistent with what I expected"? And why in the world sometimes these negative reports sound as if the researchers publishing them conferred a great favor to entire humanity just by doing their highly unproductive research? Without becoming *para*noid, we still can say that these features are quite suspicious from the point of view of the scientific objectivity. They hardly have anything to do with the much spoken objectivity of the scientific method and even less with the common sense. They do, however, have much in common with the patterns of stubbornness, obsession and some sort of a "submission to hidden curriculum" (term from [36]) and even more with an egoistic "entrapment in losing endeavors" [37].

Especially interesting is the result of the *medSNP* analyses that states, paradoxically, that negative finding reports often occur when there are a number of independent positive association reports were previously found and subsequently confirmed (Feature 5 in the Box 1). Why, really, to study a cohort of over 1000 patients and to report a negative finding without providing any explanations or rationale for the lack of association while there are at least 10 or more positive association studies on the similar (say, Caucasian) population are available? Is it because of wanting to prove "my own, my precious" point of view? Or is it the operation of another common assumption which might be called as "the need to disagree"?

Of course, these common errors and the related fallacious attitudes are, definitely, not the only source of the negative findings. Often, a negative finding might is a mere result of those errors and attitudes. Sometimes, it is a mix of a number of subjective and objective factors. For example, the contradictions in studies associating A2M (alpha-2-macroglobulin) 5bp insertion/deletion polymorphism with Alzheimer disease did made a row in biomedical press: the original positive findings of a family study were followed by several other studies that reported negative associations. The incident led the journal editors to discuss (in public) the plausibility of supplementing the results of family studies with association studies *prior* to acceptance for publication.

Para-Analysis: A2M ins/del and the Risk of Alzheimer Disease

From what we've read in this Volume, we can conclude that many factors, from accurate work to sheer data falsification can lead to a publication reporting a "positive finding". Not less many factors, from an inadequate data analysis to an apparent lack of any plausible physiological mechanisms can result in a "negative report". Sheer data falsification and Neanderthal-type of the data analyses are really the problem of those who do these things and a careful reader such as a journal editor is often able to detect this kind of problems. What we are actually interested here is to analyze the contradictions among the independent studies that are, possibly, free from these two factors and the technical errors we've mentioned. The case of the putative association of A2M 5bp ins/del polymorphism with Alzheimer does illustrate common problems that routinely arise during para-analysis of the clinical data. The most remarkable thing about the A2M example is that each new study only adds more and more contradictions to the original findings and does not clarify the issue at all. In other terms, an ortho-analysis of the heterogeneous clinical data for this polymorphism did not allow establishing any sort of a self-consistent picture and para-analysis clearly supports this conclusion.

We'll begin with a short summary of the data on the levels of function of the A2M gene and the 5bp ins/del polymorphism. Alpha-2-macroglobulin is a serum protease inhibitor that inhibits many proteases (including thrombin and collagenases) through a "bait" amino acid sequence that contains proteinase-specific cleavage sites. The A2M protein is also assumed to mediate the degradation of the major component of beta-amyloid deposits through interactions with LRP1 (low density lipoprotein-related protein 1). The beta-amyloid deposits are the key issue in the Alzheimer disease (AD) and A2M protein has been localized to senile plaques in AD. Thus, A2M gene is a likely candidate gene for the disease.

The 5bp ins/del polymorphism, although it's being called as "ex18del", is, actually, located at the 5' border of the exon (that is, in the intron area). As no "defective" protein corresponding to the 5bp del was ever detected, the variant is more likely to affect the splicing of the A2M gene. Accordingly, the primary manifestation of the polymorphism should be altered antigen levels of the A2M protein in plasma. However, it is important to notice that there were no accurate biochemical studies of the effects of the deletion on the gene function *prior* to association studies. Therefore, we can *a priori* expect unequivocal results from the association studies and the ambitious geneticists who attempt an association study of the genetic variants that lack even basic biochemical characterization do that at their own peril.

The original positive association [38] reported an increased risk for AD. The entire cohort included 310 families (679 subjects from National Institute of Mental Health, USA) with Alzheimer diagnosis assigned in accordance with the NINCDS/ADRDA and/or Khachaturian criteria. A2M genotype allele frequencies were examined in affected and unaffected subjects in the study families. Assuming an increased risk of AD for a particular variant, its frequency would be expected to be higher among affected individuals than among their unaffected relatives. For a subset of 104 families, OR was found to be 3.6 (95% CI 1.8-7.0, P=0.0003). The effects on the age of onset were investigated using Kaplan-Meier curves and ANOVA and no correlations were found. Thus, the study reports a "genetics-only" kind of association with quite high OR and CI, remarkably low P-value but without age any effects detected.

This study was almost immediately followed by several "disproval studies" (that is, by negative reports). These negative reports varied differently in the scope, comprehension and the amount of the common sense applied. A UK study used three groups to claim that there is no association [39]. By the way, genetic association studies on UK populations show, on average, a significantly larger rate of negative reports in comparison to studies on populations from other European countries and US (see Volume II). This particular study featured entirely different set of criteria for patient selection than the original study (DSM-III-R or AGECAT dementia criteria and MMSE scores) in all of the groups. The study also involved quite peculiar groups of patients and controls. Thus, in the first group the patients and the controls were apparently selected from different geographical locations (albeit within UK). The second group was a very small group of 53 cases aged over 70 years and 53 age-matched controls. The third group was an epidemiologically based cohort (age >83 years) in which the subjects were merely assessed using an obscure questionnaire without, apparently, any clinical confirmation of the AD diagnosis. Given the peculiar selection criteria of these groups, it comes as no surprise that there were no association detected and the authors of the study haste to conclude that the discrepancies with the previous study [38] may be due to "regional population differences" in AD susceptibility associated with this polymorphism or because of the linkage disequilibrium.

A larger case-control study was performed on a multi-center cohort of 2616 individuals taken from populations across the Europe and USA [40]. At least, in this study the patients were diagnosed with AD according to the same criteria (NINCDS/ADRDA). The study also presented an analysis of a similar set of the National Institute of Mental Health family samples as the one mentioned in the "positive" report. Although in the case-control settings no associations were reported (which, again, comes as no surprise given very poor analysis of the clinical data, at least as presented in the text of the paper [40]), a statistically significant association in the family sample was detected thus providing another reason for the discrepancy: the different study design (familial *vs* case-control). This particular disproval study contains a lot of talk about "power of study" but does not appear to mention any stratifications of this very large sample cohort (which would be most natural thing to do for any clinical scientist). Accordingly, the effects of the potential confounders are unknown and no reliable conclusions whatsoever can be made on the base of this negative report. In addition, the study lacks any description of the actual genotyping procedures thus precluding technical analysis of the alleged results.

Perhaps the most serious and the most substantiated was the negative report [41] performed in accordance with the family study methodology similar to that of the positive association report [38]. In this study, an overlapping set of the same family samples (143 National Institute of Mental Health sibships) was investigated and the analyses generated largely non-significant results. Moreover, the data sets demonstrated the association of AD with *APOE*E4* indicated sufficient study power to detect associations of the magnitude reported for A2M*2 while no associations with the *A2M*2* polymorphism were detected using similar techniques as in [38]. The most significant feature of this study, however, was the analysis of the plasma levels of A2M which confirmed our *a priori* expectation: although plasma A2M levels were measured only in a small sample of 14 individuals (7 "ins" homozygotes and 7 "del" homozygotes), the levels of the A2M protein were largely similar (1.80 *vs* 1.6 mg/ml). This result supports the assumption concerning the lack of the functional significance of the 5bp ins/del polymorphism and also supports the idea that the original positive findings were rather based on the linkage disequilibrium in a particular subset of the population sample [41].

Subsequent studies only further clouded the issue. Thus, a study of sporadic AD in a sample from southern Italy (132+184 subjects, age 66-78) did report a positive result: an association of 5bp ins/del with AD [42]. The direction of the association was, alas, the opposite and the 5 bp deletion appears to be a protective factor while it is the insertion (wild type) that seems to be the risk factor (OR 1.93; CI 1.2-3.11). However, this result still might make some sense since an independent study of Italian subjects also reported a significant decrease of the deletion allele in AD cases with respect to controls [43]. It might be expected, then, that in Southern Europeans the association of 5bp del with AD might, indeed, be reverted. However, a Spanish study (190+200 subjects aged 80-90) reports higher frequency of D-allele in patients older than 81 years ($p=0.001$) [44] thus again frustrating our expectations as well as contradicting to the lack of the age effects reported in the previous studies.

Summarizing the para-analysis of the A2M 5bp ins/del polymorphism as for its putative association with AD, we can remark that there was a distinct lack of the careful stratification analyses in most of the reports, both positive and negative. For example, only one report [42] presented clear percentages of the males and females in the patient and control groups. Most of the studies, including the original positive report and the three subsequent disproval reports lacked mentioning of even the gender effects, not to mention any other confounder variables. It s out of the question that the gender effects are often significant in genetic associations and that the patient groups are not infrequently biased either towards males or females. Accordingly, on the basis of the studies we briefly analyzed here we cannot conclude with certainty whether the 5bp ins/del polymorphism would be associated with AD or not. The strongest evidence *against* the association is presented in only one report [41] that combines an accurate attempt to reproduce results with reasonable biochemical evidence.

Thus, the A2M example of the para-analysis illustrates, first of all, that lack of even basic functional characterization of the variant will lead only to waste of time, a misuse of the governmental and/or private funding (just consider the costs associated with all these association studies we've mentioned) and no reliable results established. Secondly, the family study might be more sensitive instrument to detect associations. However, these associations might occur only because of the linkage disequilibrium phenomenon and thus are of very limited diagnostic value. The most important conclusion, though, is that putative genetic associations that are surrounded by too many discrepancies, overheated debate and remarkably flawed argumentation on both sides are hardly trustworthy and, therefore, indicate strongly inadequate genetic markers rather than anything else. For the sake of reference, considerably less controversy surrounds the "R681H" variant in the same gene which, thus, is more likely to be a reliable genetic marker than the perilous 5 bp deletion.

PARA-ANALYSIS: SUBSTANTIATED NEGATIVE REPORTS

Ex nihilo nihil fit (Lucretius)
"Out of nothing comes [only] nothing"

Para-analysis is not always as scandalous as it might seem when considering the case of the A2M 5bp polymorphism. The negative reports, although being often the result of inaccurate research, do have their value in the area of the genetic association studies and, for that matter, in any other area of science as well. However, in order to be of an objective value, a negative report should be performed not less accurately than a positive association report. Not only the clinical part (patient/study selection and characterization) should be accurately made but also the genetic reasoning, the technical procedures, the biochemical evidence and the data analysis should not contain any of the most common drawbacks that we've discussed. Provided this, a negative report will be more likely to reflect an objective reality and not so much the grievous reality of the Pavlovian conditioning abundant in the minds of the unscrupulous researchers who are hasty to publish something or to perish, fast...

Still, substantiated negative reports of genetic associations, though being much rarer than the hastily made ones, do exist. In the case of the A2M 5bp ins/del, the most substantiated report appears to be [41]. As another example, let's take a French study of the potential role of the ADRB2 polymorphisms for the survival in the congestive heart failure (CHF) as an example [45]. Congestive heart failure is an impairment of the ability of the heart to sustain a sufficient flow of blood through the body. The mortality rate in CHF remains high so identification of the patients at high risk of cardiovascular events is an important part of the treatment. This study presents analyses of 444 consecutive patients (365 male, 79 female) with CHF related to left ventricular systolic dysfunction. The patients underwent echocardiography, radionuclide angiography and a cardiopulmonary exercise test and were followed for approximately 3 years. During this period there were 110 cardiac-related deaths. No correlations of the CHF mortality/survival with the presence of the ADRB2 polymorphisms R16G and Q27E were established.

This study features an apparently adequate study design and selection of patients (prospective cohort consecutive CHF Caucasians: ambulatory, stable for at least 2 months, excluded if myocardial infarction *etc*), collection of detailed clinical data (echocardiography, radionuclide angiography, cardiopulmonary exercise; coronary status of the patients determined by systematic cardiac catheterization) and *a priori* study power calculations (study was powered to detect a 15% difference in the survival rate of patients with homozygotes). The selection of the polymorphisms is also adequate: these are functional polymorphisms (criterion 6, Box 6) that "down-regulate" the receptor involved in the cardiovascular function. The genotyping procedures described in the paper appear to operate correct polymorphism positions. However, among a couple of dozen of the variables, the independent predictors of survival were only percent of maximal predicted VO2 (P=0.0001), age (P=0.0001), left ventricle ejection fraction (P=0.004), creatinine (P=0.02) and atrial fibrillation (P=0.04) but not the polymorphisms.

The most remarkable feature of the study, which might suggests why it is a negative report is that the independent predictors of cardiac survival shown very small effects: the detected relative risks were close to 1.0 (peak VO2 RR=0.95; age RR=1.05; LVEF RR=0.97; creatinine RR=1.08 and atrial fibrillation RR=1.7). Small size of all the statistically significant effects suggests, in particular, presence of "unknown" confounders which were somehow missed despite reasonably well planned and performed study. This hypothesis is corroborated, in part, by absence of any correlations between the heart rate and the ADRB2 polymorphisms in the cohort. At the same time, it is well known from dozens of independent studies that these ADRB2 polymorphisms do influence the heart rate and blood pressure. A putative correlation with CHF survival for the R16G/Q27E haplotypes (at least, in univariate analysis [45]) is another indication for the presence of these confounders, since haplotype analysis is somewhat more sensitive genetic technique. The authors did take care of one of the most important confounders (myocardial infarction, hypertension, atrial fibrillation) and those unknown confounders might include family history, infection, alcohol ingestion, anemia and hyperthyroidism. In this sense, the study [45] somewhat resembles the initial results we've obtained in *The Case of IHD200*, where patients with congenital heart defects heavily skewed the entire statistics leading to a negative finding (Chapter III).

Alas, the negative reports at least as reasonably substantiated as the ones we just mentioned are infrequent and comprised less than 20% of all the hundreds of negative reports we've read during compilation of *medSNP*. The extremely low quality of the research in the negative reports does not appear to be restricted to the genetic association studies. It may be instructive to observe how many important points may be missed by the authors bent on publishing their research at all costs, even at the cost of publishing just another "negative finding" (see, for instance, an extensive discussion illustrating the point on the example of a particular meta-analysis, http://bmj.bmjjournals.com/cgi/eletters/332/7544/752).

A negative finding (as, for instance, our *IHD200* Case Study illustrated) implies that the data should be analyzed with considerably greater care and not just dismissed as being a "negative finding". Therefore, so-called "publication bias" is, in fact, a great stimulus for the researchers to analyze their data with greater care. In other words, more consistent implementation of the "publication bias" by the journal editors and by the publishers can prevent giving out some sort of laziness or lack of inquiry for a "scientific result" politely referred to as a "negative finding".

Analyses of Data Consistency and Implications for Post-Genomic Studies

The material presented in this chapter was intended to illustrate that scientific quality of a particular finding is far from being restricted to mere statistical significance. In addition to the numerical parameters reflecting statistical significance, a study should also conform to numerous criteria which we discussed at some length (see Boxes 7-11). The analytical methodologies of meta-, ortho- and para-analyses can be used for verification of the biomedical consistency of the data from various studies and allow to build a self-consistent picture of the biomedical phenomenon of interest.

With these techniques in hand, we can consider the question that we have asked in the beginning of this book which is: why the large scale association studies of the post-genomic kind (thousands of SNPs, hundreds of genes, large groups) produce so little output or even lead to conclusions that are ridiculous from the point of view of human physiology? The authors of the large-scale studies should certainly be commended for the huge efforts they put into those studies. However, these efforts alone cannot serve as a justification for not finding any tangible results (such as, at the very least, replication of the previously known associations, not mentioning elucidation of the mechanisms of disease, recommendations for disease diagnostics, prevention and treatment *etc*).

An attentive examination of a number of these publications does show many of the features we've discussed in the present chapter. Poorly designed or inappropriate questionnaires, poor clinical characterization of the patients and controls (at least, according to the data presented in the publication), lack of the awareness of the most likely potential confounders, restricting study to genetics-only, inadequate or severely limited stratifications and other deficiencies in the study methodology surface all the time in the large-scale studies. Moreover, even the positive associations are hardly tenable given the small effects and an apparent lack of reproducibility. We can continue quite for a time considering various particular examples of the post-genomic studies of the kind. For now, we can only suggest to any interested reader, when he or she would happen to see a study that reports a positive association for a only few genes out of a thousand, to ponder on the matter in view of what we have discussed in this and previous Chapters.

Chapter Summary

In this chapter, we discussed several methods of the scientific analysis to verify consistency of the data from independent biomedical studies. The comparative analyses (meta-analysis) require a set of similar studies while the integrative analyses (ortho-analyses and para-analyses) can provide a wider perspective. **Meta-analysis** is a technique that uses statistical methods to pool together the results of highly similar (homogeneous) studies. The major requirements for the meta-analysis are the **study homogeneity**, clear definitions of the **cohort sizes** and the adequate criteria for the selection of the **best evidence**. While meta-analysis mostly deals with a particular level of biological function or a particular clinical outcome, in **ortho-analysis** we attempt to look at the entire stack of the levels of function. In an ortho-analysis, the data from studies of different aspects of one phenomenon are put together and are then analyzed from various points of view. This generic definition covers numerous eventualities of which we considered **consistency across the levels of function, quantitative analyses of biochemical variables, dose-dependent effects, multivariate vs. univariate analyses** and **qualitative analyses of heterogeneous clinical studies**. The term **para-analysis** designates an analysis specifically oriented towards the identification of the potential causes of the data inconsistencies such as a "negative association" reports. These basic techniques of the data analyses allow the selection of the most reliable data on the genetic association studies from thousands of publications.

References

[1] Higgins JP, Thompson SG, Deeks JJ, Altman DG. Measuring inconsistency in meta-analyses. *BMJ* 2003;327:557-60.

[2] Burzotta F, Paciaroni K, De Stefano V, Crea F, Maseri A, Leone G, Andreotti F. G20210A prothrombin gene polymorphism and coronary ischaemic syndromes: a phenotype-specific meta-analysis of 12 034 subjects. *Heart*. 2004;90:82-6.

[3] Burzotta F, Paciaroni K, De Stefano V, Chiusolo P, Manzoli A, Casorelli I, Leone AM, Rossi E, Leone G, Maseri A, Andreotti F. Increased prevalence of the G20210A prothrombin gene variant in acute coronary syndromes without metabolic or acquired risk factors or with limited extent of disease. *Eur Heart J*. 2002;23:26-30.

[4] Wu A, Tsongalis G. Correlation of polymorphisms to coagulation and biochemical risk factors for cardiovascular diseases. *Am J Cardiol* 2001;87:1361–1366.

[5] Ye Z, Liu E, Higgins J, Keavney B, Lowe G, Collins R, Danesh J. Seven haemostatic gene polymorphisms in coronary disease: meta-analysis of 66 155 cases and 91 307 controls. *Lancet* 2006;367:651–58.

[6] Zhu M, Weedon J, Clark L. Meta-analysis of the association of platelet glycoprotein IIIa PlA1/A2 polymorphism with myocardial infarction. *Am J Card* 2000;86:1000-5.

[7] Hattersley A, McCarthy M. What makes a good genetic association study? *Lancet*. 2005;366:1315-23.

[8] Li W, Dammerman M, Smith J, Metzger S, Breslow J, Leff T. Common genetic variation in the promoter of the human apo CIII gene abolishes regulation by insulin and may contribute to hypertriglyceridemia. *J Clin Invest* 1995;96:2601-5.

[9] Waterworth D, Talmud P, Bujac S, Fisher RM, Miller GJ, Humphries SE. Contribution of apolipoprotein C-III gene variants to determination of triglyceride levels and interaction with smoking in middle-aged men. *Arterioscler Thromb Vasc Biol* 2000;20:2663-9.

[10] Waterworth DM, Talmud PJ, Luan J, Flavell DM, Byrne CD, Humphries SE, Wareham NJ. Variants in the APOC3 promoter insulin responsive element modulate insulin secretion and lipids in middle-aged men. *Biochim Biophys Acta* 2003;1637:200-6.

[11] Kang SS, Zhou J, Wong PW, Kowalisyn J, Strokosch G. Intermediate homocysteinemia: a thermolabile variant of methylenetetrahydrofolate reductase. *Am J Hum Genet* 1988;43:414-21.

[12] Christensen B, Frosst P, Lussier-Cacan S, Selhub J, Goyette P, Rosenblatt DS, Genest J Jr, Rozen R. Correlation of a common mutation in the methylenetetrahydrofolate reductase gene with plasma homocysteine in patients with premature coronary artery disease. *Arterioscler Thromb Vasc Biol* 1997;17:569-73.

[13] Ghandour H, Chen Z, Selhub J, Rozen R. Mice deficient in methylenetetrahydrofolate reductase exhibit tissue-specific distribution of folates. *J Nutr*. 2004;134:2975-8.

[14] Frosst P, Blom HJ, Milos R, Goyette P, Sheppard CA, Matthews RG, Boers GJ, den Heijer M, Kluijtmans LA, van den Heuvel LP, et al. A candidate genetic risk factor for vascular disease: a common mutation in methylenetetrahydrofolate reductase. *Nat Genet* 1995;10:111-3.

[15] McCully KS. Vascular pathology of homocysteinemia: implications for the pathogenesis of arteriosclerosis. *Am J Pathol* 1969;56:111-28.

[16] Poort SR, Rosendaal FR, Reitsma PH, Bertina RM. A common genetic variation in the 3'-untranslated region of the prothrombin gene is associated with elevated plasma prothrombin levels and an increase in venous thrombosis. *Blood* 1996;88:3698-703.

[17] Klerk M, Verhoef P, Clarke R, Blom H, Kok F, Schouten E and the MTHFR Studies Collaboration Group. MTHFR 677C/T polymorphism and risk of coronary heart disease: a meta-analysis. *JAMA* 2002;288:2023-2032.

[18] Wald D, Law M, Morris J. The dose–response relation between serum homocysteine and cardiovascular disease: implications for treatment and screening. *Eur J Card Prev Rehab* 2004; 11:250–3.

[19] Cronin S, Furie K, Kelly P. Dose-related association of MTHFR 677T allele with risk of ischemic stroke: evidence from a cumulative meta-analysis. *Stroke* 2005; 36:1581-7.

[20] Heijer M, Lewington S, Clarke R. Homocysteine, MTHFR and risk of venous thrombosis: a meta-analysis of published epidemiological studies. *J Throm Haemost* 2004; 3:292–9.

[21] Bender R, Grouven U. Logistic regression models used in medical research are poorly presented. *BMJ* 1996;313:628.

[22] Ridker PM, Miletich JP, Stampfer MJ, Goldhaber SZ, Lindpaintner K, Hennekens CH. Factor V Leiden and risks of recurrent idiopathic venous thromboembolism. *Circulation* 1995;92:2800-2.

[23] Ito T, Yasue H, Yoshimura M, Nakamura S, Nakayama M, Shimasaki Y, Harada E, Mizuno Y, Kawano H, Ogawa H. Paraoxonase gene Gln192Arg (Q192R) polymorphism is associated with coronary artery spasm. *Hum Genet* 2002;110:89-94.

[24] Bagley SC, White H, Golomb BA. Logistic regression in the medical literature: standards for use and reporting, with particular attention to one medical domain. *J Clin Epidemiol* 2001;54:979-85.

[25] Bathum L, von Bornemann Hjelmborg J, Christiansen L, Madsen JS, Skytthe A, Christensen K. Evidence for an association of methylene tetrahydrofolate reductase polymorphism C677T and an increased risk of fractures: results from a population-based Danish twin study. *Osteoporos Int* 2004;15:659-64.

[26] van der Put NM, Gabreels F, Stevens EM, Smeitink JA, Trijbels FJ, Eskes TK, van den Heuvel LP, Blom HJ. A second common mutation in the methylenetetrahydrofolate reductase gene: an additional risk factor for neural-tube defects? *Am J Hum Genet* 1998;62:1044-51.

[27] Morita H, Taguchi J, Kurihara H, Kitaoka M, Kaneda H, Kurihara Y, Maemura K, Shindo T, Minamino T, Ohno M, Yamaoki K, Ogasawara K, Aizawa T, Suzuki S, Yazaki Y. Genetic polymorphism of 5,10-methylenetetrahydrofolate reductase (MTHFR) as a risk factor for coronary artery disease. *Circulation* 1997;95:2032-6.

[28] Rodriguez-Esparragon F, Hernandez-Perera O, Rodriguez-Perez JC, Anabitarte A, Diaz-Cremades JM, Losada A, Fiuza D, Hernandez E, Yunis C, Ferrario CM. The effect of methylenetetrahydrofolate reductase C677T common variant on hypertensive risk is not solely explained by increased plasma homocysteine values. *Clin Exp Hypertens* 2003;25:209-20.

[29] Inamoto N, Katsuya T, Kokubo Y, Mannami T, Asai T, Baba S, Ogata J, Tomoike H, Ogihara T.Association of methylenetetrahydrofolate reductase gene polymorphism with carotid atherosclerosis depending on smoking status in a Japanese general population. *Stroke* 2003;34:1628-33.

[30] Strauss E, Waliszewski K, Gabriel M, Zapalski S, Pawlak AL. Increased risk of the abdominal aortic aneurysm in carriers of the MTHFR 677T allele. *J Appl Genet* 2003;44:85-93.

[31] Anello G, Gueant-Rodriguez RM, Bosco P, Gueant JL, Romano A, Namour B, Spada R, Caraci F, Pourie G, Daval JL, Ferri R. Homocysteine and methylenetetrahydrofolate reductase polymorphism in Alzheimer's disease. *Neuroreport* 2004;15:859-61.

[32] Zetterberg H, Zafiropoulos A, Spandidos DA, Rymo L, Blennow K. Gene-gene interaction between fetal MTHFR 677C>T and transcobalamin 776C>G polymorphisms in human spontaneous abortion. *Hum Reprod* 2003;18:1948-50.

[33] Curtin K, Bigler J, Slattery ML, Caan B, Potter JD, Ulrich CM. MTHFR C677T and A1298C polymorphisms: diet, estrogen, and risk of colon cancer. *Cancer Epidemiol Biomarkers Prev* 2004;13:285-92.

[34] Toffoli G, Gafa R, Russo A, Lanza G, Dolcetti R, Sartor F, Libra M, Viel A, Boiocchi M. Methylenetetrahydrofolate reductase 677 C-->T polymorphism and risk of proximal colon cancer in north Italy. *Clin Cancer Res* 2003;9:743-8.

[35] Giovannucci E, Chen J, Smith-Warner SA, Rimm EB, Fuchs CS, Palomeque C, Willett WC, Hunter DJ. Methylenetetrahydrofolate reductase, alcohol dehydrogenase, diet, and risk of colorectal adenomas. *Cancer Epidemiol Biomarkers Prev* 2003;12:970-9.

[36] Abbasi K. The four pillars of global academic medicine. *BMJ* 2004 Oct 2; 329 (http://www.bmj.com/cgi/content/full/329/7469/0-g?ijkey=5091cdf94d088fcd0e6131204ad3993e536c032b&keytype2=tf_ipsecsha)

[37] Zhang L, Baumeister RF. Your money or your self-esteem: threatened egotism promotes costly entrapment in losing endeavors. *Pers Soc Psychol Bull* 2006;32:881-93.

[38] Blacker D, Wilcox MA, Laird NM, Rodes L, Horvath SM, Go RC, Perry R, Watson B Jr, Bassett SS, McInnis MG, Albert MS, Hyman BT, Tanzi RE. Alpha-2 macroglobulin is genetically associated with Alzheimer disease. *Nat Genet* 1998 ;19:357-60.

[39] Dow DJ, Lindsey N, Cairns NJ, Brayne C, Robinson D, Huppert FA, Paykel ES, Xuereb J, Wilcock G, Whittaker JL, Rubinsztein DC. Alpha-2 macroglobulin polymorphism and Alzheimer disease risk in the UK. *Nat Genet* 1999;22:16-7.

[40] Rudrasingham V, Wavrant-De Vrieze F, Lambert JC, Chakraverty S, Kehoe P, Crook R, Amouyel P, Wu W, Rice F, Perez-Tur J, Frigard B, Morris JC, Carty S, Petersen R, Cottel D, Tunstall N, Holmans P, Lovestone S, Chartier-Harlin MC, Goate A, Hardy J, Owen MJ, Williams J. Alpha-2 macroglobulin gene and Alzheimer disease. *Nat Genet* 1999;22:17-9.

[41] Rogaeva EA, Premkumar S, Grubber J, Serneels L, Scott WK, Kawarai T, Song Y, Hill DL, Abou-Donia SM, Martin ER, Vance JJ, Yu G, Orlacchio A, Pei Y, Nishimura M, Supala A, Roberge B, Saunders AM, Roses AD, Schmechel D, Crane-Gatherum A, Sorbi S, Bruni A, Small GW, Conneally PM, Haines JL, Van Leuven F, St George-Hyslop PH, Farrer LA, Pericak-Vance MA. An alpha-2-macroglobulin insertion-deletion polymorphism in Alzheimer disease. *Nat Genet* 1999;22:19-22.

[42] Zappia M, Cittadella R, Manna I, Nicoletti G, Andreoli V, Bonavita S, Gambardella A, Quattrone A. Genetic association of alpha2-macroglobulin polymorphisms with AD in southern Italy. *Neurology* 2002;59:756-8.

[43] Nacmias B, Tedde A, Cellini E, Forleo P, Orlacchio A, Guarnieri BM, Petruzzi C, D'Andrea F, Serio A, Sorbi S. Alpha2-macroglobulin polymorphisms in Italian sporadic and familial Alzheimer's disease. *Neurosci Lett* 2001;299:9-12.

[44] Alvarez V, Alvarez R, Lahoz CH, Martinez C, Pena J, Guisasola LM, Salas-Puig J, Moris G, Uria D, Menes BB, Ribacoba R, Vidal JA, Sanchez JM, Coto E. Association between an alpha(2) macroglobulin DNA polymorphism and late-onset Alzheimer's disease. *Biochem Biophys Res Commun* 1999;264:48-50.

[45] de Groote P, Lamblin N, Helbecque N, Mouquet F, Mc Fadden E, Hermant X, Amouyel P, Dallongeville J, Bauters C. The impact of beta-adrenoreceptor gene

polymorphisms on survival in patients with congestive heart failure. *Eur J Heart Fail* 2005;7:966-73.

Chapter VI

THE FUNCTIONAL GENOMICS QUEST

INTRODUCTION

The genotype-phenotype association studies, which were the main subject of the preceding Chapters, are the major tool of investigating the various effects of common genetic polymorphisms on the human health. In brief, a genetic association study boldly attempts to correlate the two most distant levels in the hierarchy of the levels of function: the level of the clinical manifestations and the level of the genetic code (Figure 1-5). Given quite the amount of the intermediate levels and even the greater amount of the additional factors that adjust this kind of associations, genetic association studies should be carefully performed in respect to a number of points (Chapters III&V). Previously, we also maintained that a polymorphism can become a reliable genetic marker only if it is located in a gene with clear physiological function and is a functionally important polymorphism (Chapters IV&V). Accordingly, imaginative research scenarios in which tends of thousands of polymorphisms (of hardly known functional value) in thousands of genes (of barely known physiological function) are genotyped in thousands of patients (with poorly characterized clinics) will only lead to negative reports thus inducing frustration of both the highly imaginative researchers and the highly depressed readers of their reports. Does this mean that any post-genomic study of the kind is bound to be a failure?

The plain answer is, of course, no. Given the amount of details routinely neglected by the researchers attempting such a study, these failures come as no surprise. Would at least half of the details we've already discussed was implemented in at least one large-scale association study that involves many patients, many genes and many polymorphisms, that study would never be just another negative report. So, there still some hope remains that in the coming years we'll see more adequately performed association studies featuring reasonably good informational management. We return to the association studies as well as to the numerous analyses and applications of them through the *medSNP* database & expert system in the second Volume of the present book series.

Here, we would like to reiterate that the common difficulty of the researchers neglecting considerable amount of essential details is not new in biology. Neither this common difficulty is limited to the genotype-phenotype association studies. In fact, any attempt of a high-throughput study, especially when conducted without regard to the basics of physiology, will be prone to the same problem. In particular, applications of the post-genomic methodologies we consider in this chapter is often severely limited in scope. For example, functional genomics studies are often restricted mostly to intracellular processes and it is assumed that studying these processes will produce all the answers the modern biomedicine needs. However, the focus on the intracellular processes entirely neglects the concept of the internal environment, the basic physiologic concept: most of the physiological processes essential for survival take place in physiological fluids and at very intricate tissue-tissue interfaces. Accordingly, this neglect can easily make the researchers blind to the presence of multiple physiologically unsound assumptions in their research models.

In this chapter, we are interested in the post-genomic methodologies that aim at functional characterization of the genes and polymorphisms located in the vast and largely non-annotated expanse of the human genome. In view of the apparent necessity of functional characterization *prior* to genetic association studies, such techniques can be of utmost importance in post-genomic studies that involve human physiology. These experimental methodologies are collectively called as "functional genomics". Although being undoubtedly a field that includes techniques stemming from molecular cell biology and biochemistry, we decided to consider the subject of functional genomics in the volume dealing with physiology since 1. physiology is the most comprehensive field of the biomedical research (Figure 1-2) and 2. the functional and biomedical significance of the genes and the gene products is based on the significance of the physiological effects brought in by these genes and gene products *in vivo*. The viewpoints author formulates and discusses in the present chapter, far from being a result of an extensive review of literature dealing with functional genomics, rather reflect author's practical experience with the studies dealing in functional genomics.

On "Functional Genomics" and "Systems Biology"

"Do not try to run before
you learn how to walk"...
(Folk wisdom)

As of yet, there appears to be no common definition of what the "functional genomics" is. The numerous definitions vary from "high-throughput measurement of mRNA levels" to quite long and elaborated ones such as "a multidisciplinary approach to study genes, gene products and their interactions mediating physiologic responses as well as detection and characterization of aberrations in these processes that may result in diseases". Although the first definition is the most concise (and, perhaps, the most common), it is also the narrowest. Broader definitions are definitely needed. For instance, a "Programme on Integrated Approaches for Functional Genomics" (http://www.functionalgenomics.org.uk/sections/programme/index.htm) contains quite broad a definition of what the functional genomic might be and also outlines several distinct areas of the functional genomics studies: mutagenesis and gene disruption (using antisense oligonucleotides and ribozymes), gene expression profiling, proteomics (protein expression and protein-protein interactions), what is called as "structural genomics" (or "structural proteomics", the high-throughput protein structure determination, Volume V) and bioinformatics (in the form of the "database management" and "data integration"). In this chapter, we will also use the term "functional genomics" as covering each of these areas. According to this definition, the expression profiling (mRNA levels) is a functional genomics study. A proteomic analysis is a functional genomics study, too. We summarize the major branches of the functional genomics in the Figure 6-1.

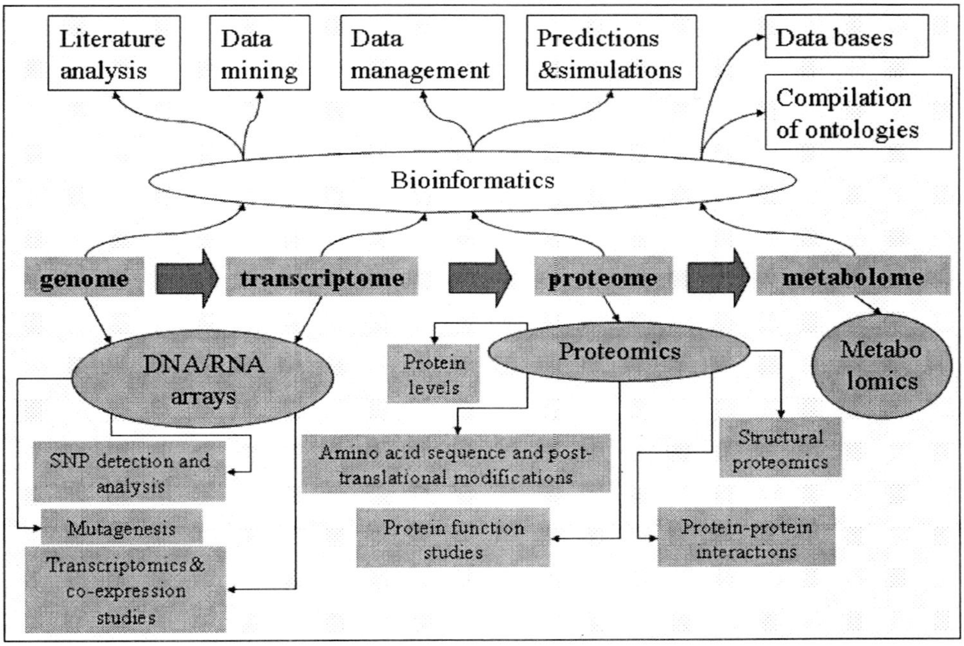

Figure 6-1. The major branches of the functional genomics. This classification is based on the post-genomic tetrad (Figure 1-3).

Another modern and popular term in respect to post-genomic studies is "system biology". While "functional genomics" implies largely an experimental data collection, "systems biology" is calimed to deal with computational models. Albeit there are no two academics that would have agreed what, actually, "system biology" is, the amount of the media hype related to this term is considerable. To read current academic publications, the magic "system biology", whatever it is or would be in the decades to come, holds the answers to the finest secrets and intricacies of the living systems and even, as some overexcited writers claim, "total redefining of what life is". This putative field of "systems biology", as its current proponents claim, "attempts to provide a systems-level understanding by systematically organizing the genomic, functional genomic and proteomic data obtained from genetic and environmental perturbations of interest and using the data to build a descriptive and mechanistic model of the biological phenomena" [1].

An intriguing definition as the above implies some sort of an integrated approach to data management. That's why we mention it in our book: since the integration of data and pieces of information is essential for any adequate physiological study [2] and the notion of homeostasis is the cornerstone of the modern physiology (Chapter I). Thus, the "system biology" is, essentially, hardly a new concept but rather a term that suggests more rigorous data integration down to the molecular and genetic levels. An input of new terms such as "system biology" and others of the kind might be necessary to support the variety in the pool of terms in modern biology. What is essential to understand in respect to this and many other equivalent terms is that they imply an integrated approach: a (parallel) data collection of the experimental information pertaining to genome, transcriptome, proteome and metabolome (Figure 1-3, Chapter I). These terms might also be helpful in order to shift the focus of studies from the reductionist (context-neglecting) approach towards an integrative (context-including) approach thus reiterating the millennia-old, pre-Aristotelian principle that stresses the priority of the whole over the sum of the parts. However, as we'll see from the further material presented in this chapter, the creeping of the reductionist shortcuts is far from being uncommon in the "functional genomics" *cum* "system biology" studies, despite the vehement rhetoric swearing allegiance to the opposite.

What we call here as a "reductionist approach" to biological research represents any study that aims at identification of some individual "key" components (genes, polymorphisms, proteins, metabolites *etc*) responsible for some particular physiological phenomenon. This kind of dissecting study is, undoubtedly, important for practical applications. However, the materials collected in the present book series can illustrate well enough that the information at only one level of function (such as "genome-only" or "proteome-only" data) only rarely can explain the specifics of the behavior of biological systems observed at the higher levels of functional hierarchy. The continuous reiteration and cross-referencing the diagram of the levels of function (the Figure 1-5) throughout this Volume serves the same purpose: to avoid reasoning attitudes that stuck at very few levels of function or at some allegedly "key" aspect out of dozens of other not less important aspects.

Generally, the term "systems biology" seems to bear strong links to cybernetics. *Cybernetics* is a comparative study of organic and artificial systems of control and communication that makes emphasis on the *functional* relations between the different parts of a system, rather than the parts themselves. This term stems from the Greek word *kybernetes* that denotes a wide range of things including "statesman", "pilot" or simply a "rudder". This generic term is as old as the term "homeostasis" in physiology. Similarly to the original Greek word *kybernetes*, the term "cybernetics" also designates quite a lot of subject matters such as artificial intelligence, complexity theory, control systems, information theory, systems engineering and what now is called as "systems biology", a field that affects to describe biological phenomena in terms of interlocking "networks of control" thus underlining the importance of the "networks" rather than some "key" elements.

Formally, such network properties as modularity, regulation through feedback, redundancy and robustness rather characterize the *complexity* of the system than indicate some sort of "key" components thus indeed allowing some researchers to abandon primitive their reductionism. At the same time, while seeking the integration of the data by using the terms and approaches directly borrowed from the science of cybernetics, it is not less important avoid any *neo-reductionism* that, instead of reducing phenomena to "key" parts, reduces them to "key" networks and "key" network properties. From the start, it is necessary to recognize that most of the biological systems cannot be fit into flexible but still quite limited concepts of cybernetics and this idea is hardly at all understood among the hot-blooded but not very cool-minded converts to systems biology.

Why do we spend already second page just to stumble around the two notions of "functional genomics" and "systems biology"? Apparently, these two terms represent "experimental" and "theoretical" branches inherent for any scientific study. Without an adequate guidance from an adequate theory, functional genomics experiments will be useless at best. So would be a theoretical branch if it is provided with false or inadequately collected experimental data, is dominated by flawed concepts or suffers from oversimplification. In their present state, both these alleged branches of the "new biology" are highly controversial subjects.

For example, let's take a look into a recent review with an intriguing title "The Agony and Ecstasy of 'OMIC' Technologies in Drug Development" [3] which was, apparently, written by a researcher from one of the giant pharmaceutical conglomerates. Supposedly, the pharmaceutical companies are very much interested in the practical applications of the functional genomics. I would like to bring attention to this article because it, at least, admits considerable problems with the current state of the *practical* applications of the post-genomic technologies and does not push on the reader the cheerfully-idiotic propaganda of the type "significant achievements were made in systems biology during last decade" *etc*. The conclusions are concise and accurate so I decided to cite here almost the entire abstract of this review: "...Advances in the 'omics' technologies such as genomics, transcriptomics, proteomics and metabonomics were touted as having the potential to revolutionize our approach to disease diagnosis, prognostication and development of novel therapeutics. However, the promise of rapid advances in medicine from the lab bench to the bedside has not manifested as of yet. Indeed, it appears that the translational applications of genomic-based research have preceded the development of both (1) conceptual framework for disease understanding and (2) effective tools that can exploit the vast amounts of data derived from these efforts... " [3].

The author then goes to explain that the post-genomic technologies, such as proteomics, can, theoretically, provide more comprehensive disease monitoring. The "omics" technologies *might* also be able to distinguish adverse effects from exposure to drugs or toxic substances. Then, the author mentions the familiar sociological pattern we mentioned several times: the frustration of over-hyped expectations with the subsequent absence of any positive results. The author states that despite the continuing investments into the "omics" technologies, there is already considerable amount of evidence that these financial infusions did not stimulate nor even approached a new era of prevention or treatment of disease which was promised.

At least, the latter statement is true when considered in view of the new drugs registered in (or, at the very least, submitted to) FDA and similar agencies worldwide. It appears that the number of these submissions has decreased significantly over the past decade which might indicate, among other things, that the research deviates from actual drug design to ultra-high expenditures into functional genomics studies that do not seem to bring much of a profit in terms of the new drugs (not mentioning the actual efficiency of those drugs). The current propagandists of the new "omics" technologies such as transcriptomics, proteomics and metabolomics also promised to enhance comprehension of the non-pathological cellular function and it was anticipated that these technologies would present major opportunities for enhanced "understanding of human disease". However, no progress that would result in ground-breaking practical applications has being made also in this area [3].

The paper we cited above is definitely worth a careful study and the interested readers might indeed do so. Here, let's ask another question: *what* can be efficiently done *at present* using post-genomic technologies such as gene expression profiling, proteome scanning and others? We can only complicate the issue of a practical application by introducing generic terms such as "functional genomics" or "system biology" and then expecting that simply naming an approach to research with a magic word of power (see Introduction) or simply by claming "data integration" will magically provide us with a full and adequate theoretical basis for research. So, what can actually be done and what can't be done using high-throughput post-genomic technologies already available on the market? The rest of this chapter as well as the following Chapters tackle the question. In particular, a Case Study at the end of this Chapter illustrates, once again, that without proper analysis of *already available* experimental data (using *already existing* concepts of physiology as analytic tools) a study based on functional genomics *cum* systems biology will only result in a waste of financial resources and time.

Aiming at Complex Diseases

The primary interest in the functional genomic studies of the human tissues is allowing more precise diagnostics and biochemical monitoring of the complex multifactor diseases. Additionally, more accurate and comprehensive monitoring might pin-point new biochemical markers. Subsequently, this monitoring might also indicate susceptibility traits (such as genetic polymorphisms) to the diseases of interest. What complicates to an enormous degree the cheerful outcome of this cheerful scenario is the enormous complexity of the human physiology.

While some sort of a perturbation technique (see, for example, a study of galactose metabolism in yeast, [4]) is often set to be as a central feature of the studies in systems biology, a scientist even with a minimum background in human physiology will quickly understand the essential difficulty of this supposedly universal method when applied to the research in human physiology. Yes, perturbation of each known essential component in a defined biochemical pathway/molecular cascade using a mutation/deletion approach or a set of environmental changes might indicate a number of physiological mechanisms. However, this approach has at least one considerable limitation: there is only a limited set of perturbations possible under lab conditions. There is even more limited set of the lab conditions, in most of the experiments. *Ergo*, in order to obtain the most adequate mathematical model of the relevant biological process, the lab conditions need to be as complex as the real conditions the biological systems under study actually operate in, if we are to avoid the reductionsm of the "key" factors. Functional genomics works, most of the time, with cells in culture, which are far out of the normal physiological context of the living tissues, not to mention organs and interactions between the organ systems (Figure 1-2). Given the complexity of the internal milieu of the physiological systems in humans, the results of experiments performed on a few disjointed cells (raised in culture) from an organism that has thousands of cell types (intricately woven together) will be difficult to interpret, especially when the experiments were done without any regard to the known particulars of human physiology. Hence, it's a likely expectation that the most productive studies in "systems biology" will be those dealing with unicellular organisms and not with human tissues. Under this condition, how can functional genomics be ever used for the studies in human physiology in the post-genomic era? And what, by the way, is the role of bioinformatics here?

Attempting to answer these questions from a cybernetic point of view, it will be necessary first to define all components of the system, then to describe regulatory relationships, interactions of proteins and biochemical pathways thus formulating the initial model. Then, this initial model based on expert knowledge can be refined using specific genetic/environmental perturbations. Although a pathway/cascade diagram *per se* cannot be a good quantitative predictor of the behavioral response of the physiological system to a certain perturbation, it definitely facilitates comprehension of at least the most prominent qualitative aspects of the physiological phenomena in question. For example, in Chapter II we have compiled the *CVD GeneScape* diagram as a result of the bioinformatical Case Study (see Figure 2-5). This *GeneScape* diagram serves as an initial crude model of CVD and, depending on particular research purpose, numerous functional genomics studies of human tissues can be designed on the base of this diagram using the basics we have outlined in the Chapter II.

For example, a researcher is interested in studying the gene-nutrient interactions in CVD to investigate susceptibility to atherosclerosis. The nutrients of interest in this study are fats. The researcher plans to include biochemical measurements (lipid profile), genotyping and, potentially, a proteomic analysis. In this case, first thing we need is to refer to the third part of the *GeneScape diagram* (Figure 2-5c) that depicts the genes involved in interactions between the bloodstream and the vessel wall. Using basic physiological information on these genes and their respective value (that would be shown in the *GeneScape diagram* as node and arc weights, see also Appendix I), we obtain an initial model of the phenomenon of interest that includes the genetic and environmental components of interest.

Another researcher is interested, say, in thrombotic complications in CVD. This task, apparently, is somewhat more difficult than the previous, since we need to take into account not only the factors that are likely to influence the size of lumen (Figure 2-5c) but, at the very least, also the factors that influence the blood viscosity (represented as coagulation cascade in the Figure 2-5b). Depending on the clinical details of the research project as well as the technical possibilities, these two parts of the diagram can be reduced to various degrees. Quite a number of other *in silico* applications of the *CVD GeneScape* diagram and *medSNP* database are considered in the Volume II of the series. For now, let's consider how methods of functional genomics might be used (and, incidentally, how they should *not* be used) for the study of complex diseases.

Aiming at Complex Diseases: The Scleroderma Example

It is sometimes assumed, perhaps somewhat foolhardily, that "system biology" might not be in any need of the information that was collected in the previous decades of physiological research and that the methods of this newborn area will allow to reconstruct the molecular mechanisms of diseases directly from the data supplied by the functional genomics. A Case Study in the next Chapter indicates the essential informational problems inherent for this kind of *de novo* approach to human physiology. Here, let's consider the plausibility of this sort of attitude on a particular example. As the example, we'll be using a particular study in functional genomics that attempts to deal away with human physiology in this manner.

The functional genomics study we consider here attempted to generate a set of candidate genes for scleroderma (multiple sclerosis) and then to reconstruct the genetic network/molecular cascade from the gene expression data [5]. Once again, the major purpose of this study is to establish genetic determinants of the disease, with subsequent association studies in mind. We compare the results the researchers have obtained as the result of their system biology endeavor with the results of the Case Study we present in this section. *The Case of Scleroderma GeneScape* was based on extensive PubMed searches with the purpose of extracting the reliable data on the genetic association studies dealing with scleroderma. The results of this Case Study represent, apparently, a short summary of the research into the genetic determinants of scleroderma made during the last decades.

Prior to comparing the results, something needs to be said about whether these two approaches to study can be compared at all and in what way. The most important difference is that the functional genomics studies the processes characteristic for a particular type of cells thus almost entirely neglecting the complex reality of gene regulation within the multicellular organism. The data we collected from the genetic association studies are free from this restriction and imply a variety of physiological mechanisms for different cell types.

Further, the functional genomics study [5] deals with a carefully planned expression data analysis and its primary result is, apparently, a small network of mutually regulated genes which might, in theory, represent a molecular cascade or a point of interaction of several molecular cascades (something like "metabolic context", Chapter VII). Again, this gene network (Figure 6-2) was constructed for a particular cell type (skin fibroblasts). As it is not known whether skin fibrosis is primary or secondary phenomenon in the disease etiology (see Box 12), there is no warranty whatsoever that the genes involved in this particular gene network would contain reliable genetic markers or even some common genetic polymorphisms of biomedical significance. On the contrary, *The Case of Scleroderma GeneScape* involves epidemiological data (genetic association studies) that directly addressed the question of the genetic determinants/modifiers of the disease. The putative molecular cascades were then reconstructed using this GeneScape diagram and an expertise in human physiology and molecular cell biology (Figure 6-3). Thus, from the beginning, we can expect quite a number of discrepancies between these two sets of data.

We'll start with presenting and discussing the results of the functional genomics study. The study seems to feature an unusually accurate study design and selection of patients (twin pairs, differential expression analysis). As there is great inter-individual variability of biochemical variables (influenced by both genetic and environmental factors), it was necessary to reduce the amount of statistic noise. For this, the authors of [5] selected candidate genes by identifying differentially expressed genes from microarray data analysis of skin fibroblasts of 16 pairs of affected and unaffected SSc twins. Each pair represents an SSc patient and a normal individual.

The major assumptions of the authors were as follows: 1) the definition of the systemic sclerosis is limited to one of its most obvious manifestations: the skin fibrosis; 2) if the expression levels of the genes between monozygotic patients and their normal twins showed no significant differences, but expression levels between dizygotic patients and their normal counterparts showed differences, then these differences are more likely to reflect the genetic predispositions; 3) four modes of comparisons were used: lesional skin of dizygotic patients and their normal twins, nonlesional skin of dizygotic patients and their twins, lesional skin of monozygotic patients and dizygotic normals and non-lesional skin of monozygotic patients and dizygotic normals. As the result of the study a number of genes, potentially involved in the etiology were identified and a quantitative model of the gene network was built as a set of linear equations. Then, RNA interference was used to perturb the network. The genes of this gene network included various types of collagens, TGF-beta, CTGF, a SMAD protein (Figure 6-2).

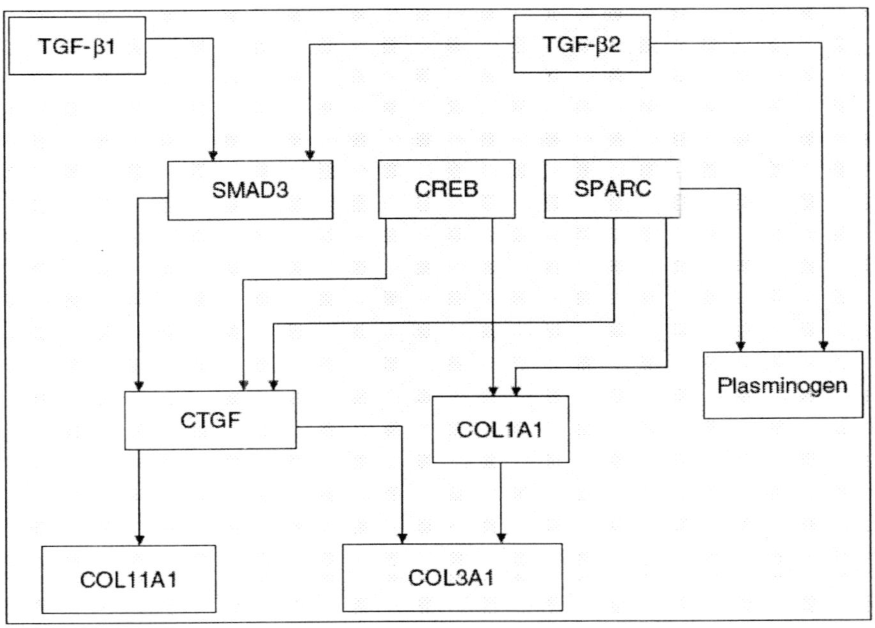

Figure 6-2. A putative gene network for scleroderma using functional genomics *cum* systems biology (after [5]).

To credit the authors, it is quite a remarkable and quite an exceptional study of the problem. Comparisons of the expression patterns in monozygotic/dyzigotic twin pairs are, no doubt, an important feature that seems to support the validity of the results. The major difficulty with the biological significance of the results of this study, however, is that none of the genes the authors put forward as potential genetic determinants of the scleroderma ever figured in a reliable association study. Thus, none of the genes of the putative gene network from [5] is present in the GeneScape summary for scleroderma (presented in the Figure 6-4) nor in our attempt to reconstruct the most general outlines of the related molecular cascades (Figure 6-3). On the contrary, the published data on the genetic associations, analyzed in the course of *The Case of Scleroderma GeneScape,* rather suggest only negative genetic associations for the genes in the "gene network" (Figure 6-2) proposed in the functional genomics study [5]. Although collagens and TGFb-related genes are, apparently, implicated in scleroderma [6], no reliable genetic studies reporting positive associations were ever published. Moreover, the available association studies on collagen and TGF-beta polymorphisms almost unanimously report lack of associations with scleroderma.

Box 12. Scleroderma: A synopsis of the clinical data.

Systemic sclerosis (scleroderma, multiple sclerosis) is a chronic and rare disease (incidence rates of 1/50000 and lower) characterized, in particular, by excessive deposits of collagen in various tissues of the human body. Mortality and morbidity are substantial and are directly related to the extent of the visceral involvement and fibrotic alteration. The most evident symptoms are chronic inflammation, hardening of the skin and associated scarring, calcinosis; swelling of the extremities often precedes the onset of skin problems. Visceral involvement with fibrosis is present in most patients. Diagnosis of the disease is made on the base of these and a number of other clinical symptoms, presence of specific autoantibodies and, occasionally, by histological studies of the biopsies.

The seriousness of the disease is determined by the level of visceral involvement (that is, literally, beneath the skin) and by the total area covered by the disease. Less serious cases tend to be localized and do not appear to involve the internal bodily functions. There are three major forms of scleroderma: diffuse, limited and morphea. Diffuse and limited scleroderma are both systemic disease, whereas only the morphea type is localized mostly to the skin. Diffuse scleroderma is the most severe form of the disease that often involves musculoskeletal, pulmonary, renal and other complications. It features a rapid onset and is more life threatening. The limited form is much rarer and much milder: it has a slow onset and progression, skin hardening is usually localized to hands and face with internal organ involvement being less severe.

There are no obvious causes for systemic sclerosis. The overproduction of collagen *per se* is thought to be a result of the autoimmune dysfunction: T-cells accumulate in the skin and are thought to secrete various cytokines and other proteins that stimulate collagen deposition [6]. Although the most apparent component in systemic sclerosis is persistent activation of the expression of the collagen genes, abnormalities in at least 3 types of cells are intimately involved in the development of the clinical and pathologic manifestations of the disease: fibroblasts, endothelial cells and leukocytes. Damage to the endothelium is an early abnormality in the development of scleroderma which is likely to be the result of both collagen accumulation and direct alterations caused by cytokines and platelet adhesion. Increased endothelin and decreased vasodilation has also been documented.

Since there is no known cure for the disease, the current therapeutic approaches concentrate on treatment of the symptoms: reducing inflammation with nonsteroidal anti-inflammatory drugs, prescribing specific drugs for the specific complications (in the cases of diffused scleroderma) and using drugs affecting immune system. (After [6], see also International Scleroderma Network http://www.sclero.org/ and Wikipedia http://en.wikipedia.org/).

If we give some further thought to this discrepancy between the results of the functional genomics study (Figure 6-2) and the summary of the results of the genetic association studies (Figures 6-3 and 6-4), we might be able to conclude, once again, that physiological expertise is paramount for the functional genomics studies, if they are to produce biologically and medicinally significant results. A study in functional genomics that, apparently, neglects the fundamentals of the physiological process of interest, cannot be expected to produce results of any biomedical significance.

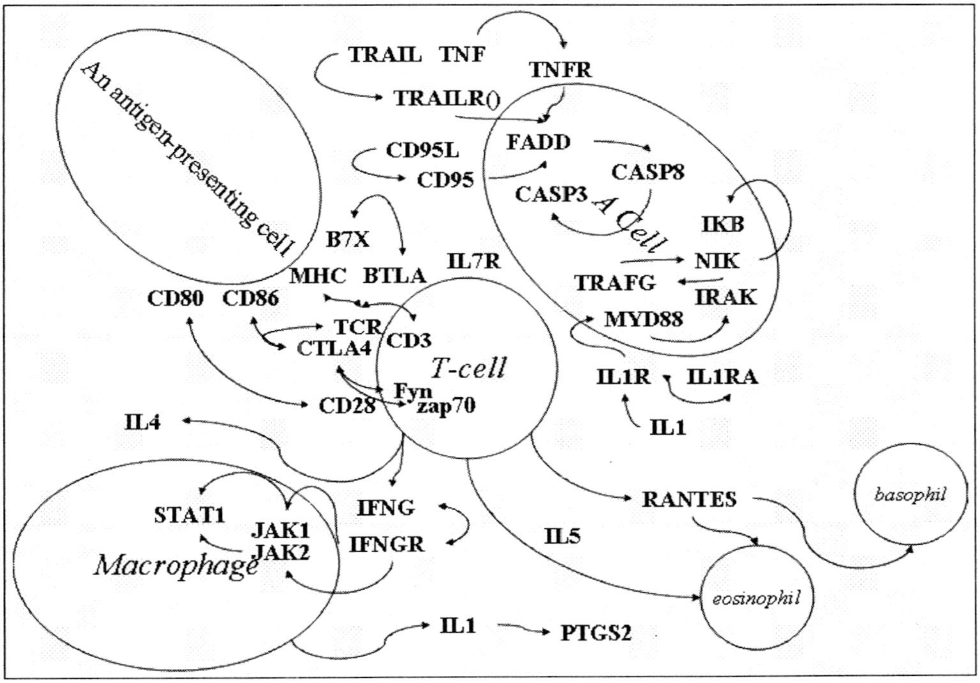

Figure 6-3. An brief summary of molecular cascades potentially involved the etiology of systemic sclerosis. The diagram was compiled on the base of the inflammation genes in the *Scleroderma GeneScape* and some expertise in molecular cascades. Actually, the genes indicated in the diagram can belong to different cell types from different tissues. The diagram thus represents a disease-oriented cross-section of the human molecular cascades and merely provides most general functional context for each of the genes mentioned.

Indeed, the most apparent and almost universal clinical feature of systemic sclerosis is the progressive fibrosis of the skin. This, however, is only the most apparent symptom of the disease. It should not be forgotten that multiple internal organs are always involved even if this involvement is subclinical [6]. Moreover, the swelling of extremities (so-called "Raynaud syndrome") often takes precedence in the clinical manifestations of scleroderma. Thus, contrary to the integral approach that "system biology" seems to claim, we still see thriving of a *neo-reductionist* shortcut: the host of clinical symptoms was reduced to one "key" symptom (skin fibrosis), the entire problem of multiple sclerosis (that, most likely, includes a number of tissues and many cell types, see Box 12, Figures 6-3 and 6-4) was reduced to only one cell type (skin fibroblasts) *etc*.

Our putative reconstruction of the general outlines of the molecular cascades involved in scleroderma (Figure 6-3) underlines the importance of the inflammation processes and the relevant genes in the etiology of the systemic sclerosis. Inflammation is the first response of the immune system to infection or irritation characterised by redness, heat, swelling, pain and visceral dysfunctions common in multiple sclerosis. The cellular component of inflammation involves the movement of lymphocytes from blood vessels into the inflamed tissue. The cellular kinesis of lymphocytes has an extremely important role in inflammation: guided by cytokines such as interleukins and interferons, they filter out into the inflamed tissue, act as phagocytes and release further cytokines (such as IL1 and TNF) that activate endothelial cells to upregulate immune cell receptors (such as selectins) *etc*. In respect to the systemic sclerosis, it is important to remark that, in particular, interleukins are involved in systemic inflammation essential for the systemic sclerosis [6].

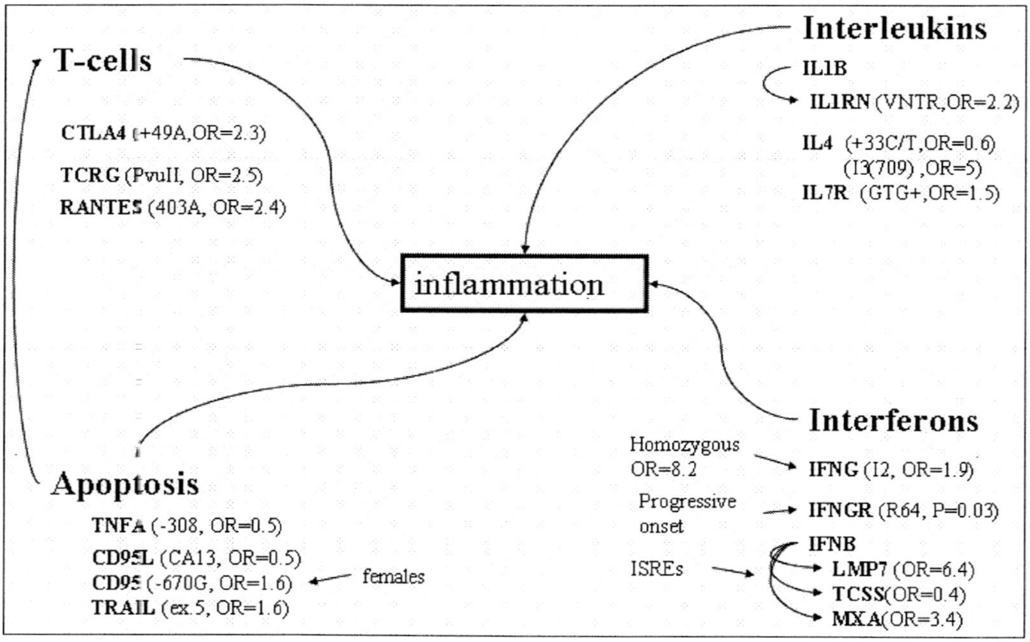

Figure 6-4. Systemic sclerosis and genetic association studies. This figure presents the inflammation-related genes: genes involved in apoptosis and lymphocyte fuction, interferons and interleukins. The processes behind these subheadings are well-known to be considerably interrelated (T-cells are activated by some interleukins and secret some other interleukins, they also secret/activated by interferons *etc*) so the diagram omits the relevant links and simply enumerates the genes involved. Also, the data for the Caucasian and Asian populations were pooled together.

Pin-pointing the molecular pathways potentially implicated in scleroderma is far from being a novel research problem. For example, an excellent review summarizes the data from about two decades of biomedical research in the area [6]. Despite that contribution of the genetic factors to the disease etiology is quite equivocal, this contribution is suggested by familial clustering of the disease, high frequency of autoimmune disorders in family members and differences among ethnic groups. Although the concordance of systemic sclerosis among identical twins did not differ significantly from the concordance of disease in dizygotic twins, the concordance for the presence of specific autoantibodies was substantially higher [7]. The genetic factors implicated are, potentially involved in the production of specific autoantibodies, collagen synthesis and processing, balance between immunoregulatory T-cell populations and effects of various products of inflammatory cells (including interleukins, interferons and TNF). In the clinical science of scleroderma, it is also well-standing notion that fibrosis, even though it is the most obvious feature of the disease, is likely to be only the final step in the pathogenic pathway of systemic sclerosis [6,8].

Accurately digesting the materials cited above, we will be able see that the astounding discrepancy between the results of the two studies we present here is far from being something unexpected. Since the chronic inflammation, an essential physiological feature of scleroderma, was neglected in the functional genomics study, the discrepancy is, indeed, of no surprise at all. Chronic inflammation involves quite a number of different cell types and most of those mentioned in the Figure 6-3. From this comparison we can also infer how important it is for a functional genomics study to select not just any cell type (or, at best, the most obvious one) but the cell types that are the most likely to be fundamental in the pathology of interest. We will return to the matter of selecting the adequate cell types for functional genomics experiments in another Case Study at the end of this Chapter (*The Case of Estrogens and Venous Thrombosis*). Here, we present

THE CASE OF SCLERODERMA GENESCAPE,

in which we compile a diagram of the putative genetic determinants of the scleroderma on the base of the published association studies.

Introduction

Scleroderma (systemic sclerosis) is a chronic and a rare disease (Box 12). Although the most apparent clinical features of systemic sclerosis are related to the progressive thickening and fibrosis of the skin, multiple internal organs are always involved in some degree even when this involvement is not clinically apparent [6]. The common pathologic changes include severe tissue fibrosis, chronic inflammation characterized by production of specific autoantibodies as well as impared lymphokine production and microvascular disease characterized by intimal proliferation and obliteration of the lumen of small arteries and arterioles.

In some cases, environmental agents seem to play more important role than the genetic factors. For example, a study reported a remarkably low concordance in the development of systemic sclerosis among homozygous twins, indicating that the heritability component of the disease was very low leading to conclusion that the most important risks are likely to be environmental [7]. In particular, organic solvent exposure, certain pesticides, hair dyes and industrial fumes might among these environmental risk factors. Generally, the disease is not inherited in a classical Mendelian pattern although, as the results of the present Case Study suggest, genetic factors do seem to contribute to its development and the extent of the clinical manifestations. The present Case Study aimed at collection and systematic analysis of the data on the genetic association studies relevant to the etiology and the course of scleroderma.

Methods

We used largely the same bioinformatic techniques of the data retrieval and analysis as in the Chapter II (*The Case of A GeneScape for CVD*). Again, we made use of the *medCrawler* script, the phenotype-related keywords were "scelroderma", "sclerodermia", "multiple sclerosis", "systemic sclerosis". The association studies extracted with the script were further analyzed using the study verification criteria presented in the preceding Chapters.

Results

Totally, there were over 150 of the association studies dealing with systemic sclerosis. Analysis of these studies according to the selection criteria (see Chapters III, V) has allowed identification of 36 studies that conform to the criteria. These studies provided information on polymorphisms in 25 genes that were associated with the disease in at least one of association study.

The essential peculiarity of the entire array of the data is that there were no genes/polymorphisms for which associations with scleroderma were confirmed in numerous independently made association studies. In other words, presently available (2006) genetic associations for each of the genes/polymorphisms were confirmed (in some cases, were *not* confirmed) by no more than 2..3 independent groups of researchers.

The genes identified belong to numerous functional groups, from genes involved in the immune system to metalloproteinase and a highly pleiotorpic protein kinase. A significant proportion of the genes identified (15 out of 25) were apparently related to the inflammation processes and included genes involved in apoptosis, T-cell function, interleukins and interferon-related genes (Figure 6-4). Multiple sclerosis is an autoimmune disease so it is only natural that many of the genes involved in the functioning of the immune system (in particular, inflammation-related genes) were found as the result of this Case Study. Here, we consider the results mostly for the genes in these four groups, while the entire results of this Case Study are being deposited into the *medSNP* database.

The GeneScape diagram is not a representation of any sort of a gene network or molecular cascade. It simply enumerates the genes with polymorphisms associated with attenuation of risk for systemic sclerosis under various conditions. In this diagram, we simply piled together the data from association studies performed on various Caucasian and Asian populations and which were found to be reliable (at least, according to the criteria we mention). A putative reconstruction of the relevant molecular cascades, based on the GeneScape diagram (Figure 6-4) and author's expertise in cell biology (however limited) is presented in the Figure 6-3.

AIMING AT COMPLEX DISEASES: CVD

Man can learn nothing unless he proceeds from the known to unknown.

This saying is attributed to Claude Bernard (1813-1878), famous French physiologist who was one of the first to introduce the idea of the "internal environment". Albeit being said more than a century ago, the statement bears direct relation to the role of physiological knowledge in functional genomics studies. Thus, the case with scleroderma we just considered has several peculiarities that need to be reiterated. Albeit being a multifactor disease, scleroderma is quite rare. How crucial is the genetic component in the disease etiology is not clear, although certain genetic polymorphisms are likely to modify the course and severity of scleroderma. The association studies dealing with scleroderma are quite few (a hundred, in comparison to the many thousands of the studies dealing with CVD). Nevertheless, the pathological mechanisms behind this very particular disease are very likely to involve a number of different cell types. Thus, even in such a relatively "simple" case, a functional genomics study faces many challenges, often hardly surmountable when taken without due consideration of the relevant physiological processes. Such is the situation in the case of a rare and relatively "simple" disease. What about a common and much more complex disease such as CVD? Could the functional genomics be of any use at all in this case? And, rephrasing Prof. Bernard's statement, how reasonable it would be *to proceed* with functional genomics without any regard to what is already *known* in human physiology?

In view of the experience we might've obtained while reading the above scleroderma example, the perspectives for functional genomics to study more complex diseases can appear bleak: a more complex disease would involve even greater variety of the cell types and more complex interactions on the levels of tissue and organ systems, not just processes in particular types of cells. However, this is the case where the *purpose-oriented* approach to bioinformatics can be of great help. In other words, attempting any sort of the functional genomics study to pinpoint "genetic determinants of CVD" will, indeed, be a ridiculous task even to consider. Still, when this kind of know-all pretensions is set aside, a sane research purpose is formulated and an adequate bioinformatic strategy is figured out, then we can have some hope that the efforts in functional genomics will not be in vain.

A few examples we present below were, again, assessed by comparison to the corresponding GeneScape diagram obtained as the result of analysis of already available association studies (*CVD GeneScape*, Figure 2-5). In order to be of any practical use (let alone establishing fundamentals of the corresponding physiological processes), a functional genomics approach, be it proteomics or gene expression profiling, needs to be properly applied for a particular purpose. The onus here is that the purpose should be neither too general nor ultra-narrow. Of course, it is up to the researchers to strike the appropriate balance between these two extremes.

For example, the decade-old efforts of several research groups to establish some general "human cardiac proteome" [9,10] might have being pioneering and allowed to list over 150 proteins. Of course, the human proteomes are not compiled only for the sake of establishing potential genetic determinants for the future association studies. Nevertheless, it is quite instructive to observe that out of 150 proteins of the putative human cardiac proteome (see the web references in [9,10]), only APOAI can be found in our GeneScape diagram (Figure 2-5c). None of the entries for "human endothelial proteome" features in the *CVD GeneScape* despite the fact that the Figure 2-5c lists at least a dozen of proteins/genes relevant to the state of the endothelium wall and corresponding pathologies.

As we appropriately narrow the scope and shape the research purpose more adequately, more meaningful results get a chance to surface. For example, in the following study [11], the researchers used gene expression profiling in a CVD-related study. Their purpose was to see how functional genomics can be used to map the genes involved in the blood pressure regulation. The map of these genes can then be used for understanding the genetics behind the chronic elevation of blood pressure in hypertension. Far from denying the previous decades of research in human physiology, the researchers first formulated the known physiological mechanisms of the process of interest. This careful attitude has allowed to select adequate cell types for the functional genomics study and to obtain a number of physiologically plausible results.

Thus, basic physiology states that blood pressure is regulated by sympathetic, parasympathetic nerve activity and humoral mechanisms that further interact to control cardiac output and vascular resistance. In addition, the central integration of afferent neural and humoral inputs from the periphery also modulates autonomic nerve activity and thereby contributes to regulation of the blood pressure. Given the central role of the neural system in the regulation of the blood pressure, the cells selected for gene profiling were those of hypothalamus and brainstem regions. The authors were also very careful to avoid detection of the secondary effects in gene expression by using hypothalamus and brainstem areas of neonatal (pre-hypertensive) rats [11]. Stimulation by angiotensin II was used to alter the gene expression.

Authors did mention that reproducibility of the results of the functional genomics studies is a common problem. However, the results presented in [11] are of definite biomedical interest. Thus, over 40 genes were identified with the expression observably affected in at least 30% of the experiments. It cannot be denied that most of these genes would look, to a medical geneticist, rather like a list of genes predisposing to various cancers than the genes related to physiology of hypertension. Indeed, B-cell leukemia/lymphoma 2 (BCL2), epidermal growth factor, glutathione synthetase, microtubule-associated protein 2 (MTAP2) *etc* are hardly strong candidate genes to be involved in the hypertension. Indeed, these genes are involved in the most fundamental cellular and organism processes so their expression is affected by an exceptionally wide range of physiological conditions, not only after treatment of the cells with angiotensin. However, a number of the genes identified in this functional genomics study represent a direct or, at least, quite close a hit when we look at our *CVD GeneScape* (Figure 2-5). These would be natriuretic peptide precursor type A (NPPA), insulin receptor substrate 2 (IRS2), interleukin 1A, gamma-adducine and a few others.

Another example deals with using proteomics to discover novel biochemical markers of the cardiac allograft rejection [12]. This study, again, features clearly defined purpose, an accurate clinical work and a realistic usage of functional genomics (proteomics, in this case). In brief, endomyocardial biopsy remains the most reliable method of detecting rejection following cardiac transplantation. The procedure is invasive, associated with risks for the patient, labor-intensive and costly. Therefore, it would be of considerable benefit to have a reliable noninvasive biochemical method to detect acute rejection, preferably as a sort of a specific blood test. Within the framework of this purpose, proteomics can be used to detect which proteins are upregulated or downregulated in the heart tissue during rejection. These proteins can become novel blood markers of the cardiac graft rejection.

In order to identify these proteins, sequential cardiac biopsies from several patients were analyzed by 2D electrophoresis. To detect allograft rejection, over 100 of the endomyocardial biopsies were reviewed independently by two expert histopathologists. Over 100 proteins were found to be upregulated 2-fold and 5-fold during rejection and an initial panel of 13 proteins showing a 2-5-fold change in response to rejection in all the patients was selected. Two of these (RB-crystallin and tropomyosin) were considered as potentially interesting markers and were subsequently measured by ELISA in the sera of 17 patients followed for 3 months after their transplants. Mean levels of RB-crystallin and tropomyosin were significantly higher in sera associated with biopsies showing all grades of rejection compared to no-rejection biopsies ($P<0.03$, see details in [12]).

As we can see, precision in the formulation of the research purpose and awareness of the basics of physiology can be crucial to obtain results of at least some biomedical significance even in the case of very complex diseases such as CVD. Proceeding from known to unknown is the natural approach of developing human knowledge and it's folly to ignore this Law of Human Nature known to humans for millennia. On the contrary, jumping-to-conclusions approach that neglects even the known fundamentals of physiology will lead to nowhere. These two principles (precise enough research purpose and awareness of the basics of physiology) are, of course, by no means restricted to the few studies we just considered. Consistent implementation of these principles within any biomedical study that involves post-genomic technologies such as functional genomics can be almost a warranty of obtaining non-trivial and physiologically plausible results.

Identification of the Target Genes: Transcriptomics

Even from the very few examples we've considered so far in this chapter, it is clear that one of the important tasks functional genomics aims at is establishing valid candidate genes for certain pathologies. The gene expression profiling (transcriptomics) and the protein levels detection (proteomics) are the two major and relatively independent branches of the functional genomics that can be used for the tasks of the kind (Figure 6-1). Previously, we have considered both of the techniques using particular studies as examples. We have seen what pitfalls a functional genomics study of an improper design and a deficient information management will face. We have also discussed how at least some of these pitfalls can be avoided.

In respect to transcriptomics, the most important point to be kept in mind is that the gene expression is regulated at multiple levels. Mere identification of the mRNA levels is next to useless, especially when attempted to be done by disregarding the complexity of the mechanisms of the gene regulation. Even in a particular cell type, there are highly coordinated (and, often, cell-specific) layers of control that variously mediate the implementation of the information encoded in the genome. In particular, post-transcriptional levels of regulation (such as transcript turnover and translational control) comprise an integral part of gene expression and in their apparent complexity might even rival the transcriptional control. Accordingly, the transcriptomic attempts of the discovery of the "target" genes should take into account these fundamental intricacies of the cellular regulation of the gene function. For further details on these levels of regulation, the interested reader may refer to an excellent review [13].

Through decade-long studies in classical molecular biology a fundamental fact was established: activity of a gene (and, therefore, the protein production) is regulated at several levels such as transcriptional control (the rates of the DNA to mRNA transcription), post-transcriptional control (the rates of the mRNA translation and decay) and post-translational control (biochemical modifications of the protein). Etiology of a disease can involve a disturbance or loss of regulation at any of these levels and by no means is restricted exclusively to the level of the transcriptional control (as it is assumed in most of the published studies in transcriptomics). The apparent bias of the researchers towards transcriptional control might be rationalized in terms of historical and technical reasons but, nevertheless, post-transcriptional regulation (processing, export, localization, turnover and translation of mRNAs) cannot be neglected, especially in a large scale and a high-throughput study.

For example, *mRNA turnover* can be regulated by several different mechanisms. Decay rates of mRNA appear to be encoded by specific sequence patterns (located in the 3'UTR of the mRNA) which are recognized by a number of cellular RNA-binding proteins. This contention quickly becomes a basic fact of molecular biology and it has direct repercussions for the transcriptomics. Thus, although studies in the gene expression profiling claim to investigate the transcriptional control of the genes, they, in the most of the cases, just measure momentary levels of specific mRNAs using PCR-based techniques. The levels measured are, actually, the steady-state mRNA levels that reflect production, stability and decay of the mRNA transcripts [13]. Transcripts encoding core metabolic proteins might have longer half-lives, whereas transcripts encoding transcription factors or members of the ribosome-biogenesis machinery are significantly less stable [14]. Therefore, for a transcriptomic study to be more biologically adequate, it should be complemented by a large-scale, high-throughput determination of the turnover/stability of the mRNAs involved. Usage of the finer timescales of the transcriptomics analysis is also advisable.

Translational control represents another level of post-transcriptional control. It occurs during the translation on ribosome and employs both general cellular mechanisms and transcript-specific ones. Rate-limiting step for the protein synthesis is the translation initiation which is also the main target for translational control. While the effect of the general mechanisms can be taken into account by collecting sufficient statistics, the transcript-specific regulation poses additional challenges for the studies in transcriptomics. Transcript-specific regulation involves RNA binding proteins that recognize the control elements encoded as specific sequence patterns in the UTRs of the mRNA. The regulation of translation appears to be especially important under conditions that require abrupt and quantified changes in protein levels required for cell survival (stress, apoptosis, cell growth *etc*). Experimentally, the effects related to translation control can be measured by fractionation of the transcripts based on the number of associated ribosomes. This "translational profiling" should, in general, supplement any study in transcriptomics [13].

Since there are mRNA-specific mechanisms of the post-transcriptional regulation, it is clear that the straightforward attempts of gene profiling aiming at identification of the genes important for a disease are very likely to fail or to produce equivocal results at best. Such attempts might be considered as another indication of *neo-reductionism* that is already on the rise within the field that is supposedly "comprehensive" and "integrative". From psychological point of view, these attempts can be nothing more than a manifestation of the infantile instinct to reduce multilevel complexity to a few oversimplified dogmas. In other words, very few of the transcriptomic studies take into account the effects related to post-transcriptional control. Usually, the researchers, apparently, strictly adhere to assumption that results of biomedical significance can be magically produced merely by waving the wand of measuring mRNA levels at different time intervals and using a few specific stimulae. And it is quite good a study if it uses some rationale from the physiology and/or cell biology to select an adequate cell types and combination of the stimulae. Most of the contemporary transcriptomic studies lack even this rationale. Therefore, it comes as no surprise that it is exceedingly difficult to infer any biomedical conclusions from the results of the most of the published transcriptomic studies.

Most of the transcriptomic studies oriented at identification of the target genes using the "standard" scheme of 2, 6, 24 and 48 hrs measurements for a particular stimulant. The major difficulty with the transcriptomics studies of the sort is a profound lack of correlation among the results obtained in different time frameworks: only rarely the set of genes obtained, say, after 2hrs after stimulation will correlate in any way with the set of genes obtained after 6-8 hrs. This might be considered as "an extremely complex reality of the cellular regulation of the gene expression", quoting anonymously a distinguished reviewer. However, this is also indicative of what we call here, in this book, as an extremely simple reality described as the "elephant-in-the-dark" syndrome (Introduction). In science lingo, this famous fable implies that some key factors were missed from the experiment and the results are, most likely, scientifically unsound. Often, this omission is the result of the inadequate selection of the cell type and of stimulation regime which is far from any realistic assessment of the physiology involved (as it was with the unfortunate examples we mentioned previously in this chapter as well as with the transcriptomics experiment we mention in *The Case of Estrogens and Venous Thrombosis*). Sometimes, the possible reasons take to much time to ponder... We present additional analyses of the problem in the Volume III of the present book series ("Molecular and Cellular Biology").

Despite that the foregoing points and examples point to a number of problems inherent to the current transcriptomic studies, this discussion in no way implies that studies dealing in transcriptomics are of very little promise in post-genomic biology. As we continue to reiterate throughout this volume (no doubt, in order to bore the readers to death), post-genomic technologies, *when* combined with an adequate expertise in physiology, hold indeed a great promise for the development of biology and medicine. Take, for instance, the Allen Brain Atlas: genome-wide database of gene expression in the mouse brain (http://www.brain-map.org/). This atlas is based on the RNA *in situ* hybridization performed on the consecutive slices of the mouse brain. While, again, non-genetic factors known to affect gene expression (age, gender and others) were, apparently, neglected in this excellent pilot study, this study is, for the most part, free from reducing the physiological processes to only one cell type. Bioinformatic analyses even of this snapshot of the data can be very helpful to direct further research in the area [15] and this database can be gainfully employed by post-genomic neuroscience (Volume VI) to study the brain at significantly increased level of cellular and molecular details.

IDENTIFICATION OF THE TARGET GENES: PROTEOMICS

Proteomics might be one of the most promising branches of functional genomics in respect to studies of the complex diseases such as CVD. Indeed, given the numerous occasions on which we referred to the diagram of the levels of function (Figure 1-5), we might have, finally, registered that it would be more productive to use some intermediate levels in the hierarchy of the levels of function *along with* analysis of the clinical manifestations and the genetic data. Unlike mRNA transcripts, proteins are the actual actors in the cells and bodily fluids. Proteins mediate most of the actual physiological responses. The levels of certain proteins, being what we call "biochemical variables", can be helpful intermediaries in studying the genotype-phenotype associations (Chapters II&III). The changes in the protein levels detected by methods of proteomics can, perhaps, allow to arrive at more certain and reliable conclusions concerning the target genes as well as concerning novel biomarkers of human disease.

Following the definition of functional genomics we use (Figure 6-1), the general idea behind proteomics is a high-throughput determination of the levels of (particular) proteins. However, proteomics is not restricted to this task and can also involve biochemical and biophysical characterization of particular proteins, including post-translational modifications and protein-protein interactions. Proteomics also faces a number of challenges which have to be surmounted before any actual clinical application can be based on the data from proteomic profiling. The additional layers of functional complexity, the vast range of protein abundance and the compartmentalization of proteins, highly dynamic nature of the proteome and a number of other factors indicate considerable differences between quite straight-forward of genome sequencing and the task of the human proteome characterization [16]. Given thousands of various cell types involved in hundreds of tissues, along with unique proteome for each cell type in each tissue (not mentioning cellular organelles), it's clear that any sort of, shall we say, threestomachic studies (meaning, of course, "three-stomachs-of-a-study") that aim at determination of all proteomes for all human cell types, their protein constituents, levels *etc* are hardly feasible. Therefore, it's rather accurate crafting of the purpose-oriented studies will be crucial for the advancements in the clinical applications of proteomics.

Before we proceed with considering basics, trivialities and important details concerning proteomics, let's make a brief feasibility assessment concerning the use of proteomic markers in the studies of CVD. Again, relying on the published data of association studies, we can assess the feasibility of some current proteomic markers using a particular list of proteomic markers and our *CVD GeneScape* diagram (Figure 2-5). This particular list of markers has nothing special about it (http://www.superarray.com/gene_array_product/HTML/OHS-037.html) and it was selected simply because it deals with a particular microarray for the cardiovascular studies available on market. This particular list, apparently, over-represents cell adhesion molecules (numerous integrin subunits), while it under-represents, for instance, the proteins of the blood coagulation cascade. Also, the entire idea of comparing an arbitrarily selected list with the results of association studies do seem to be preposterous so this comparison of ours does not bear any kind of promise nor does it imply any conclusions of any crucial importance. However, it is still intriguing to notice that at least 20% of the proteins mentioned in the list (such as ICAM1, ITGB3, SELE, SELP, SAA1 *etc*) also featuring in the *CVD GeneScape*. The 20% correspondence between what might seem to be two non-related lists of genes/proteins is definitely worth noting, especially in respect to the problem of identification of the target genes. This correspondence is considerably higher than, say, the one between the scleroderma *GeneScape* and the functional genomics study we previously discussed.

As it is the case with all biomedical studies, in order to obtain results of any clinical significance, it's necessary to exclude as much of the statistical noise as possible. Proteomic studies are no exception to the principle. Whether the purpose would be identification of the target genes or biomarker discovery, noise reduction of the *in vivo* experiments can be achieved by selecting individuals from a highly homogeneous cohort (in animal studies, for instance, that will involve specially bred lines of animals). *In vitro*, noise reduction is achieved, as we already mentioned more than once, by careful consideration of the physiological effects of interest and by not less careful selection of the tissue and cell types for the high-throughput experiments. Let's consider two examples of application of both of these strategies in proteomics.

First example deals with an attempt of discovering a biomarker for liver cirrhosis [17]. While bilirubin test is the common and reliable marker of the hepatic function in general, there are no established biochemical markers specific for cirrhosis. The gold standard for diagnostics of cirrhosis is histological examination of the liver biopsy. This invasive procedure is, apparently, associated with considerable risks for the patient and a biochemical marker that detects early stages of cirrhosis would be of great help to clinicians. Using proteomic approach within an animal study aimed at discovery of such a biomarker (rats in laboratory conditions) allowed exclusion of much of the informational noise which is otherwise unavoidable in the study of the human subjects [17]. Although the results are, no doubt, preliminary (since, in brief, rats are *not* humans), they are still worth considering, at least for the sake of observing an accurate methodology of the proteomic study.

The authors of the study [17] used a three-step approach that included (1) protein profiling using SELDI (surface-enhanced laser desorption/ionization); (2) selection of a statistically significant set of serum biomarkers using a simplistic AI algorithm and (3) peptide sequencing of the selected biomarkers. Three groups of the rats participated in the experiment: normal, with thioacetamide-induced liver cirrhosis and with liver fibrosis induced by bile duct ligation. Scanning of the serum protein profiles indicated that 3495 Da and 8201 Da proteins were down-regulated while 4186 Da and 4902 Da proteins were up-regulated. Further analyses suggested that the 3495 Da protein was consistently down-regulated in most of the cirrhotic samples. Peptide sequencing suggested that sequence of this

3495 Da shares some similarity with histidine-rich glycoproteins. While biological significance of this putative protein in the liver cirrhosis is far from being obvious, the study suggested, at least, what seems to be a perspective biomarker [17]. Clearly, this positive finding was achieved, in particular, through careful planning of the physiological part of the study and through accurate informational management of the entire research.

Another example we would like to mention here deals with interferon-inducible proteins [18]. This study illustrates an appropriate selection of the tissues and cell types as well as an accurate informational management of the entire project. It also illustrates that studies of greatest consequence in functional genomics/systems biology would feature both of the major branches of the functional genomics combined within the framework of an adequately designed clinical study.

Interferon-related responses of the human systems are important for the antiviral defense. The study [18] aimed at examination of the liver responses to interferons that might be related to inhibition of virus replication. Albeit the purpose is far from being clearly and precisely defined, the results of this study of functional genomics should certainly be drawn to the reader's attention as an example of an accurate and, for the most part, reliable study although it was based on somewhat awkward assumptions.

In brief, the authors aimed at studying some abstract "liver-specific response" to interferon treatment stating that interferon treatment is, in particular, an essential component of the current therapies against hepatitis C virus (HCV). Therefore, the researchers have selected a liver-derived cell line (human hepatoma cell line Huh7) and compared the proteomics profiles of these cells in the presence and in the absence of the interferon treatment.

There are two assumptions behind this study that need to be questioned. First, using an abnormal cell line (hepatoma-derived cells) might not be the best way to study molecular mechanisms neither antiviral defense in general nor HCV defense in particular. Secondly, the authors, apparently, thought that they were studying some specific response while, in fact, they were studying an extraordinarily general physiological mechanism. Thus, "liver-specific response to interferon treatment" might seem to be well defined purpose when we forget the central function of the liver in the human physiology. When we recall, however, that the liver is the center of protein synthesis for the entire body, we might conclude that, actually, it would be of great interest to study response of the liver cells to almost any treatment, not only interferons or, say, estrogens (which we discuss, by the way, in *The Case of Estrogens and Venous Thrombosis* right after this section). Despite these two assumptions, the study was definitely a success and can serve as a positive example, especially when considered against the bulk of downright flawed and largely useless efforts advertised nowadays under the exciting title of "functional genomics".

In their proteomic analysis, the authors identified >1360 proteins affected by interferon (IFN) treatment. Among these, 54 were IFN-induced by more than 2-fold (15 known and 39 novel proteins) and 24 were IFN-repressed by more than 2-fold. These IFN-regulated proteins represented multiple cellular functions including antiviral defense, immune response, cell metabolism, signal transduction, cell growth and cellular organization. Then, the authors attempted to investigate the cellular pathways with which the proteins they identified were associated.

What appeals most in this study (at least, to the author of the book you now read) is that it has demonstrated a strong convergence of the results from proteomics with the data from the classical research into molecular biology and physiology of the antiviral cell defense. Thus, decades-long research discovered several well-known IFN-induced proteins (such as the protein kinase PKR, 2'-5' oligo-adenylate synthetase, interferon regulatory factors (IRFs) *etc*). In the study we mention [18], the authors seemed to identified all of the classical IFN-induced proteins except IRFs. Further analysis with categories using so-called gene-ontology classifications has indicated that the proteins are unevenly distributed among the various predefined categories dealing with cell processes. Proper informational management and careful analysis of the results of the results produced by functional genomics experiments did seem to produce a wealth of data that have biological significance and, potentially, a number of biomedical applications.

According to the gene ontology classification, 11 of the 54 induced proteins were labeled as "involved in the immune response". Indeed, previous research abundantly indicated primary roles of interferon in the immune function. In physiological studies, interferon treatment has been also shown to reduce levels of cellular metabolism. Accordingly, the authors observed that interferon suppressed expression of the genes involved in metabolic pathways including fatty-acid metabolism (fatty acid synthase, long-chain fatty-acid-coenzyme A ligase 4, fatty-acid-binding proteins) and aminoacid metabolism (ornithine aminotransferase, sarcosine dehydrogenase). Previously, interferons were reported to inhibit cell growth and the authors apparently register induction of PKR (PKR reportedly downregulates cell growth and proliferation). Further analyses has shown a number of correlations with already known physiological effects of interferon and suggested several novel correlations of the cellular pathways, including IFN signaling. Multiple known components of the IFN signaling pathways were identified (such as STAT1, NMI in the JAK-STAT pathway, Mx1, PASK) and, in particular, related to the G-protein signaling (PLCD1, GNB1, GPS1 *etc*). Finally, the authors have attempted to compile a general cellular circuitry involved, much like we did in *The Case of Scleroderma GeneScape* (Figure 6-3), using the results of their functional genomics study along with the expertise in cell biology (see Figure 6-5).

These few examples aimed at illustrating a simple truth. The truth here is that purpose-oriented approach in functional genomics, be it guided even by quite general expertise in physiology, is a likely way to scientific success in the functional genomics studies. On the contrary, the *neo-reductionist* studies that feature one-sided physiological views (or none whatsoever) and three-stomachic attitudes achieve only the purposes that they really aim for (that is, extortion of money from federal budget, embezzlement of the funds from biotech companies *etc*). Accordingly, it's up to the reader to choose his own path in the area generally called as "scientific career"... But that is, of course, a long-term involvement. At present, we are more interested in presenting to the reader the Case Study we have mentioned several times throughout the present Volume. Here it is:

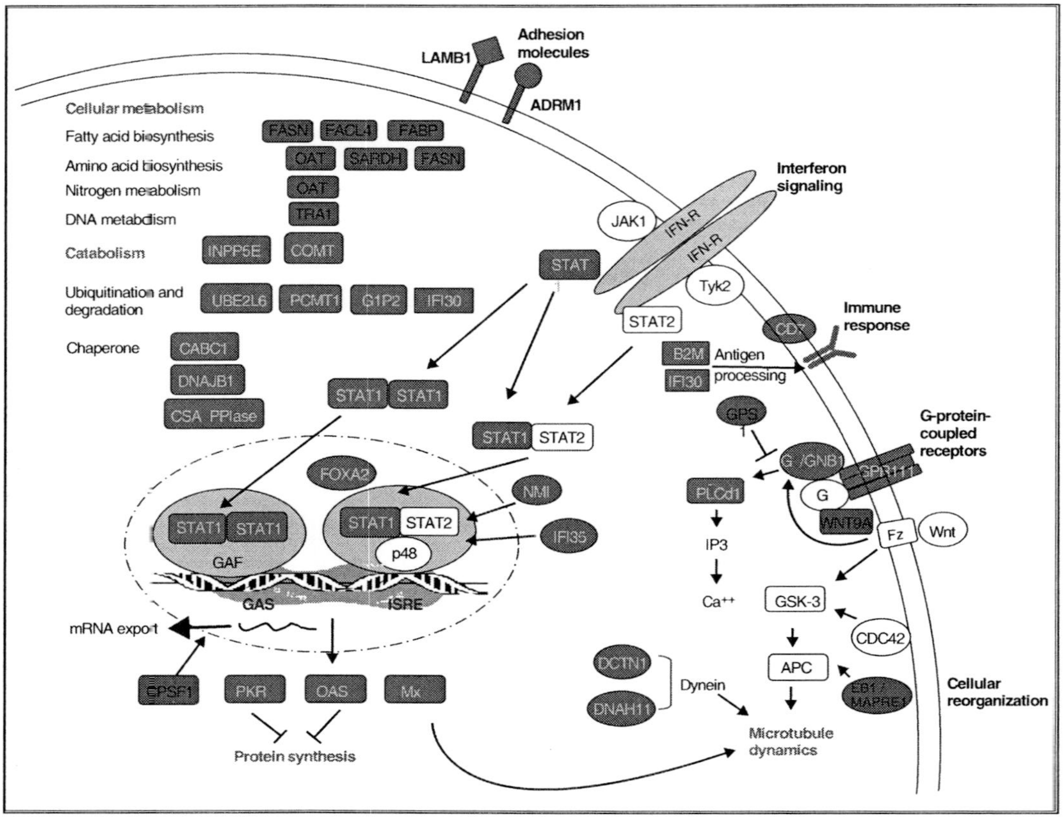

Figure 6-5. Combining functional genomics with expertise in physiology can lead to results of biological significance (after [18], the figure adopted from the free resource http://genomebiology.com/2004/5/8/R54).

THE CASE OF ESTROGENS AND VENOUS THROMBOSIS

> *"Estrogens and oral contraceptives are known to be risks for the VTE (venous thromboembolism). However, the precise physiological mechanism for this connection is unclear. A number of contradictory hypotheses were put forward. Prior to investigating the link using post-genomic approach, your laboratory needs a viable physiological hypothesis..."*

In this Case Study, we won't consider numerous proposals that were formulated to explain away the link between the estrogens and VTE. The main problem with these semi-hypotheses is that they lack an integrative outlook of the levels of function essential for post-genomic era. Therefore, in order to formulate consistent hypotheses, we need to start from fundamentals of molecular cell biology, biochemistry and physiology without, of course, forgetting to use common sense. An elaborated discussion of the results is available further in this chapter; here we present the thinking algorithm, the basic outline of the hypothesis and a summary of the results of *in silico* transcriptomics (performed by searching a literature database).

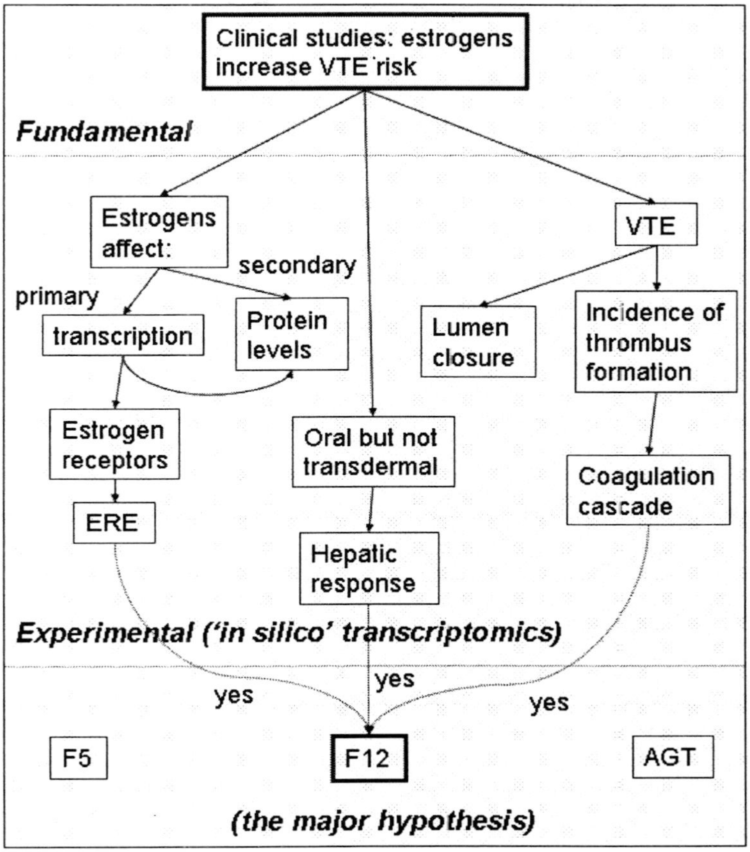

Figure 6-6. The deduction method and the VTE-estrogen link.

A diagram of the reasoning is presented in the Figure 6-6. It summarizes the three independent lines of thinking which were essential in arriving to logically consistent and experimentally verifiable results. These three lines of logic provide the general outline of *any* hypothesis aiming at the estrogen-VTE link. The diagram divided into "fundamental" and "experimental" parts. Thus, the fact that estrogens act mostly, if not exclusively, by altering gene transcription through estrogen receptors; the mechanisms of thrombus formation and the distinct hepactic response upon oral estrogen delivery are fundamental concepts firmly established in the minds of the most of professional biomedical scientists.

The "experimental" part refers to the experimental data that might be less clear. In particular, F12 protein, previously thought to be non-essential for the blood coagulation, comes as the major hypothesis. We must remind ourselves that, as with everything in biology, "non-essential" under one set of conditions becomes "vital" when the conditions are changed. Accordingly, relying on a hoary tradition (such as the fallacious belief that F12, present in the makeup of the coagulation cascade of many mammals and humans, is just a "non-essential protein") is hardly a practical attitude. The general considerations summarized in the diagram, appear to be confirmed by experimental results in the animal studies. That the three independent lines of thinking unequivocally converge at the F12 protein also serves as a logical support for the major hypothesis.

ESTROGENS AND VENOUS THROMBOSIS: AN EXTENDED DISCUSSION

Evidence of the link between estrogen-based contraceptives and VTE is derived from numerous observational and randomized studies (for example, [19-21]). This epidemiological studies point to a 2..6-fold increase in VTE risk associated with the estrogen intake. The estrogen-induced VTE does not appear to be restricted to Caucasian populations [22] and quite a number of meta-analyses and reviews were made on the subject (for instance, [23,24]). The accompanying *Case of Estrogens and Venous Thrombosis* presented a novel set of physiological hypotheses that link estrogens and VTE. A related Case Study in the Volume II of the present book series employs another bioinformatic methodology: genome-wide sequence analysis.

Having said what was said about the drawbacks of transcriptomics and advantages of the proteomics in respect to large-scale studies and to the identification of the target genes in the previous sections, we still maintain that in some cases transcriptomics even at its simplest (PCR of mRNA transcripts) still can be of greater use than proteomic studies. The problem we consider in this section is likely to be one of such cases. At least, investigating the relationship between the estrogens and venous thrombosis presented in *The Case of Estrogens and Venous Thrombosis*, we used a bioinformatic technique we called by the term as funny as "*in silico* gene expression profiling". In this section, we discuss the details of this Case Study and, in particular, the scientific thinking behind it.

First of all, it should be noticed that what we called as "*in silico* gene expression profiling" operates with apparently non-uniform cell types. After what we have stated more than once about careful selection of the cell types for the functional genomics studies, this might seem to be a blatant violation of the author's own principles. However, in this Case Study we imply *transcriptomic approach* to study rather than transcriptomics *per se*. Secondly, if we would take a closer look at the data we present in the Tables 6-1 and 6-2, we will notice that those were *not* functional genomics studies. For the most part, these were classical 'case studies' in molecular biology that dealt with transfection experiments of the promoters of certain genes. This sort of data is considerably more reliable and is, certainly, less dependent on the cell type than functional genomics is. In view of these two points, there is no warranty whatsoever as for the conclusions presented in the Case Study will be valid. Nevertheless, despite the lack of any warranties, the literature search provided enough indications to formulate, at the very least, an initial model which is 1. physiologically plausible 2. fits in with a set of various biomedical and biochemical data and 3. can be directly verified by experimental studies. With this in mind, let's consider various aspects of the problem summarized in the Figure 6-6 in greater detail.

Following the left branch of the Figure 6-6, estrogens regulate the growth and differentiation of cells of the female and male reproductive systems, have important effects on brain development, autoimmune diseases, and bone metabolism. In order to be systematic in our analysis, let's make use of the diagram of the levels of function (Figure 1-5). Considering the diagram, we conclude that an external factor such as a chemical substance can affect genome, gene expression and protein level or protein activity. Mutagenic activities of estrogens are hardly worth discussing thus the influence of estrogens on the gene expression and protein levels are to be considered. Transcriptome is the next hierarchical level of function after the genome and it is well-known that the potent physiological stimulators such as estrogens are not mere inhibitors or cofactors of some proteins but rather act on a grander scale: by binding to specific receptors which, in turn, affect transcription of hundreds of genes. Although it might be imagined that estrogens (or by-products of their biotransformation) can act as protein inhibitors in picomolar to nanomolar range, it is clear that estrogens and related substances are not likely to flow freely in the blood inhibiting all the proteins around. Rather, they transported by albumin and/or by SHBG (sex hormone-binding globulin) to the target cells [25]. What is considerably more important is that direct and reliable data on the inhibitory actions of estrogens on particular proteins are next to none. Therefore, proceeding from known to unknown, we assume that physiological action of estrogens is exerted primarily through binding to estrogen receptors followed by alterations in the gene transcription which is, probably, the universal mechanism of *any* estrogen-related physiological effects.

Further, we need to consider general physiology of the thrombus formation (the right branch in the Figure 6-6). In brief, there are two major physiological mechanisms through which obstruction of the blood flow can occur: the higher incidence of the thrombus formation and the closure of the lumen (either due to lower flexibility, distension or atherosclerosis of the blood vessels, Chapter II). Since estrogen therapies tend to lead to somewhat improved lipoprofile and the effects of the estrogens on the blood vessels are quite moderate, the higher incidence of the thrombus formation appears to be the most likely general mechanism underlying the VTE-estrogen link. Thrombus formation depends on the formation/dissolution of the fibrin clots, platelet-vessel interaction, platelet-platelet aggregation and other mechanisms we have summarized in the *CVD GeneScape* (Figure 2-5). As we also remember, the molecular cascade essential for the thrombus formation is the blood coagulation cascade. This molecular cascade contains both pro-coagulant and anti-coagulant proteins and corresponding changes in the levels of these proteins are known to influence hemostasis to a considerable degree.

Next, we need to consider the physiological context of the link between estrogens and VTE (the central branch in the Figure 6-6). Oral estrogen significantly increases levels of acute-phase proteins and procoagulant factors while transdermal estrogen does not have these adverse effects [26]. It is also clear that administration of estrogen-base hormone therapies via the transdermal route avoids hepatic or "first-pass" metabolism. Although patches may be associated with local skin reactions [27], there is very little evidence that transdermal estrogen stimulation ever caused higher risk of VTE. Evidence of the kind is anecdotal, self-contradictory and appears to belong mostly to popular media rather than any kind of scientific publication (see, for instance, a recent one at http://news.yahoo.com/s/ap/20060217/-ap_on_he_me/birth_control_patch). At the same time, lower VTE risk using transdermal estrogens was confirmed by dozens of independently made studies. In PubMed alone, there are at least a hundred of publications (such as [26-32] and many others) that clearly state the basic fact: transdermal application of estrogens bypasses hepatic response and, incidentally, appears to lower VTE risk.

Bringing together considerations from these three independent lines of thinking (estrogen action through gene transcription, estrogens and thrombus formation, the hepatic response), we have an outline of the *general hypothesis* that is likely to cover most of the potential mechanisms of the estrogen-VTE link. Elaborating logic along these three lines, we can come to more detailed hypotheses.

At present, at least three cellular mechanism of the estrogen action were postulated. The first mechanism is the "classical" model which involves binding of the ligand-activated estrogen receptors to specific DNA patterns known as estrogen-responsive elements (EREs). The binding then stimulates transcription initiation through interactions with a number of co-activator/co-repressor molecules. According to the second mechanism, ligand-activated estrogen receptors influence activities other transcription factors (such as AP1, NFkB or SP-1). The third mechanism is largely hypothetical and it is called as "non-genomic pathway". In this case, ligand-activated estrogen receptors would activate cytosolic pathways of signal transduction. While each of these mechanisms can, under various sets of conditions, be more important, it is clear that the ERE-mediated mechanism of the estrogen action is the most common and the most obvious mechanism that directly affects the gene expression.

Numerous studies in molecular cell biology show that ERE-mediated action of estrogens involves intracellular receptors mainly located in the cytoplasm of the target cells [25]. These receptors act to regulate mRNA levels of certain genes by controlling gene transcription and by regulating the stability of mRNAs. The annotated part of the human genome appears to contain only two genes coding for the estrogen receptors (ESRA and ESRB). Transcription activation or repression is brought about by the interaction of the receptors with other proteins termed coregulators, which can activate or repress gene transcription. Upon binding an estrogen molecule, the receptors dimerize and get translocated to the nucleus where they bind to estrogen response elements. Generally, the site of steroid receptor binding on the DNA is known as a "hormone response element". Although structural similarities of the DNA-binding domains of glucocortiocoid, estrogen, androgen and progesterone receptors suggest that they all might bind to the same hormone response element, there a particular "consensus" sequence which is more specific for estrogen receptors. This consensus sequence is known as "estrogen response element": 5'-AGGTCAnnnTGACCT-3', where 'n' denotes any nucleotide. These response elements also act as recognition sites of specific transcription factors. ER interacts with DNA by direct binding to sequences related to the consensus sequences of the kind, by tethering to other proteins bound at AP-1 and SP-1 sites and through interaction with genes containing isolated GGTCA half-sites [25, 33].

The above information, in the framework of our general hypothesis, suggests at least three ways of investigating the problem further: a literature search, a gene database search and a sequence database search. Here, we consider the results of the first two approaches while the genome-wide sequence search, which can also be termed as an '*in silico* gene expression profiling' is dealt with in the related Case Study in the Volume II.

The results of the gene database searches were rather equivocal. At first, it seemed plausible to search the estrogen-related data in such excellent databases as SWISSPROT (http://excpasy.ch) and GENE (http://www.ncbi.nlm.nih.gov/entrez/query.fcgi?-db=gene) to extract all of the human genes that might be related, in some way, to estrogens. After all, all of the genes discussed in the present Case Study are well known and annotated, so it seemed plausible that a database search would provide a useful feedback. However, SWISSPROT search produced only a few hits while searching the GENE database lead to over 500 hits with only few of the actually relevant genes (the relevance was judged *a posteriori* according to our analysis of the literature presented below).

The published data we've extracted directly from PubMed dealt with the protein levels attenuated by estrogen stimulation and with influence of estrogens on the levels of certain mRNAs. Although data both on human and animal cells were analyzed, it is clear that the results of the human studies are the most relevant. It is well known that levels of many proteins are attenuated upon stimulation of the cells with estrogen. The data on protein levels from independently made studies often appear to be contradictory, especially in respect to the implications for thrombosis. This apparent contradiction results from not making a distinction between *primary and secondary events* following estrogen stimulation. The primary events are, clearly, 1. the direct estrogen influence on the gene transcription (analyzed below); 2. the change in the levels of the proteins corresponding to the affected genes and 3. potential effects brought about through the non-classical mechanisms of the estrogen action. Without making this essential distinction, the data on the protein levels indeed will be largely inconsistent.

For example, potential pro-thrombotic effects of oral hormone therapy include increased plasma factors VII and IX, C-reactive protein and decreased antithrombin, PROC, PROS and TFPI, while potential protective effects include decreased blood pressure, lipids, fibrinogen and plasminogen activator inhibitor levels. However, the overall balance of the prothrombotic and protective varies greatly from patient to patient and from study to study, whatever the reasons for this variance might be [34]. Sometimes, it is not clear whether loss of endogenous ovarian steroid production or the effects of the orally administered contraceptive steroids that caused significant changes in hemostatic factors (antithrombin III, protein S, protein C, plasminogen, tissue-type plasminogen activator, plasminogen activator inhibitor 1, histidine-rich glycoprotein, and VII, VIII, X, XII coagulation factors). The changes in the protein levels induced by estrogen treatment were even claimed to be within "normal ranges" [35]. A randomized double-blind, placebo-controlled study suggests favorable changes of factor VII activity and fibrinogen and also lowered plasma levels of PAI-1 [36] while another study have not shown anything of the kind *etc*. Obviously, by studying protein levels alone it would be hardly possible to come any closer to the physiology behind the estrogen-VTE interaction.

Without making the distinction between the primary and the secondary effects, the situation gets further complicated by the fact that proteins of the coagulation/clotting cascade themselves can affect gene transcription. For instance, thrombin can up-regulate synthesis of the tissue factor pathway inhibitor mRNA through a cyclooxygenase-2-dependent, EGF receptor-independent mechanism [37]. Activated protein C up-regulates cyclooxygenase-2 expression in endothelial cells via binding to endothelial cell protein C receptor and activation of protease-activated receptor-1 [38]. In human keratinocytes, FVIIa/TF complex appears to affect transcription of quite a number of genes including transcription regulators c-fos, egr-1, ETR101, BTEB2, c-myc, fra-1, tristetraproline, growth factors amphiregulin, hbEGF, CTGF, and FGF-5, proinflammatory cytokines IL-1B, IL-8, LIF, and MIP2A, collagenases 1,3 *etc* [39]. Further ramifications of these secondary effects can entirely blur physiological picture. Therefore, we are obliged to apply the Occam's Razor ("*Entia non sunt multiplicanda praeter necessitatem*") by choosing the simplest hypothesis: the one that deals with the analysis of the primary effects.

The primary effects are the changes in mRNA levels stimulated by estrogens. The literature was sought for reliable studies reporting altered mRNA levels and/or promoter activity of the genes upon estrogen stimulation. The resulting set of the animal and human studies provided a number of interesting clues (see Tables 6-1 and 6-2). *Animal studies* suggested a number of genes: (Egr-1) is an immediate-early gene induced by 17 beta-estradiol [40]; coagulation factor V mRNA levels increased via estrogen receptor [41], 6-fold stimulation of FXII gene transcription [42], regulation of PAI-1 promoter activity by estrogen receptors [43] *etc* (Table 6-1).

Table 6-1. The genes and protein with mRNA levels affected upon estrogen stimulation: animal studies

Gene	mRNA of	Animals	Cells/Tissue	Remark	Reference
EGR1	Early growth response-1	Rodents	Uterus	(+) E2-responsive region	[40]
eNOS	Nitric oxide synthase	Bovine	Endothelium	(+), human ERalpha used	[44]
F12	Factor XII	Rats	Liver	(+), 6-fold	[42]
F5	Factor V	Mice	bone marrow	(+) through ERalpha	[41]
PAI1	Plasminogen activator inhibitor	Bovine	Aortic endothelium	(+-) human PAI-1; differential and opposing regulation	[43]
VEGF	Vascular endothelial growth factor	Rats	endothelium	(+-) 5' UTR and 3' UTR	[45]

There were considerably more studies dealing with *human* cells and tissues [46-56]. These results, irrespectively of the type of the study and the cells used, were piled up in the Table 6-2. An apparent feature of both the animal and the human data is that the genes identified are mostly up-regulated by estrogens. It is quite understandable taking into account that estrogens directly affect gene transcription through the estrogen receptors. Some of the genes were up-regulated both in animal and human studies (egr-1, F12, eNOS).

The relevance and significance of the data collected in the tables to the clinical data and physiology can be further considered in terms of estrogen biotransformations, lipoprotein profiles, vasodilatation and vasoconstriction, prostaglandin synthesis and, of course, the effects on the coagulation cascade. The two major products of an estrogen's **biotransformation** are the sulfo-conjugates and the hydroxylated derivatives. The two major genes that produce the enzymes catalyzing these biotransformations were identified as a result of our *in silico* gene expression profiling: SULT2A1 and CYP1B1. A major role of SULT2A1 and other sulfotransferases is the metabolism of estrogens and other steroids into more hydrophilic sulfate conjugates that are easier excreted. Cytochrome P450 1B1 (CYP1B1) is a key enzyme in the estrogen/steroid hydroxylation.

Table 6-2. The genes and protein with mRNA levels affected upon estrogen stimulation: human studies.

Gene	mRNA of	Cells/Tissue	Remark	Reference
AGT	Angiotensinogen	HepG(2)	(+) promoter activity	[46]
APOA1	ApoA-I	HepG2	(+) 3-fold, promoter activity	[47]
BMP6	Bone morphogenic protein 6	Osteoblasts	(+) mRNA 3-fold, protein 6-fold	[48]
CCND2	Cyclin D2	Keratinocytes	(+) both mRNA and protein	[49]
COX2	Cyclooxygenase-2	microvascular endothelium	(+) mRNA 2-fold, protein 2.8-fold, 1.5 increase in prostaglandin synthesis	[50]
CYP1B1	Aryl hydrocarbon hydroxylase	MCF-7	(+) 7-fold, putative ERE	[51]
EGR1	Early growth response-1	MCF-7	(+) E2-responsive region	[40]
eNOS	Endothelial nitric oxide synthase	endothelium	(+)1.7-fold promoter activity	[52]
F12	Factor XII	HepG2	(+) 2-fold	[53]
F12	Factor XII	-	(+) promoter activity	[54]
HDLR	HDL receptor	MCF-7	(+) promoter activity	[55]
IGFBP1	Insulin-like growth factor-binding protein 1	HepG2	(+) 2-fold	[53]
PI-9	Proteinase inhibitor 9	HepG2	(+) 15-fold	[53]
SULT2A1	Dehydroepiandrosterone sulfotransferase	Various tissues	(+) 2.6-fold promoter activity	[56]

The clinical data suggest that use of the estrogen therapy and oral contraceptives might lead to improved **lipoprotein profile**. Transcription activation of APOA1 (major component of HDL) and of the HDL receptor does correlate with the beneficial changes in the lipoprofile. Endothelial NO-synthase is the key enzyme in production of the nitric oxide which is a major endothelium-related **vasodilatator**, while angiotensinogen is a precursor of the angiotensin II, a major **vasoconstrictor**. Angiotensin II might exert a prothrombotic influence through platelet aggregation upon vasoconstriction.

COX2 is the major enzyme in **prostaglandin synthesis** and prostaglandins are related both to vasodilatation/vasoconstriction and to thrombotic tendencies. COX2 (PTGS2, prostaglandin-endoperoxide synthase 2) produces prostaglandin H2 which is the major precursor of prostaglandin E2, prostacyclin and thromboxanes. Prostaglandin E2 and prostacyclin dilate are vasodilatators while thromboxanes, synthesized in platelets, are potent vasoconstrictors which also induce platelet aggregation. Thus, a first glance on the metabolic pathway of the prostaglindin synthesis suggests that activation of the COX-2 can both increase or decrease prothrombotic tendencies, depending on quite a number of secondary factors. Since we aim at the analysis of the primary effects, we will now turn to the most important hypothesis: the direct activation of the genes in the coagulation cascade.

According to the data in the Tables, **the coagulation cascade** can be affected through at least three pathways: F5, PAI-1 and F12. Factor V is essential for prothrombin activation by the factor Xa and clinical data do suggest that lack of F5 activity leads to hemorrhagic disease while higher activity was associated with thrombosis. The study in mice suggests that estrogens activate F5 through an ERalpha mechanism [41]. PAI-1 (plasminogen activator inhibitor 1) acts at a major control point in the regulation of fibrinolysis. Lack of PAI-1 activity is characterized by hemorrhage while higher activity is associated with thrombosis. The study on bovine cells indicated a differential and opposing regulation of PAI-1 [43] upon estrogen stimulation suggesting an equivocal link to prothrombotic effects.

Finally, the activation of the F12 represents, perhaps, the most significant of the hypotheses addressing the estrogen-VTE link. Analysis of the pro-coagulant parts of the coagulation cascade shows that there are two branches of the clotting initiation: the extrinsic and the intrinsic (see Figure 7-5a in the Chapter VII). It is important to notice that F7 is the entry point for the extrinsic clotting pathway (tissue factor pathway) while F12 is the entry point for the intrinsic pathway (vascular injury pathway). Traditionally, blood coagulation is thought to be initiated mostly by factor VII in complex with the tissue factor (see Appendix I) and some physiologists hardly ever believed F12 to play an important role for coagulation *in vivo* (despite the numerous biochemical data *in vitro*). A recent animal study challenged this long-standing but, apparently, not entirely adequate assumption [57]. Although F12-deficient mice did not experience spontaneous or excessive injury-related bleeding under the laboratory conditions, they were, in fact, protected against collagen- and epinephrine-induced thromboembolism under the laboratory conditions. Infusion of human F12 into F12(-) mice restored injury-induced thrombus formation. The results of the study [57] show F12 to be essential for thrombus formation *in vivo*.

Thus, *The Case of Estrogens and Venous Thrombosis* has allowed to establish the necessary theoretical framework and produced several hypotheses (transcription activation of AGT, COX2, F5 and F12 activation by estrogens) that can explain the link between thrombosis and venous thrombosis. According to the results of this Case Study, promoter activation of F12 appears to be the primary influence on the blood clotting/coagulation cascade. Activation of F12 by estrogens both in animal and human studies [42,53,54] does suggest a direct influence of the estrogens on the function of the clotting cascade.

Further corroboration of this hypothesis follows from the results of the analysis of the gene structure performed in the parallel Case Study of the Volume II, that deals with prediction and analysis of the potential ERE sites. The role of F12 in thromboembolism is not direct (*i.e.*, F12 acts as prothrombotic factor mostly upon a vascular injury) and this can explain, in part, variability of the patient responses to the estrogen treatment (for instance, in terms of the extent of vascular microinjuries unique to each patient, extent of atherosclerosis, inflammation *etc*). The numerous changes in the levels of the clotting factors in plasma that were seen in many clinical studies can be explained in terms of the secondary effects we already mentioned.

The results of the *Case of Estrogens and Venous Thrombosis* also allow us to make several important recommendations for the experimental studies of the estrogen-VTE link. The experimental studies will include cellular studies on the elucidation and corroboration of the cellular mechanism of the F12 activation through ER alpha/beta receptors; functional genomics (see the following section); clinical studies that can include systematic analysis of the differences between the transdermal and other methods of the estrogen delivery in comparison with the oral estrogens in relation to the VTE effects.

ESTROGENS AND VENOUS THROMBOSIS: SUGGESTIONS FOR FUNCTIONAL GENOMICS STUDIES

The Case of Estrogens and Venous Thrombosis suggested several hypotheses that can be verified using, in particular, methods of the functional genomics. Direct mRNA measurement via PCR is one of the basic methods to explore the processes of gene transcription. The most generic approach to the problem would be using functional genomics (gene expression profiling) which determines the concentration profiles for all of the transcribed mRNAs. Proteomics can be used as a secondary tool corroborating that the levels of the clotting factors of interest indeed change upon the estrogen stimulation.

The important questions concerning functional genomics experimentation include selection of the tissues and cells, of the estrogens and of the regime of the stimulation. Without paying regard to these most important factors, a functional genomics study is not likely to produce any results of biomedical significance. The major hypothesis to which we arrived in the Case Study (activation of F12 transcription through hepatic metabolism) suggests that using cells from the liver will be essential to obtain any significant results. The major difficulty here is that liver biopsy is far from being a common procedure in the clinical practice and it is performed mostly in grave cases of the liver disease. Using other kinds of cells is likely to produce inadequate results in an estrogen-VTE study, as indicated by the following two examples.

There were only a few functional genomics studies that deal with estrogen action on the human cells and their results are largely inconclusive since they vary greatly from study to study and from estrogen to estrogen. For example, a global gene expression analysis of estrogen receptor transcription factor cross talk suggested several dozens of the genes that are likely to be affected by estrogens through a non-classical (namely, the AP-1) pathway [58]. Taking aside the major findings reported in the study [58], what interests us more for the purposes of the present discussion is that the Supplementary Materials for the paper [58] document alteration of the mRNA levels followed estrogen (E2) stimulation after 6hrs and 24hrs. These supplementary data are of primary interest in our case and they are remarkable in illustrating what will be the result of a functional genomics study aiming at the VTE-estrogen link if it is made without regard to basic physiology. In fact, such a study was made (unpublished data) and the results were close to those described in the [58], at least in their total lack of conclusiveness. In the study [58] the researchers used MCF-7 cell line (breast cancer cells in culture) while in the failed and unpublished study of the VTE-estrogen link the researchers used normal endothelial cells. While the authors of the [58] did manage to come to substantiated conclusions concerning the AP-1 activation, the supplementary data on the expression after estrogen stimulation are quite illustrative.

Thus, if we take the data from [58] as an example, we'll see that 6hrs after stimulation with E2 estrogen, levels of at least 10 mRNAs were upregulated more than 1.5-fold, after 24hrs also about 10 mRNAs were upregulated. Between mRNAs activated after 6hrs and after 24hrs there was only one common hit: mRNA of a hypothetical "serine protease 23". None of the proteins encoded by mRNAs upregulated in both cases had any relation to the thrombus formation whatsoever, except protein C mRNA (X02750) which was 1.3-fold upregulated. The same situation was with the downregulated mRNAs: none of the mRNAs observably (at least 1.5-fold) downregulated either after 6hrs or 24hrs had any relation to blood coagulation or any other molecular known cascade involved in the thromboembolism. These results illustrate that selection of the tissue and the regime of stimulation will be crucial in order to arrive to the results of biological significance for the strogen-VTE link (given that technical performance of the experiments would be adequate). It's also clear that selection of the conditions for this kind of the functional genomics studies cannot be adequately made without regarding the physiological context of the problem under study: in our case, the one outlined in *The Case of Estrogens and Venous Thrombosis*.

Chapter Summary

The results of any high-throughout study of any kind, be it a **genetic association study** or a study in functional genomics are prone to the neglect of the essential physiological details. In this chapter, we considered post-genomic methodologies that aim at functional characterization of the genes and the gene products. These experimental methodologies are collectively called as **functional genomics**. The gene expression profiling (**transcriptomics**) and the protein levels detection (**proteomics**) are the two major branches of the functional genomics. While functional genomics largely implies experimental data collection, the newly arising branch of **systems biology** claims to deal with computational models based on the data from the high-throughput experiments. Both functional genomics and systems biology are assumed to aim at elimination of the **reductionism** from biological studies. However, without precision in the **formulation of the research purpose** and **awareness of the basics of physiology** this task can hardly be achieved. Neither any biomedical applications of the functional genomics will be possible without employing a clearly formulated **purpose-oriented approach** based on sound physiology. On the example of a particular problem of biomedical research (the study of relationship between the **estrogens and venous thrombosis**) we considered how bioinformatic methods can help to avoid common pitfalls in functional genomics studies.

REFERENCES

[1] Aggarwal K, Lee K. Functional genomics and proteomics as a foundation for systems biology. *Brief Funct Genomic Proteomic* 2003;2:175-84.

[2] Unglaub D, Silverthorn P. *Human Physiology: An Integrated Approach.* Prentice Hall, 1st ed (1997)

[3] Bilello J. The agony and ecstasy of "OMIC" technologies in drug development. *Curr Mol Med* 2005;5:39-52.

[4] Ideker T, Thorsson V, Ranish J et al. Integrated genomic and proteomic analyses of a systematically perturbed metabolic network. *Science* 2001;292:929–934.

[5] Xiong M, Feghali-Bostwick CA, Arnett FC, Zhou X. A systems biology approach to genetic studies of complex diseases. *FEBS Lett* 2005;579:5325-32.

[6] Jimenez S, Derk C. Following the molecular pathways toward an understanding of the pathogenesis of systemic sclerosis. *Ann Intern Med* 2004;140:37-50.

[7] Feghali-Bostwick C, Medsger TA Jr, Wright TM. Analysis of systemic sclerosis in twins reveals low concordance for disease and high concordance for the presence of antinuclear antibodies. *Arthritis Rheum* 2003;48:1956-63.

[8] Rychly J, Nebe B. Therapeutic strategies in autoimmune diseases by interfering with leukocyte endothelium interaction. *Curr Pharm Des.* 2006;12(29):3799-806.

[9] Evans G, Wheeler C, Corbett J, Dunn M. Construction of HSC-2DPAGE: a two-dimensional gel electrophoresis database of heart proteins. *Electrophoresis.* 1997;18:471-9. http://www.doc.ic.ac.uk/vip/hsc-2dpage/

[10] Pleissner K, Sander S, Oswald H, Regitz-Zagrosek V, Fleck E. The construction of the World Wide Web-accessible myocardial two-dimensional gel electrophoresis protein database "HEART-2DPAGE": a practical approach. *Electrophoresis* 1996;17:1386-92. http://userpage.chemie.fu-berlin.de/~pleiss/

[11] Veerasingham S, Sellers K, Raizada M. Functional genomics as an emerging strategy for the investigation of central mechanisms in experimental hypertension. *Prog Biophys Mol Biol* 2004;84:107-23.

[12] Borozdenkova S, Westbrook J, Patel V, Wait R, Bolad I, Burke M, Bell AD, Banner N, Dunn M, Rose M. Use of proteomics to discover novel markers of cardiac allograft rejection. *J Proteome Res* 2004;3:282-8.

[13] Mata J, Marguerat S, Bahler J. Post-transcriptional control of gene expression: a genome-wide perspective. *Trends Biochem Sci* 2005;30:5-14.

[14] Yang E, van Nimwegen E, Zavolan M, Rajewsky N, Schroeder M, Magnasco M, Darnell J. Decay rates of human mRNAs: correlation with functional characteristics and sequence attributes. *Genome Res* 2003;13:1863-72.

[15] Lein ES, Hawrylycz MJ, Ao N, Ayres M, Bensinger A, Bernard A, Boe AF, Boguski M et al. Genome-wide atlas of gene expression in the adult mouse brain. *Nature* 2006 Dec6; doi:10.1038/nature05453

[16] Hanash S. Building a Foundation for the Human Proteome: The Role of the Human Proteome Organization. *J Proteome Res* 2004;3:197-199

[17] Xu X, Leow C, Lu X, Zhang X, Liu J, Wong W, Asperger A, Deininger S, Eastwood Leung H. Molecular classification of liver cirrhosis in a rat model by proteomics and bioinformatics. *Proteomics* 2004;4:3235-45.

[18] Yan W, Lee H, Yi EC, Reiss D, Shannon P, Kwieciszewski BK, Coito C, Li X, Keller A et al. System-based proteomic analysis of the interferon response in human liver cells. *Genome Biol* 2004;5:R54

[19] Gomes M, Deitcher S. Risk of venous thromboembolic disease associated with hormonal contraceptives and hormone replacement therapy: a clinical review. *Arch Intern Med* 2004;164:1965-76.

[20] Hulley S, Furberg C, Barrett-Connor E, Cauley J, Grady D, Haskell W, Knopp R, Lowery M, Satterfield S, Schrott H, Vittinghoff E, Hunninghake D; HERS Research Group. Noncardiovascular disease outcomes during 6.8 years of hormone therapy: Heart and Estrogen/progestin Replacement Study follow-up (HERS II). *JAMA* 2002 ;288:58-66.

[21] Lidegaard O, Edstrom B, Kreiner S. Oral contraceptives and venous thromboembolism. A case-control study. *Contraception* 1998;57:291-301.

[22] Venous thromboembolic disease and combined oral contraceptives: results of international multicentre case-control study. World Health Organization Collaborative Study of Cardiovascular Disease and Steroid Hormone Contraception. *Lancet* 1995;346:1575-82.

[23] Farquhar C, Marjoribanks J, Lethaby A, Lamberts Q, Suckling JA; Cochrane HT Study Group. Long term hormone therapy for perimenopausal and postmenopausal women. *Cochrane Database Syst Rev* 2005;3:CD004143

[24] Prentice R, Langer R, Stefanick M, Howard B, Pettinger M, Anderson G et al; Women's Health Initiative Investigators. Combined postmenopausal hormone therapy and cardiovascular disease: toward resolving the discrepancy between observational studies and the Women's Health Initiative clinical trial. *Am J Epidemiol* 2005;162:404-14.

[25] Nussey S, Whitehead S. *Endocrinology: an integrated approach.* BIOS Scientific publishers, 2001.

[26] Menon D, Vongpatanasin W. Effects of transdermal estrogen replacement therapy on cardiovascular risk factors. *Treat Endocrinol* 2006;5:37-51.

[27] Samsioe G. Transdermal hormone therapy: gels and patches. *Climacteric* 2004;7:347-56.

[28] Ansbacher R. The pharmacokinetics and efficacy of different estrogens are not equivalent. *Am J Obstet Gynecol* 2001;184:255-63.

[29] Crook D. The metabolic consequences of treating postmenopausal women with non-oral hormone replacement therapy. *Br J Obstet Gynaecol* 1997;104(S)16:4-13.

[30] O'Connell MB. Pharmacokinetic and pharmacologic variation between different estrogen products. *J Clin Pharmacol* 1995;35(9):18S-24S.

[31] Ockrim J, Lalani el-N, Kakkar A, Abel P. Transdermal estradiol therapy for prostate cancer reduces thrombophilic activation and protects against thromboembolism. *J Urol* 2005;174:527-33;

[32] Douketis J. Hormone replacement therapy and risk for venous thromboembolism: what's new and how do these findings influence clinical practice? *Curr Opin Hematol* 2005;12:395-400.

[33] Lehninger A, Nelson D, Cox M. *Principles of biochemistry* (2nd ed). Worth Publishers Inc., NY, 1997, p. 779

[34] Lowe G. Hormone replacement therapy: prothrombotic vs. protective effects. *Pathophysiol Haemost Thromb* 2002;32:329-32.

[35] Samsioe G. Coagulation and anticoagulation effects of contraceptive steroids. *Am J Obstet Gynecol* 1994;170:1523-7.

[36] Borgfeldt C, Li C, Samsioe G. Low-dose oral combination of 17beta-estradiol and norethisterone acetate in postmenopausal women decreases factor VII, fibrinogen, antithrombin and plasminogen activator inhibitor-1. *Climacteric* 2004;7:78-85.

[37] Neaud V, Duplantier JG, Mazzocco C, Kisiel W, Rosenbaum J. Thrombin up-regulates tissue factor pathway inhibitor-2 synthesis through a cyclooxygenase-2-dependent, epidermal growth factor receptor-independent mechanism. *J Biol Chem* 2004;279:5200-6.

[38] Brueckmann M, Horn S, Lang S, Fukudome K, Schulze Nahrup A, Hoffmann U, Kaden JJ, Borggrefe M, Haase K, Huhle G. Recombinant human activated protein C upregulates cyclooxygenase-2 expression in endothelial cells via binding to endothelial cell protein C receptor and activation of protease-activated receptor-1. *Thromb Haemost* 2005;93:743-50.

[39] Camerer E, Gjernes E, Wiiger M, Pringle S, Prydz H. Binding of factor VIIa to tissue factor on keratinocytes induces gene expression. *J Biol Chem* 2000;275:6580-5.

[40] Chen C, Lee W, Safe S. Egr-1 is activated by 17beta-estradiol in MCF-7 cells by mitogen-activated protein kinase-dependent phosphorylation of ELK-1. *J Cell Biochem* 2004;93:1063-74.

[41] Moverare S, Skrtic S, Lindberg M, Dahlman-Wright K, Ohlsson C. Estrogen increases coagulation factor V mRNA levels via both estrogen receptor-alpha and -beta in murine bone marrow/bone. *Eur J Endocrinol* 2004;151:259-63.

[42] Farsetti A, Misiti S, Citarella F, Felici A, Andreoli M, Fantoni A, Sacchi A, Pontecorvi A. Molecular basis of estrogen regulation of Hageman factor XII gene expression. *Endocrinology* 1995;136:5076-83.

[43] Smith L, Coats S, Qin H, Petrie M, Covington J, Su M, Eren M, Vaughan D. Differential and opposing regulation of PAI-1 promoter activity by estrogen receptor alpha and estrogen receptor beta in endothelial cells. *Circ Res* 2004; 95:269-75.

[44] Sumi D, Ignarro L. Estrogen-related receptor alpha 1 up-regulates endothelial nitric oxide synthase expression. *Proc Natl Acad Sci USA* 2003;100:14451-6.

[45] Hyder S, Nawaz Z, Chiappetta C, Stancel G. Identification of functional estrogen response elements in the gene coding for the potent angiogenic factor vascular endothelial growth factor. *Cancer Res* 2000;60:3183-90.

[46] Zhao YY, Sun KL, Ashok K. Analysis of a cis-acting element involved in regulation by estrogen of human angiotensinogen gene expression. *Sheng Wu Hua Xue Yu Sheng Wu Wu Li Xue Bao (Shanghai)* 1998;30:445-448.

[47] Zhang X, Jiao J, Bhavnani B, Tam S. Regulation of human apolipoprotein A-I gene expression by equine estrogens. *J Lipid Res* 2001;42:1789-800.

[48] Rickard D, Hofbauer L, Bonde S, Gori F, Spelsberg T, Riggs B. Bone morphogenetic protein-6 production in human osteoblastic cell lines. Selective regulation by estrogen. *J Clin Invest* 1998;101:413-22.

[49] Kanda N, Watanabe S. 17beta-estradiol stimulates the growth of human keratinocytes by inducing cyclin D2 expression. *J Invest Dermatol* 2004;123:319-28.

[50] Tamura M, Deb S, Sebastian S, Okamura K, Bulun S. Estrogen up-regulates cyclooxygenase-2 via estrogen receptor in human uterine microvascular endothelial cells. *Fertil Steril* 2004;81:1351-6.

[51] Tsuchiya Y, Nakajima M, Kyo S, Kanaya T, Inoue M, Yokoi T. Human CYP1B1 is regulated by estradiol via estrogen receptor. *Cancer Res* 2004;64:3119-25.

[52] Kleinert H, Wallerath T, Euchenhofer C, Ihrig-Biedert I, Li H, Forstermann U. Estrogens increase transcription of the human endothelial NO synthase gene: analysis of the transcription factors involved. *Hypertension* 1998;31:582-8.

[53] Krieg A, Krieg S, Ahn B, Shapiro D. Interplay between estrogen response element sequence and ligands controls in vivo binding of estrogen receptor to regulated genes. *J Biol Chem* 2004;279:5025-34.

[54] Citarella F, Misiti S, Felici A, Farsetti A, Pontecorvi A, Fantoni A. Estrogen induction and contact phase activation of human factor XII. *Steroids* 1996;61:270-6.

[55] Lopez D, Sanchez M, Shea-Eaton W, McLean M. Estrogen activates the high-density lipoprotein receptor gene via binding to estrogen response elements and interaction with sterol regulatory element binding protein-1A. *Endocrinology* 2002;143:2155-68.

[56] Seely J, Amigh K, Suzuki T, Mayhew B, Sasano H, Giguere V, Laganiere J, Carr B, Rainey W. Transcriptional regulation of dehydroepiandrosterone sulfotransferase (SULT2A1) by estrogen-related receptor alpha. *Endocrinology* 2005;146:3605-13.

[57] Renne T, Pozgajova M, Gruner S, Schuh K, Pauer HU, Burfeind P, Gailani D, Nieswandt B. Defective thrombus formation in mice lacking coagulation factor XII. *J Exp Med* 2005;202:271-81.

[58] DeNardo D, Kim H, Hilsenbeck S, Cuba V, Tsimelzon A, Brown P. Global gene expression analysis of estrogen receptor transcription factor cross talk in breast cancer: identification of estrogen-induced/activator protein-1-dependent genes. *Mol Endocrinol* 2005;19:362-78.

Chapter VII

METABOLIC PATHWAYS AND MOLECULAR CASCADES

INTRODUCTION

The post-genomic tetrad that we refer to throughout this book (Figure 1-3) is a general scheme. As any other general scheme, it needs to be elaborated and completed in order to provide more adequate view of the corresponding realities. The two important additions we will be interested in this and the following chapter are the "interactome" and the "reactome" areas (Figure 7-1). An "interactome" of a cell is the entire array of protein-protein interactions that occur within this kind of cell. The reactome, apparently, is the entire array of all the chemical reactions. Although both notions were defined here in respect to a particular cell, they can be widened to include interactomes/reactomes of specific organs or even species. Accordingly, there would be two additional fields of functional genomics: "interactomics" and "reactomics".

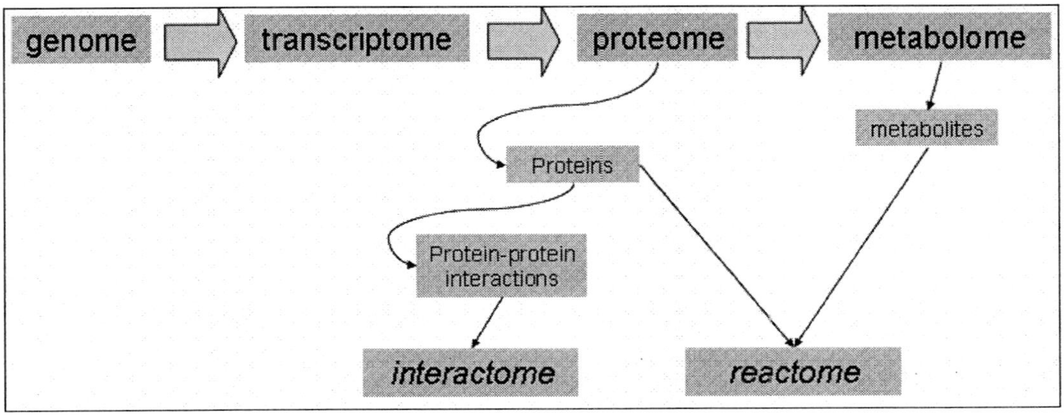

Figure 7-1. An updated version of the post-genomic tetrad.

The elaborated diagram reflects more complex biological reality. Thus, proteomes of two biological systems might be very close. However, the subtle differences in the amino acid sequences which emerge on the stage of the case of a cross-species comparisons and number of other factors can make the interactomes of these two systems differ to a greater extent than can be predicted on the base of mere contents of the proteome. The interactome and reactome of a particular biological system determine the molecular cascades and the metabolic pathways that will be present in this system, thus defining the range of responses of the biological system to the external and internal events.

For example, comparing two different systems from two different species (say, some animal cells in culture and human cells in culture), the differences in the amino acid sequence, being they even relatively few, will influence specificity and stability of the protein-protein interactions as well as binding of specific metabolites by specific proteins. Accordingly, there will be considerably more differences among the "interactomes" and "reactomes" of these two systems than there are among their proteomes and metabolomes. Some of these differences can be crucial for survival.

For example, whatever was the actual cause of the terrible side-effects seen in a recent UK drug trial (New Scientist 17 Mar 2006, see the news item at http://www.newscientist.com /article/dn8863-catastrophic-immune-response-may-have-caused-drug-trial-horror.html), it is still interesting to observe how the scientists involved reacted to this unfortunate event. Thus, one scientist stated that CD28 (one of the proteins involved) is "highly conserved across species" and then concluded that it is "very strange" that it turned out to be idiosyncratic case in humans. The company, obviously, insisted that absence of the side effects during the animal testing suggests that the disastrous side-effects in humans (six men in the critical care unit) were completely unexpected.

However, there is nothing "strange", "unexpected" and even less unusual when the naïve reductionism is set aside and the problem is looked upon from a post-genomic point of view. A post-genomic point of view inherently operates with the notions of proteomes *and* interactomes. Despite the fact that protein-protein interactions are often formed by a significant number of amino acid residues of the proteins involved, there are some amino acid residues (which might not be necessarily "highly conserved", see Volumes III-V of the present book series) that are more important for the formation and/or stability of the particular protein-protein interfaces. Accordingly, a change of even one amino acid residue in otherwise "highly conserved" protein might significantly modify or even abolish a number of important protein-protein interactions as well as might introduce some undesirable protein-protein interactions. When there are some metabolites involved, the changes affected by even a single amino acid substitution will influence not only the interactome but also the reactome thus affecting the difference in function of the molecular cascades involved.

Accordingly, the studies of the actual interactomes and reactomes involved in a particular physiological process are not only of fundamental significance but also might have direct repercussions for the biomedical applications. The onus here is, clearly, how these two phenomena can be studied or predicted within the framework of the post-genomic research. Functional genomics will be, apparently, the chief experimental instrument of the post-genomic biology (Figure 6-1). One of the most important perspective applications of this instrument is reconstruction and elucidation of the molecular cascades and the metabolic pathways in particular organelles, cells, organs and organisms.

MOLECULAR CASCADES, METABOLIC PATHWAYS AND METABOLIC CONTEXTS

In this section, we consider several important concepts that need to be remembered when dealing with any sort of interactome or reactome study. These concepts are those pertaining to the structure, the system-specific occurrence, the system-specific implementations of the interactome/reactome as well as to the multiple biological functions of the proteins and other compounds involved.

The very first thing that needs to be digested in the respect to interactomes and reactomes is that both the interactome and the reactome of a particular biological system have **definite structure**. Far from being an unspecific and highly confusing meshes of interactions (as it is often presented in the contemporary papers on functional genomics), both the interactome and the reactome contain specific units of interactions that have important functional value. It is the decades-long research in physiology, cell biology and biochemistry that allowed us to establish with great certainty the presence and the basic properties for many of these functional units such as "blood coagulation cascade", "glycolytic pathway" and hundreds of others. Referring to interactome, we might call these units as "molecular cascades" (Figure 1-2). When reactome is under consideration, these units can be called as "metabolic pathways". A molecular cascade, generally, is a wider concept than a metabolic pathway. That is, all of the metabolic pathways are molecular cascades (i.e., cascades comprised of interacting molecules) while not all of the molecular cascades are involved in metabolism *per se* (depending, of course, on how one would define metabolism). The difference can be illustrated by the current use of the professional lingo. Thus, calling the blood coagulation cascade as a "molecular cascade" is not likely to raise many objections. When, however, a less informed bioinformatician would risk calling the coagulation cascade a "metabolic pathway", that statement usually raises quite many an eyebrow, especially if this slip of the tongue happens before biomedical audience.

Secondly, there is **system-specific occurrence** of the molecular cascades. A particular molecular cascade might occur in any cell and in any organism. Glycolysis, for example, is one of the most common metabolic pathways. It occurs in any of the known cellular organisms. At the same time, occurrence of many other molecular cascades is restricted only to certain types of organisms (the blood coagulation cascade can occur only in organisms that, at the very least, have blood). The essential thing to understand here is that there are quite a few molecular cascades which can be uniformly found in any cell of any organism. This statement needs to be elaborated.

In accordance with the variation among the organ systems across species, widely known from the studies in comparative anatomy and physiology, there are numerous molecular cascades that occur only in certain classes of organisms or might even occur exclusively in one particular organism. For example, cross-species comparisons of the immune phenomena show different components of the immune system present in different animals [1]. While protozoans do not appear to have any of the components that we now came to call as "immune system", metazoans feature some sort of a histocompatibility system and allogeneity. While immunocytes appear to be present in all metazoans, allograft immunity is not likely to be present in ribbon worms but is present in segmented worms *etc* [1].

Moreover, nature allows us to fnd an overwhelming number of highly specialized physiological mechanisms and, correspondently, of the highly specialized molecular cascades in particular organisms. Only to cite briefly the most apparent marvelous phenomena of the kind would take a huge volume of the size of the *Who's Who* directory. We'll use three of the most common examples. For instance, there are special valves and other anatomical features in the neck of giraffe (*Giraffa camelopardalis*) that prevent otherwise inevitable damage to the brain tissue would the giraffe like to drink from a water source below. In this most peculiar animal, there is also a lift-like system that allows the food to travel through the same long neck. Another example is the special secretory lobes in the hind section of bombardier beetles (*Brachinini, Ozaenini etc*, over 500 species). These lobes mix hydroquinone and hydrogen peroxide with the enzymes and other substances when the beetle senses danger. As soon as these substances are combined, quinone at the temperature of over 100 C is ejected at the source of potential danger. To round up, the yellow-green lights of the fireflies (*Lampyridae*, over 2000 particular species) produced during their communication are formed by a specific molecular cascade that involves, in particular, luciferase enzyme. System-specific occurrence of the molecular cascades in the biological systems is an extremely important issue to consider when somebody attempts to make any generalizations at the systems level.

Thirdly, despite the fact that there are some common molecular cascades that, potentially, occur in any organism, there are always **differences between the particular manifestations (implementations)** of the molecular cascades in various species and even cell types. The interactome perspective in post-genomic biology suggests that presence of all or most of the molecules participating in a particular cascade/pathway does not at all imply that this cascade/pathway would even be present in a particular organism. For example, a pathway known as "Cori cycle" involves utilization of lactate produced by anaerobic glycolysis in muscle for hepatic gluconeogenesis. The lactate formed as a result of short-term but intense muscular work is released into blood. To prevent acidosis, the liver enzymes recycle the lactate (through pyruvic acid and glucose-6-phosphate) back to glucose which is, in part, stored as glycogen in liver cells and, in part, released back into the blood stream and absorbed by muscles and other organs. Thus, the pathway known as "Cori cycle", in the original sense of the term, includes interactions of at least three organ systems: muscle, liver and cardiovascular system. In no way this pathway is restricted to a particular cell type.

Here comes the one of the most common difficulties inherent for post-genomic era: that of excessive and unsubstantiated generalization. Supposing that most of the molecular components involved in the Cori cycle (for instance, according to Reactome database there are over 40 of them, see http://www.reactome.org/cgi-bin/eventbrowser?DB=gk_current&ID=70268&) would be found in some other organism after genome sequencing and all the related predictions and inferences. How *physiologically* reasonable would be to infer that this pathway, originally discovered in higher mammals and that involves an interaction of at least three organ systems, is also present in unicellular organisms (as the electronic inferences, available at the web-address above, seem to suggest)? There is no unequivocal answer to this and other questions of the kind and it is hardly arguable that mere comparisons of the genomes, proteomes *etc* across species would provide any. Answering this sort of question requires accurately planned, adequately managed and physiologically certifiable experimental studies. Alternatively, accurately designed and physiologically certifiable bioinformatic analyses are required.

Of course, many organisms share common metabolic pathways such as glycolysis or citric acid cycle. From an engineer's point of view, which is well supported by biochemical studies, the similarity of the metabolic pathways in a cross-species comparison suggest them to be the common elements of the metabolic design. From physiological point of view, however, a notion of some general metabolism, universal for organisms can be confusing. With the exception of a few systems (such as glycolysis), a physiological study of a particular organism is not so generic in its approach as a biochemical study tends to be. Assuming existence of some universal and always invariable blocks of metabolic design that invariably occur in any organism can lead to various confusions or even health problems. For example, there is a current controversy over whether hepactic lipase activity is pro- or anti-atherogenic. Thus, in various animal models the enzyme appears to act differently according to the particular organism. Albeit some of these differences might be related to other conditions [2], one of the reasons for this seeming contradiction is the lack of regard of the differences in the details of particular implementations of what is assumed to be a common metabolic pathway. CD28 protein, mentioned in the introduction to this chapter, can be used as another of many hundreds of possible examples.

Fourthly, there are **multiple biological functions of proteins involved** in a particular molecular cascade/metabolic pathway. That is, many of the proteins involved in a particular biological process and in the related molecular cascade can be involved in a number of other, entirely different processes and the related molecular cascades. These other processes can manifest themselves under a specific happenstance. For example, while discussing the link between VTE and estrogens (see the previous chapter), we mentioned that proteins of the blood coagulation cascade can influence gene transcription. These additional functions have, apparently, little to do with the major functions of the proteins as components of the coagulation cascade.

Another example: apoptosis and glycolysis are processes supported by molecular cascades that seem to be remarkably similar among various species. These two processes are also considered to be entirely non-related. However, a number of glycolytic enzymes are more than just components of the glycolytic pathway [3]. Depending on a particular organism or cell type, certain glycolytic enzymes can have a number of additional, non-glycolytic functions, such as transcriptional regulation (hexokinase, lactate dehydrogenase A, glyceraldehyde-3-phosphate dehydrogenase, enolase), stimulation of cell motility (glucose-6-phosphate isomerase), regulation of apoptosis (glucokinase, glyceraldehyde-3-phosphate dehydrogenase, hexokinase) *etc*. There is also a possibility that interconversion between the glycolytic and other biological functions of the glycolytic enzymes might occur as a result of phosphorylation and/or other chemical modifications of the multi-subunit glycolytic enzymes [4].

In addition to these inherent peculiarities of the molecular cascades (namely, the structuring of the interactome/reactome, the system-specific occurrence, the system-specific implementations and the multiple biological functions of proteins involved), there are further complications related to the interactomes and the proteomes that can greatly reduce the actual value of post-genomic studies when ignored, whether formally or effectively. Purpose-oriented approach to post-genomic studies requires elaboration of the informational models adequate for the research problem in hand. An adequate model should not be too general. Neither should it feature excessive details irrelevant for the research task of interest.

From this point of view, both the concepts of molecular cascade and metabolic pathway, valuable though they are for biochemistry and for molecular biology, can be considered as being too general and lacking essential details, especially in the case when we are interested in a particular problem of physiological or biomedical research. Indiscriminate and mechanical operation of the terms like "molecular cascade" and "metabolic pathway" current in professional media leads to a danger another neo-reductionism of the sort "first, we'll describe all the pathways and then we'll infer the cascades and then the physiological effects". Apparently, this sort of neo-reductionism attitudes is incompatible with any practical application.

In other words, there is a distinct possibility that by the time the actual knowledge of the molecular cascades will reach the precision that will allow actual practical applications, the entire game will not worth the time spent nor the money wasted, since practical applications are not interested at all in these sorts of futile generalizations. Alternatively, the practical problems formulated within the framework of the purpose-oriented research will be solved much earlier than such precise knowledge of the molecular cascades will be achieved. In this perspective, we would like to introduce here the concept of a "metabolic context" that allows more efficient management of the informational complexity of particular metabolic pathways. Apparently, this term refers to reactome and the parallel term for the interactome would be "molecular context". We implicitly introduce the term in the following Case Study that deals with reconstruction of metabolic context of a particular enzyme/gene.

THE CASE OF A METABOLIC CONTEXT RECONSTRUCTION (MTHFR)

"The levels of plasma folates are known to influence predisposition and course of a wide range of disorders. What could be the related molecular mechanisms? How they could be retrieved/described/elucidated using bioinformatic tools?"

In the scope of the purpose-oriented bioinformatics, the most reliable strategy lies, of course, in the efficient use of the MEDLINE searches. Even having minimum background in physiology, a few searches in MEDLINE or any other abstract database with keywords like "folate", "plasma", "molecular" or "gene" will allow to find that the activity of the enzyme called MTHFR (methylenetetrahydrofolate reductase) is one of the crucial determinants of the folate plasma levels. However, in a physiological system an enzyme rarely acts alone and is always a part of a metabolic pathway. At first glance, these pathways can be easily extracted using standard bioinformatic resources such as KEGG (http://www.genome.jp) or BIOCYC (http://biocyc.org). The question with the large databases of the kind is how complete information there would be and how regularly it is updated.

In KEGG, for example, MTHFR is listed as belonging to "one carbon pool by folate" and "methane metabolism" metabolic pathways. The outline of the "one carbon pool" pathway is presented in the Figure 7-2. Given a couple of dozen of the enzymes and about 10 of the metabolites, the resulting picture does look impressive and complex. However, how could we abstract the essential information from this complex picture? Can we use this picture to explain how, for example, the folate levels are related to the cardiovascular or oncological etiology?

It is clear that a representation of the metabolic pathway such as shown in the Figure 7-2 still requires some additional work to extract the necessary information. Moreover, there is always a risk that this particular representation might be incomplete. Therefore, to answer the questions of the biomedical significance such as the one we asked in this Case Study, we need some other kind of the data representation. Not less important, we need to verify the data for their completeness and to ensure that all of the currently known essential details were included. With the help of KEGG, BIOCYC and MEDLINE searches, we have compiled a specific representation of the MTHFR-related metabolism that we call "metabolic context of MTHFR" (Figure 7-3).

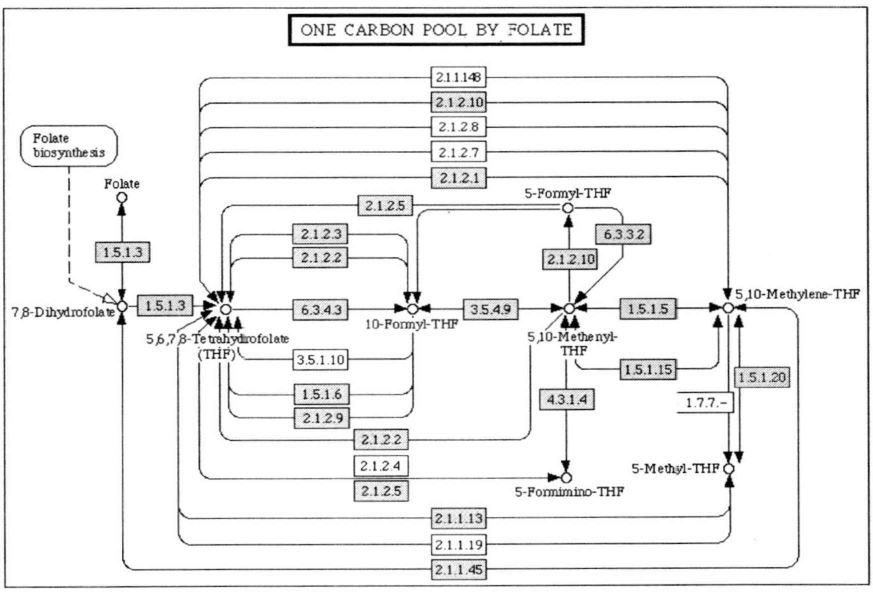

Figure 7-2. One carbon pathway by folate. The figure was generated using the DBGET server of KEGG (http://www.genome.jp/dbget-bin/show_pathway?hsa00670+4524).

The major difference between the two representations is the enhanced simplicity of the metabolic context. In comparison to the one-carbon pool folate pathway (Figure 7-2), the diagram of the metabolic context of MTHFR (Figure 7-3) mentions only the substances and the enzymes involving interactions with the immediate substrate/products of MTHFR. Therefore, the metabolic context can outline the most important metabolic effects induced by a change in the MTHFR activity. As the reader, perhaps, remembers, we used the metabolic context of MTHFR during ortho-analysis of the levels of function (Chapter V).

The other difference between a metabolic context and a metabolic pathway is completeness: a metabolic context is not necessarily limited to a particular metabolic pathway. Thus, the metabolic context of the MTHFR reaction combines the data from distinct metabolic pathways (according to BIOCYC, these are biosynthesis of pyrimidine deoxyribonucleotides; glycine biosynthesis; glycine degradation; methionine biosynthesis; S-adenosylmethionine biosynthesis and reductive acetyl coenzyme A pathway). We discuss other implications and applications of the metabolic context further in this chapter.

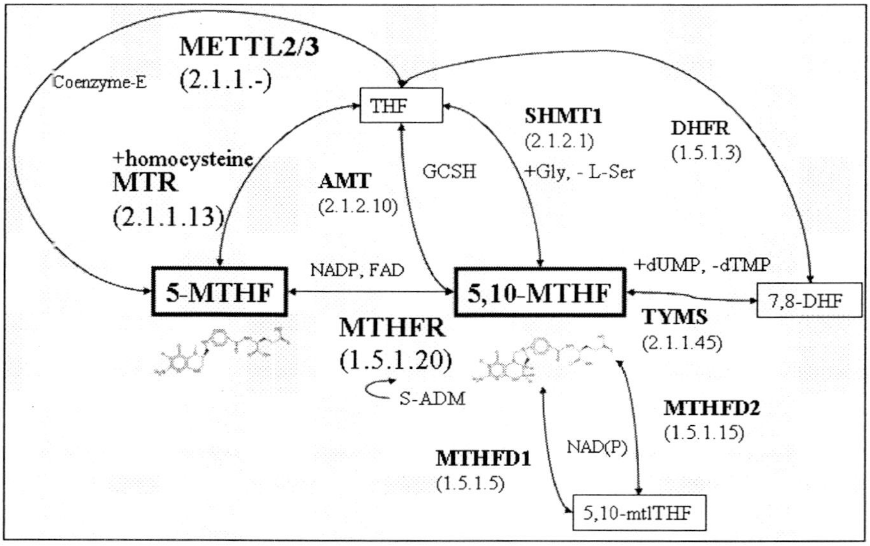

Figure 7-3. Metabolic context of MTHFR. 5,10-methylenetetrahydrofolate reductase (MTHFR) catalyzes the interconversion of 5,10-methylenetetrahydrofolate and 5-methyltetrahydrofolate. MTHFR is the key rate-limiting enzyme of this conversion. MTHFR enzymes requires NAD(P), FAD and is allosterically regulated by S-adenosylmethionine. This diagram of the "metabolic context" of MTHFR includes data from the following metabolic pathways: de novo biosynthesis of pyrimidine deoxyribonucleotides; glycine biosynthesis; glycine degradation; methionine biosynthesis; S-adenosylmethionine biosynthesis and reductive acetyl coenzyme A pathway. Each enzyme is annotated with the gene name and E.C. number. Compound names are as follows: S-ADM (or "SAM"), S-adenosylmethionine; FAD, flavin adenine dinucleotide cofactor; NADP, nicotinamide adenine dinucleotide phosphate cofactor; dUMP, deoxyurdine phosphate; dTMP, deoxythymidylate; 7,8-DHF, 7,8-dihydrofolate; 5,10-mtlTHF, 5,10-methenyltetrahydrofolate; THF, tetrahydrafolate; GCSH, glycine cleavage system H protein;

Metabolic Context Analysis and Molecular Mechanisms of Human Disease

As it was indicated in the foregoing Case Study, a metabolic context of a particular metabolite, reaction or gene can be an efficient way of representing the data on metabolic pathways, especially if we would like to consider questions of biomedical importance. It is quite understandable in view of the fact that metabolic context is not limited to a single metabolic pathway but includes all the data currently available for the known pathways. In terms of physical chemistry, a metabolic context of a compound reflects the system of kinetic equations for that compound (see the next section). Since no reliable information on the protein-protein interactions of the MTHFR protein is currently available, the metabolic context of MTHFR represents, thus, complete description of the most immediate metabolic effects the MTHFR activity might have (the word "complete" here and in any other place in the book implies the *currently available* information). MTHFR catalyzes the reaction of interconversion of 5,10-methylenetetrahydrofolate/5-methyltetrahydrofolate. In respect to this reaction, the metabolic context in the Figure 7-3 is, apparently, a first-order metabolic context. That is, it mentions largely the substances and reactions in which both the substances (5,10-methylenetetrahydrofolate and 5-methyltetrahydrofolate) are directly involved. In respect to MTHFR protein/gene, it is a second-order metabolic context. That is, it mentions not only the substances that directly interact with MTHFR enzyme but also substances and enzymes that interact with the primary MTHFR substrates.

The biomedical implications of the information presented even in a first-order metabolic context are, nevertheless, of considerable biological significance. Let's take a particular research question as an example. Among cardiovascular researchers there is well-established notion that levels of plasma folate(s) and homocysteine(s) are closely related. Although "folate" denotes a certain substance (folic acid), it is also used as a general term for the substances with a pteridine ring joined to both paminobenzoic acid and glutamic acid. Biomarkers of the folate status include red blood cell folate levels, serum homocysteine, serum SAH concentrations and DNA hypomethylation. The major forms of "folate" in plasma are 5,10-methylenetetrahydrofolate and 5-methyltetrahydrofolate. Similarly, although the word "homocysteine" stands for a particular aminoacid ($NH_2(COOH)CH\text{-}CH_2\text{-}CH_2\text{-}SH$), in physiology this same term often refers to the circulating homocysteine pool which includes homocysteine and its derivatives (homocystine and homocysteine-cysteine disulfides) [5].

Both folate and homocysteine influence the cardiovascular risk: folate appears to be a protective factor while homocysteine is a risk factor. What could be the related molecular mechanism(s) involved in the interaction between the levels of folate and homocysteine? However long we would look at the "one carbon pathway by folate" (Figure 7-2) and browse the related database entries, it won't be easy to come to any certain conclusions. A single look at the metabolic context of MTHFR (Figure 7-3) almost instantly suggests a number of plausible answers: homocysteine is involved into the transfer of the methyl-group from the 5-methyltetrahydrofolate to form L-methionine. This reaction (catalyzed by methionine synthase E.C.2.1.1.13, gene MTR) appears to be the only link between the homocysteine and folates. A deficiency in the activity of the MTR enzyme will affect both homocysteine and folates. A deficiency in the MTHFR activity will lead to dissipation of the folate pool and, consequently, to lower levels of all forms of folate (in particular, 5-methyltetrahydrofolate). As there is less 5-methyltetrahydrofolate available, less homocysteine will be methylated. Therefore, lower MTHFR activity is associated with lower folate levels and with higher homocysteine levels. This conclusion we based on the analysis of the metabolic context of MTHFR was proven by numerous epidemiological studies.

Although the above logical deduction based on the metabolic context is hardly an original one, it still illustrates the immediate usefulness of the metabolic context as a method to represent and analyze information on metabolic pathways. This method is especially important for answering particular questions arising in the course of biomedical research. There are many other applications. First of all, metabolic contexts serve as direct link to quantitative modeling of the metabolic reactions (see below). Then, metabolic contexts are very helpful for the ortho-analysis of the levels of function (Box 9) and during ortho-analysis of the heterogeneous clinical data (Box 10). The ortho-analyses, as we remember, are one of the major means to ensure consistency of a particular biomedical study or a group of studies (Chapter V). If metabolic contexts of particular reactions/genes can be established experimentally, this allows more consistent and more reliable reconstruction of the relevant metabolic pathways than some sort of a high-through-put-off study. Combining metabolic contexts for certain reactions, genes or metabolites can be an efficient way of analyzing the molecular mechanisms of human diseases at any level of detail. In the rest of the present section, we will analyze the metabolic contexts for the homocysteine(s) and the related substances. Then, we will look for correspondences between the inferences from the metabolic contexts and what became known from the available epidemiological studies of homocysteine effects, including the genetic association studies. In a way, this analysis can be considered as a further elaboration of the ortho-analysis of the MTHFR presented earlier (Box 9).

Ortho-analysis of the clinical data for the MTHFR 677 C>T polymorphism (Box 11) indicates that lowered MTHFR activity is associated with a wide range of human diseases including cardiovascular diseases, hypertension, cancers, obstetrical complications and osteoporotic fractures. Numerous epidemiological studies have illustrated the correlations between lower MTHFR activity, lower folate/high homocysteine and the adjusted risks of the diseases. Here, we are interested in the following question: is it possible to rationalize the risk adjustments in terms of metabolic contexts? In other words, would it be possible to trace certain adjustments of a disease risk to certain changes in the folate/homocysteine metabolism? We attempt this kind of deduction while analyzing the metabolic context of homocysteine (Figure 7-4). In our analyses we assumed that: **1.** Either homocysteine or some of its immediate products can directly affect the disease-related physiological changes; **2.** These changes are brought about exclusively through the metabolic context of homocysteine and the immediately adjacent metabolic chains; **3.** We limited the length of the metabolic chains to 3 links (that is, the direct effects would have a substance that is not further than 3 reactions from the center of the metabolic context of homocysteine---*i.e.*, the homocysteine itself); **4.** The compounds assumed to have the direct influence on pathology were regarded in respect to their specificity of the related physiological effects.

Beforehand, we ought to admit that the action of the homocysteine through the metabolic context might not be the only cause of the physiological effects exerted on the body by higher levels of homocysteine. For example, homocysteine appears to decrease gene transcription of APOAI and increase expression of CYP7A1 thus increasing the risk of atherosclerosis [6], possibly through the PPAR-alpha mechanism. It is not entirely clear what would be the role of the metabolic context of homocysteine in this case. However, as we'll see from the discussion below, the metabolic context allows us to perform bioinformatic analysis of the systemic effects of the higher homocysteine concentrations ("systemic" here means "not restricted to a particular cell type or a tissue").

According to the metabolic context of homocysteine (Figure 7-4), the major products of homocysteine transformations are L-methionine, cystathionine and SAH (S-adenosylhomocysteine). L-methionine is one of the 20 most common aminoacids that make up all of the proteins in the body. Therefore, physiological effects of the altered levels of methionine are extremely hard to deduce from the metabolic model. Cystathionine is a very specific semi-product involved in the synthesis and degradation of the sulfur-containing amino acids cysteine, homocysteine and methionine and cannot be a "target substance" of the analysis we aim at. The only lead that seems to remain is S-adenosylhomocysteine (SAH).

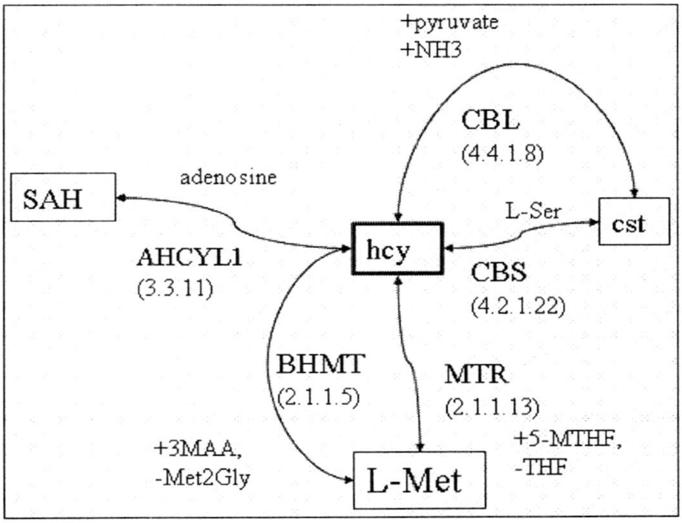

Figure 7-4. The metabolic context of homocysteine. Abbreviations: hcy, homocysteine; cst, cystathionine; SAH, S-adenosylhomocysteine; 3MAA, trimethylammonioacetate; Met2Gly, dimethylglycine; 5-MTHF, 5-methyltetrahydrofolate; THF, tetrahydrofolate.

SAH and it's major product SAM (S-adenosylmethionine) are involved in a number of reactions that can directly affect the physiological changes related, in particular, to cardiovascular disease. The metabolic context of SAH/SAM is quite large and includes almost 50 reactions (not shown), so here we simply summarize the results of the analysis of these metabolic contexts relevant to carviovascular disease(s). The routes of the metabolic influence that we consider to be the links between the metabolic effects of homocysteine and cardiovascular pathologies are synthesis/degradation of catecholamines and nicotine metabolism.

These two compounds, SAH and SAM, are directly involved in the **synthesis of catecholamines**. The most common catecholamines are neurotransmitters epinephrine ("adrenaline"), norepinephrine ("noradrenaline") and dopamine. Epinephrine is CNS neurotransmitter which can also act as a hormone. Norepinephrine mainly acts as a sympathetic PNS (peripheral) neurotransmitter. Dopamine also acts on the SNS. Generally, catecholamines affect physiological changes that prepare the body for short bursts of physical activity (such as so-called "fight-or-flight response") and high levels of catecholamine levels in blood plasma are associated with higher degree of stress. The most typical physiological effects of catecholamines include increased heart rate and blood pressure. Immunomodulatory effects of the blood catecholamines are reasonably well documented [7] and catecholamines are likely to be involved in inflammation processes; it was hypothesized that catecholamines can cause vascular injury thus aggravating atherosclerosis. Catecholamine-induced aggravation of aortic and coronary atherosclerosis was shown in animal studies [8].

Qualitative analysis of the metabolic contexts of the SAH/SAM shows that both substances are involved in the interconversion between noradrenaline and adrenaline. These reaction is catalyzed by phenylethanolamine *N*-methyltransferase (PNMT, E.C.2.1.1.28). The increase in the levels of SAH and SAM is likely to be associated with a shift of metabolic equilibrium towards higher noradrenaline and adrenaline. Both substances are also involved in the dopamine transformation to 3-methoxytyramine catalyzed by catechol O-methyltransferase (COMT, E.C.2.1.1.6). Moreover, SAH is well-known to be a competitive inhibitor of COMT enzyme and COMT-inhibitors are, in turn, known to raise the levels of the catecholamines.

Summarizing, analysis of the metabolic contexts of MTHFR, homocysteine and SAH indicates that lower MTHFR activity results in higher homocysteine levels. Through the subsequent shifts of metabolic equilibrium, higher homocysteine is likely to lead to higher catecholamine concentrations. Higher catecholamine levels are directly associated with increased blood pressure, inflammation and atherosclerosis--- the common cardiovascular risks.

The involvement of the SAH/SAM into **nicotine metabolism** might account for some of the effects smoking has on CVD risk. Smoking is a recognized CVD risk (Figure 2-3). However, the relationships between the smoking and CVD etiology/course are not always unequivocal. In particular, it is known that levels of homocysteine might have something to do with patient-specific response to smoking. According to the metabolic context of SAH/SAM, the interconversions between the SAH and SAM can involve biotransformations of nicotine (indolethylamine N-methyltransferase, gene INMT, E.C.2.1.1.49) and of the nicotinamide (nicotinamide N-methyltransferase, gene NNMT, E.C.2.1.1.1). SAM is also a known cofactor of the NNMT enzyme.

How these deductions do compare with epidemiological studies? Although this sort of comparison attempts to bring together levels of function quite distant from one another (Figure 1-5), there are definite correspondences between these metabolic deductions of ours and the biomedical studies of the effects of higher levels of homocysteine. Thus, acute MI or death due to coronary disease was associated with higher homocysteine after adjustments for diabetes, hypertension, total/HDL cholesterol and other factors (OR=3.4, 95% CI 1.3-8.8, P=0.01) [9]. A study of the Framingham cohort showed an increasing prevalence of carotid-artery stenosis directly proportional to homocysteine concentrations after adjustment for gender, age, HDL cholesterol, blood pressure and smoking (P< 0.001) [10]. Among normotensives, the prevalence of atherosclerosis went from 22% in the lowest quintile of homocysteine values up to 53% in the fifth quintile (OR=4.1, P<0.05). Adjustments for age, gender, systolic blood pressure, cholesterol and smoking have shown even stronger correlation (OR=6.4, P<0.01) [11]. In the framework of the metabolic model, stenosis is associated with the higher catecholamies that induce vascular injury and stimulate the inflammation component of atherosclerosis.

The second most important metabolic deduction (higher blood pressure associated with the higher homocysteine due to higher adrenaline and other catecholamines) also seems to be confirmed in biomedical studies. The median homocysteine value was 11.5 micromol/L for hypertension cases and 9.9 for control subjects ($P<0.001$). After adjustments for other factors, homocysteine still remained significantly associated with systolic hypertension ($P=0.02$) [11]. The 677T variant of MTHFR gene, associated with higher homocysteine was as mild but significant a risk factor for essential hypertension in Caucasians (OR=1.6; 95% CI:1-2.4;P=0.03) [12]. Given that homocysteine seems to act at the fundamental level of the basic human physiology (catecholamine synthesis and degradation), it is likely that that these associations will not be limited to any particular human race. For instance, the association of hyperhomocysteinaemia and hypertension was confirmed in a Sri Lankan population [13].

How the deductions from the metabolic contexts would compare with the studies of the physiological effects of homocysteine-induced pathology? It was stated that possible mechanisms by which elevated homocysteine levels lead to the development and progression of vascular disease include effects on platelets, clotting factors and endothelium [14]. Albeit effects on the platelets and blood clotting do seem to be reproducible, they are, most likely, to be the secondary effects of the homocysteine action. In the framework of the metabolic model, they are far from being apparent and would be hard to deduce using only the analysis of metabolic changes. However, the fact that homocysteine induces endothelial damage in higher primates and in human cells in culture is known for several decades [15-17].

There are several important remarks that need to be made to conclude the analysis we presented. First, after completion of the analysis of the metabolic contexts of the homocysteine and SAH/SAM the respected reader just read, the author decided to look in the MEDLINE for the studies that would directly associate higher levels of homocysteine with the higher catecholamines. No experimental studies were apparently found. It came as a surprise that, nevertheless, a hypothesis of the pathophysiological effects of homocysteine that provides predictions compatible with our analysis of the metabolic contexts was explicitly formulated some time ago [18]. The main component in the surprise, however, was the fact that this hypothesis, most likely to be the correct one, was largely ignored by the contemporary learned establishment.

Second, the metabolic contexts we analyzed in the present section, were, in part, derived from the data based on the annotations of the human genome sequence. These data are known to be incomplete since less than 50% of the human genes were annotated with at least some degree of credibility. Nevertheless, we can observe how analysis of the metabolic contexts can, at the very least, produce a rational for further studies in human physiology and post-genomic studies. We, of course, are quite aware of the fact that these ideas might seem to be "weird", "unlikely" or even "ridiculous" to more narrow-minded scholars that attach themselves to biomedicine and physiology.

Third, it is the matter of the biological mechanisms that link the folate levels with pathologies other than CVD. For example, in the ortho-analysis of MTHFR (Box 10) we associated cancers with the lower folate status. However, this is rather a conventional than a causational attribution. Within the framework of the metabolic model we presented here, it appears that the pathophysiological action of the lower folate levels related to cancer can be achieved through the same chain of the metabolic events (namely, folate-homocystein-SAH/SAM). Albeit a number of quite contradictory hypotheses were proposed for the link between the low folate and cancer etiology (such as uracil misincorporation, decreased DNA methylation, methylation-regulated gene expression *etc*), it can hardly be denied that lower folate levels correspond to abnormally lowered DNA methylation: lower folate leads to higher homocysteine which in turn leads to higher SAH; SAH is known to be a strong inhibitor of SAM-dependent DNA methylation [19]. As DNA methylation affects expression of many genes (in both directions), it is clear that the lower folate levels *will* affect the cancer risk. However, it cannot be said in which direction and both lower and higher risks of cancer at lower folate can be expected.

Fourth, there is a distinct possibility that metabolic contexts might be not sufficient for explanation of all the known physiological effects of the lower MTHFR activity/higher homocysteine levels. For example, there are effects of the higher homocysteine on the blood clotting/platelets or the effects of homocysteine levels on the gene transcription that we've already mentioned. These effects are not apparent in the scope of the metabolic model *per se*. Indeed, at least theoretically, homocysteine would influence gene expression, inhibit or activate certain proteins from pathways other than "standard" metabolic pathways involving homocysteine molecule *etc*. The same can be said about the products of the homocysteine's biotransformations. Here, the experimental post-genomic studies of the reactomes and metabolomes will be crucial. For any sort of such study to be successful, a strong physiological background and purpose-oriented approach to bioinformatics will be required.

Mathematical Modeling of Metabolic Pathways and Molecular Cascades

The metabolic contexts we just considered represent qualitative models of the metabolism of particular compounds. Subsequent qualitative analyses of the metabolic contexts allowed us to arrive at physiologically plausible conclusions that can be used as starting hypotheses in the ongoing physiological research. It is clear that the value of the analysis of metabolic contexts would be greater if it would also allow making some quantitative estimates by taking into account the actual rates of the individual enzymes, the actual ranges of the compound concentrations and other factors.

Metabolic contexts are functional representations of the metabolism which are considerably more suited for particular purposes of biomedical research. This suitability, however, in no way abrogates the usefulness of the larger functional blocks such as metabolic pathways. Both metabolic pathways and metabolic contexts are qualitative models. They show the relationships between the various substances (most of all, the small-molecule metabolites) indicating which reaction spends a particular metabolite and in which reaction this or that metabolite is produced. Metabolic pathway also indicates along which route(s) a particular substance can be converted into another.

A qualitative model of the kind indicates, at the very least, the components and their essential interactions. As such, it can be well represented by a block diagram (see, for instance, the figures depicting the metabolic contexts discussed in the previous sections). A qualitative model of the kind can also serve as the basis for elaboration of the quantitative model: the one which will operate with concentrations of the compounds and with the actual conversion rates.

The transition from the qualitative model to the quantitative model is achieved by using the method of the *classical chemical kinetics* (Box 13). What needs to be noticed is that a "metabolic context" is considerably more practical concept to make this transition than a "metabolic pathway" is.

Box 13. The method of the classical chemical kinetics

Chemical kinetics and thermodynamics are the two major branches of the physical chemistry. Chemical kinetics deals, apparently, with the rates of the chemical reactions. Being first formulated as a quantitative science in the middle of 19th century, chemical kinetics is a common instrument in chemistry, biochemistry, enzymology and quite a number of other disciplines.

The basic unit in chemical kinetics is a *chemical reaction*. Each reaction is associated with a differential equation (also generally known as a "*rate equation*") that encodes the quantitative aspects of the reaction. In simplest terms, if there is, for example, a reaction in which the reaction rate depends on one compound (1st-order reaction):

$$A \Rightarrow B$$

then the differential rate equation is:

$$\frac{d[A]}{dt} = -k[A],$$

where "[A]" denotes the concentration of the compound A and "k" is the *rate constant*. When we have a reaction with the rate depending on the concentration of the two compounds (2nd-order reaction):

$$A + B \Rightarrow C$$

the differential rate equation for the compound A is:

$$\frac{d[A]}{dt} = -k[A][B]$$

When there are several reactions that use a number of common compounds, the differential equation of each compound is formulated in the similar manner by adding or subtracting corresponding members in the right part of the equation (see homocysteine example in the text).

Being rather straightforward, the approach of chemical kinetics still has a number of caveats, few of which we mention here. Firstly, whether the actual reaction is first-order or second-order does not follow from the apparent stoichiometry of the reaction and often needs to be determined through an experimental study of the reaction kinetics. Secondly, the rate constant depends on the temperature and, generally, the reaction rates are higher at higher temperatures. The dependence of the rate constant on the temperature is described by the Arrhenius equation:

$$k = A_0 e^{-\frac{E_a}{RT}},$$

where "E_a" is a reaction-specific constant called "activation energy". Thirdly, the rate equations of the chemical kinetics we've just considered were formulated for homogeneous and isotropic phases (such as well-mixed gases or fluids) and reflect, apparently, an ideal case. In most of the cases, especially in biological systems, the chemical reactions tend to occur at the surfaces or at phase interfaces. Fourthly, almost any biological reaction is supported and controlled by a certain catalyst, a substance that accelerates the reaction but remains unchanged. In the case of the overwhelming majority of the metabolic reactions, the catalysts are specific proteins known as *enzymes*.

Thus, traditional definitions of the metabolic pathways that feature in the textbooks on biochemistry and public databases do not provide sufficient means for quantitative modeling (largely because these definitions focus on *subsets* of networks without consideration of the network-wide interactions [20]). The same cannot be said about the metabolic contexts. Indeed, after careful consideration of the basic tenets of the chemical kinetics we briefly summarized in the Box 13 (and, perhaps, after vaguely recalling the appropriate university course the reader might have had), we can immediately notice that a metabolic context for a particular compound directly translates into the differential rate equation for this compound. Combining the differential equations for several compounds (or, in our terms, combining metabolic contexts) allows us to build, at least in theory, a quantitative kinetic model of any complexity.

Let's consider a particular example of how a metabolic context for a particular compound serves as the basis for formulating the rate equation for the compound. We'll take the metabolic context of homocysteine (Figure 7-4). First, we transform the data encoded in the metabolic context into a series of chemical equations. According to the metabolic context we specified in the Figure, there are 5 reactions involving homocysteine (hcy):

1. hcy + L-serine = cystathionine + H_2O (CBS)
2. cystathionine + H_2O = ammonia + pyruvate + hcy (CBL)
3. hcy + 5-MTHF = L-methionine + THF (MTR)
4. hcy + 3-MAA = L-methionine + Met2Gly (BHMT)
5. hcy + adenosine = SAH+H_2O (AHCYL1)

Each of these reactions is catalyzed by a certain enzyme, indicated in parentheses, so we mark the rate constants with the enzyme/gene IDs. Assuming that each of the reactions is reversible, there will be two rate constants: one for the forward ("f") and one for the backward ("b") reaction. The compilation of the rate equation then proceeds reaction by reaction as follows. Let's take the reaction (1) as an example. In this reaction, homocysteine is spent in the forward reaction (constant k^f_{CBS}) and is gained in the backward reaction (constant k^b_{CBS}). The corresponding members on the rate equation will be $-k^f_{CBS}$[hcy][Ser] and $+k^b_{CBS}$[cst]. The latter member is a pseudo-1st order reaction and it can be written in the unimolecular form (since the water is in excess). Similarly, we formulate the rest of the members, regroup and write the final rate equation for the metabolic context of homocysteine:

$$\frac{d[hcy]}{dt} = -[\mathrm{hcy}](k^f_{CBS}[\mathrm{Ser}]+k^b_{CBL}[\mathrm{pyr}]+k^f_{MTR}[\mathrm{5MTHF}]+k^f_{BHMT}[\mathrm{3MAA}]+k^f_{AHCYL1}[\mathrm{ads}])+$$
$$+[\mathrm{Met}](k^b_{MTR}[\mathrm{THF}]+k^b_{BHMT}[\mathrm{Met2Gly}])+k^b_{CBS}[\mathrm{cst}]+k^b_{AHCLY1}[\mathrm{SAH}]$$

Albeit the equation is somewhat long and looks quite messy, that's not the main problem with its application. Apart from possible typoz, there are at least four major difficulties in practical usage of this equation in its present form. First of all, of course, the equations for the rest of the compounds in the system we consider need to be written. For this, we'll need the metabolic contexts for each of the compounds involved and then we'll have to decide which of the concentrations will be treated as variables and which as parameters of the kinetic system. Further, we'll have to define the initial conditions (concentrations) for all participating substances. Second, we should know precisely the kinetic order of each reaction involved. In the above rate equation, we assumed some of the reactions to be bimolecular and that may be not so, depending on the initial conditions. Third, we also need to know the detailed molecular mechanism of the reaction. Especially in the case of enzymatic catalysis, each of the 5 reactions above might need to be divided in sub-processes (such as diffusion of susbtrate(s), binding of substrate-1 to enzyme, binding of another substrate to the substrate-1-enzyme complex *etc*) and then the rate-limiting sub-process has to be defined. Fourth, all the rate constants need to be unequivocally defined or experimentally determined. The most likely obstacle here is that the constants we might extract from literature will inevitably be determined under different conditions and thus might be mutually incompatible. Nevertheless, this example conveys the general idea of how the bridge from the qualitative metabolic contexts to the quantitative rate equations is built.

So, the above example was worth rather of a lecture course in a good old university. Practically, any researcher interested in the kinetic problems in chemistry or biology is involved not so much in writing huge equations of the kind but rather in developing programs or in using already developed programs. However, in addition to the general caveats which are inherently associated with the method of the chemical kinetics as a method of physical chemistry (Box 13), there are also practical considerations that significantly limit straight-forward application of those programs, especially in relation to biological problems such as quantitative pathway modeling.

Provided that we might, in some way, account for heterogeneity, compartmentalization and for mass transfer, there is still one most important limitation of the method. The modern programs, such as, for instance, KINTECUS (see www.kintecus.com) have, apparently, no limit in the number of compounds and reactions that can be involved. However, years of practical expertise in modeling actual kinetic tasks do suggest that credible math models of various kinetic systems include no more than 50..100 reactions and a comparable number of compounds. The most accurate kinetic models currently known usually contain less than 10..20 compounds. The point here is that a valid kinetic model (that is, the set of compounds, reactions, rate equations, constants *etc*) is built on the base of the experimentally derived kinetic curves. When modeling is done after experimental kinetic curves, there is a definite limit of accuracy that can be actually reached by introducing new reactions and compounds. When this limit is reached, further addition of the compounds and reactions is pointless since there is no valid experimental basis to justify these additions. This problem of experimental justification or *falsifiability* of the kinetic models is especially important to be held in mind when dealing with biological tasks such as modeling of the metabolic pathways, signal cascades or molecular cascades.

The next important point to be remembered is that for building the quantitative models of metabolism it is rather metabolic contexts that should be used and not so much the formalized entities currently known as "metabolic pathways". This statement becomes entirely clear if we recall that a metabolic context of a compound directly translates into the rate equation for this compound. Combining metabolic contexts for two specific compounds produces the metabolic context of the corresponding reaction. At the same time, what is called as "metabolic pathway" often represents much extraneous information and cannot be directly used for building the quantitative models. If, for example, we compare the metabolic context of the MTHFR reaction (Figure 7-3) with the "one carbon metabolic pathway" (Figure 7-2), we will notice that the metabolic pathway contains a greater amount of extraneous information (such as alternative enzyme routes) and still lacks a few important details (the key compounds such as homocysteine that, formally, do not belong to the "one carbon metabolic pathway").

Actual studies that successfully employ the chemical kinetics for the analysis of the metabolic pathways do tend to use approximations closer to what we call as "metabolic contexts" than to "metabolic pathways". A metabolic context for a particular compound might not necessarily be exhaustive (that is, include all of the known reactions) but, at the very least, it has to be crafted in accordance with the purposes of the particular study. For example, if we consider a reasonably successful quantitative model of the "one carbon metabolism" [21], it does operate with what looks rather like combined metabolic contexts of the various forms of folate and homocysteine than the bulky metabolic pathway (see Figure 1 in the paper [21]). This model, in particular, not only produced physiologically reasonable correlations between the major compounds involved (which, by the way, are quite apparent from the qualitative analysis of the metabolic contexts), but also allowed to produce some non-trivial predictions (such as inverse association between folate and homocysteine being strongest at very low folate levels, potential influences on the DNA methylation, respective changes in the levels of SAH and SAM *etc* [21]).

The final *caveat* is the matter of the kinetic modeling of the protein-protein interactions. Metabolic pathways are networks of substances/compounds and the related enzymes that support corresponding biochemical reactions among the compound involved. A metabolic pathway is a standalone functional unit of the reactome. Mathematical modeling of the metabolic pathways (metabolic contexts) is, apparently, quite close to the framework provided by the tenets of the classical chemical kinetics (Box 13), even taking into account the points we mentioned discussing the rate equation for the homocysteine (see above). However, it is the modeling of the protein-protein interaction networks rather than of the metabolic pathways that is of the major interest for post-genomic biology.

Formally, the networks of the protein-protein interactions still can fit into the framework of the classical chemical kinetics by considering a "chemical reaction" that describes association-dissociation equilibrium between various proteins/subunits (though, in this "chemical reaction", no covalent bonds are altered). However, using the concept of chemical reaction, borrowed from *general chemistry*, and studying it by applying computational techniques borrowed from *physical chemistry* and doing all that in order to describe protein-protein interactions in *biology* looks awfully like attempting to fit old shoes of a child to the adult (though both wear *some* shoes, of course). The similarity to this shoe analogy grows considerably if we turn our attention not to the process of building the models, but to the actual *results* of such modeling activities. That is, to the profound absence of any significant results.

Apart from several minor points, the major difficulty with quantitative models of the protein-protein networks is almost total lack of the data on experimental kinetics that could be immediately used to confirm or to disprove the results of the modeling. Because of the problem of justification of the kinetic models that we've mentioned earlier, actual *in vitro* studies that deduce kinetics of the protein-protein associations rarely involve more than 4..8 of various subunits or their complexes. When a scientist, who works in biology or computer science (and who, typically, lacks any expertise in chemical kinetics) attempts to model a system that (1) involves a several dozen compounds and more than a hundred reactions and (2) the reactions reflecting the *protein-protein interactions*, not metabolic processes, then the credibility of the results drops sharply. It drops to zero level when the actual limitations of the chemical kinetics as a method (Box 13) are deliberately neglected. In other words, this is the very same problem that is raised by so-called "falsifiability" of the scientific studies (Introduction). Translating this criterion for the purposes of our present discussion, we can say that you can model as much as many reactions as you like but this model will not be able to prove or disprove anything.

Neverthelss, because of the high demand, the modeling of the molecular cascades and, in particular, of the molecular cascades that involve extensive networks of the protein-protein interactions still draws some attention of post-genomic researchers. Their generic reasoning goes, perhaps, like this: "OK, due to the difficulties you've mentioned, we cannot model kinetics on a quantitative level *precisely*. Would it be possible to model, at least, some *qualitative* effects with the help of a quantitative model, be it inaccurate as it is?".

Indeed, what interests an honest biologist most is not how to write a huge equation and to sell it to the federal government under disguise of a "research grant", but, rather, what are the mechanisms behind the real biological processes. Why, for instance, blood continues to carry multiple clotting factors which then "suddenly" begin activate each other at crazy rates when a vascular injury occurs? Or, in a completely different area of biology, why do the birds "suddenly" form a flock which often behaves as a separate living entity? There are numerous linear and non-linear processes of the kind in biology and human physiology at any level of function so the studies of these kinds of "emergent behaviors" [22] are of considerable interest in post-genomic era. Apparently, processes of the kind do not immediately follow from the properties of the individual elements but rather "emerge" when these elements are put together in a proper manner.

From this point of view, attempting to model the complex networks of protein-protein interactions might not be entirely a fool's errand. If a quantitative model, however flawed it would be, adequately reproduces a wide variety of the behaviors known for the particular system (say, a molecular cascade), then this model still might be usefully employed not only in studies of general interest but also for particular tasks of biomedical research. Given this scenario, what we need further are examples of actual success of modeling of particular molecular cascades that would produce non-trivial qualitative predictions.

The mere cheerfulness of the scenario, however, is in no way a basis for the successful math modeling in the area of molecular cascades (nor in any other area, for that matter). For example, the epidermal growth factor (EGF) signaling cascade has been used repeatedly as a test system for the mathematical modeling of the molecular cascades. EGF signaling is a good example to consider *status quo* in the area, since it is the best-studied and, allegedly, the best-modeled system. Even in the case of this "best-modeled" system, the current math models (based, by the way, on the improper usage of the math apparatus of the classical chemical kinetics, Box 13) still have a wide range of profound difficulties and do not successfully reproduce even the essential behaviors known for this particular molecular cascade (not to mention making any non-trivial predictions). Often, reproduction of this or other known behavior of the EGF cascade is achieved only after extensive tempering with the model parameters and this "fudging" (using Einstein's terminology) is hardly a practice of high scientific standards.

The first difficulty with the current EGF models is the lack of comparison with the experimental data. That is, the comparisons are made not to experimental kinetic curves but rather to some qualitative behaviors such as a sequence of certain events, for instance. Albeit, as we already said, it won't be fair to brush off this approach immediately as a sheer nonsense, it still has significant drawback: namely, *theoretical* modeling without direct *experimental* data.

The second difficulty is the size of the kinetic system. As we already remarked in passing, there are no obvious limits to the number of compounds and reactions involved in a kinetic system. However, century-long expertise accumulated in the course of actual modeling of various kinetic systems does suggest optimal limits for both the number of compounds and reactions: no more than 50..100 of both. At the same time, the EGF modeling systems tend to include about 100 compounds participating in 150 reactions [23] and the task of modeling kinetic system of that size is outside the boundaries of the practical applicability of the method of chemical kinetics.

The third difficulty is the definition and the relative values of the rate constants used in the model. Often, it appears that the values of the kinetic constants are set not so much in accordance with actual experimental data but either in accordance with arbitrary whims of particular individuals or in a way that would allow merely to mimic certain behavior of interest. For example, the author directly witnessed how "corrections" of the kind exceed two or three orders of magnitude (such as changing a reaction constant from $3*10^7$ to $8*10^5$ *etc).* Fudging with parameters in this manner is hardly scientific at all.

Apart from this quite disturbing issue of using arbitrary values of constants, in studies of the sort the reactions are pooled together without any regard to the relative numeric values of these constants. There might seem to be nothing in the basic tenets of the chemical kinetics to prohibit this (Box 13). Once again, however, practical experience in chemical kinetics suggests that when in the same kinetic system some constants are on the order of 10^7, other constants are about 0.1 and yet some others on the order of 10^{-7}, it is clear that there will be dominant effects of some few constants almost under any set of the initial conditions. Accordingly, the "models" of the kind are, shall we say, highly degenerate ("degenerate" in the sense the word is used in quantum mechanics) and are likely to be dominated by a few "strong" constants rather than to represent any sort of a realistic picture of the complex biological process.

In author's opinion, just these three major difficulties with the current math models of EGF and other molecular cascades are quite enough to make any reader endowed with common sense to be extremely cautious when he or she would see the usual happy trumpeting in a high impact journal that cheerfully reports "outstanding results of the mathematical modeling" of a molecular cascade. In the case of encountering such an event, I would advise the interested reader to take into account the points we've discussed in the present section. In particular, the three items we discussed in the preceding paragraphs are the most fundamental since they indicate the severe clashes between "modeling" of the kind and the results of actual experimental studies of the kinetic systems, be they biological or chemical.

The problems with quantitative modeling of the protein-protein interaction networks (and with the modeling of EGF signaling, in particular) are definitely not restricted to the list of these three items (namely, lack of comparison with experiment, system size and the numerical values of the kinetic constants). One of the best reviews of the EGF modeling attempts directly states that the current quantitative models of the EGF signaling are very far from being fit to be used for biomedical studies [23]. In particular, the authors cite compartmentalization, spatiality, endosomal trafficking, complexity of signalosome complexes and diffusion as the putative challenges in the realistic modeling of the EGF signaling. Indeed, incomplete (sometimes, even incorrect) information on the molecular cascade, over-simplification of the protein-protein interaction networks, neglect of the feedback loops and of the spatial aspects might have some role in the failures of the current models of the EGF signaling. At the same time, it hardly can be denied that the fundamental issues concerning the entire *range of applicability* of the chemical kinetics, which are not even considered in those "modeling" papers, are the most crucial in determining success or failure of any quantitative model.

Thus, as it is the case with the Newton equations that are clearly not enough for the quantitative modeling of the large biomolecular systems (Volume V of the book series), the quantitative modeling of the networks of the protein-protein interactions also requires introduction of quite a number of assumptions in addition to the basic tenets of the chemical kinetics. Practically, these assumptions are often arbitrary and undermine the very basics of the chemical kinetics rendering the entire modeling activity perhaps pleasurable but entirely meaningless. At the same time, quantitative modeling of the metabolic pathways still might be of theoretical and practical use provided that it is performed with a due regard to the limitations of the methodologies borrowed from the classical physical chemistry.

NETWORK APPROACHES TO MOLECULAR CASCADES

So far, we have mostly considered the modeling and possible practical applications of the metabolic pathways. In particular, we have introduced the concept of metabolic context that allows an effective usage of the information on the metabolic pathways that comprise the reactome. On a number of occasions, we've also illustrated applications of this concept. Now, it seems proper to turn our attention to the interactome and to the related concept of the molecular cascades.

We consider metabolic pathways as larger functional units of the reactome. Formally, we can also refer to a metabolic pathway as being a kind of a molecular cascade (as metabolic pathways do consist from a certain set of interacting molecules). However, involvement of several enzymes in a metabolic pathway does not necessarily imply that there will be direct protein-protein interactions among them. Interactions between the enzymes in a metabolic pathway are largely mediated by small-molecule metabolites and the protein-protein interactions are, more than likely, have only secondary roles in the performing of biological function (such as stability of the multimeric enzymes, for instance). At the same time, the term "molecular cascade", as we use it here, implies the primary role of the protein-protein interactions in the biological function.

Often, it is the interactome and the protein-protein interactions that are considered to be of primary importance in post-genomic studies. Many thousands of the functional networks of the protein-protein interactions (*i.e.*, molecular cascades such as EGF signaling, coagulation cascade and others) interlock on multiple levels and form the astoundingly complex and yet precisely functional web of interactions essential for the adequate and versatile behaviors of the physiological systems. Since much interest is invested in the interactome studies, it is an honest question to ask: would the high-throughput studies of the functional genomics type be able, at the very least, to produce adequate qualitative models of already known molecular cascades? If so, would it be possible to establish and to elucidate novel molecular cascades, most of which are not even known in contemporary biology? Apparently, these questions are not restricted to experimental studies but also include *in silico* studies which are the primary area of bioinformatics.

It's important to stress here that not simply *some* networks of protein-protein interactions are of interest for the human physiology and medicine but the *functional* networks of protein-protein interactions. There are two kinds of extremism that should be carefully avoided to make any progress in the study of the functional networks of the protein-protein interactions. The first extreme is the reductionist attitude driven by the development by the experimental 'case studies' in molecular biology during the last century [24]. A reductionist attempts to explain complex phenomena by merely reducing it to the "key" component parts. Surely, the proper dissection into parts (*"analysis"*) is essential for successful scientific research. However, biological systems are clearly much more than the sum of their parts. Even the small boy who took the fly to pieces in our Introduction was considerably puzzled when the fly "disappeared".

The second extreme is, apparently, the neo-reductionism (using the term we've coined earlier). For example, a post-genomic enterprise known as "systems biology" heavily borrows from cybernetics and attempts to predict the emergent properties of complex biological processes by borrowing ready-made models from classical cybernetics and from math branches such as graph theory. No doubt, data integration (*"synthesis"*) is also essential for the productive scientific research. Data integration includes defining the parts of the system, mapping their interactions and the ways they can be put together to operate smoothly. However, here is the danger that some quite general but, in the long run, an oversimplified concept will be taken as the fundamental principle with no regard to any other point of view, including the actual success or failure of its application.

This is what seems to be happening with the concept of the "biological networks". Prior to discussing any particular details, let's illustrate the foremost difficulty that's very likely to plague both experimental and theoretical studies of the functional aspects of interactome. We'll do so by using the analogy of a computer class/library (Box 14). Indeed, if we are to analyze the "function" of each user in a computer class, what really counts is *what* each particular user is doing and *how* he/she interacts with the rest of the world through the Internet. Understanding of this function will be impossible if we neglect the stream of the long-range interactions through the Internet or, say, the data on the social occupation and interests of each particular user. At the same time, high-throughput but highly superficial "mapping" of all interactions between all "entities" in the computer class who are available at some particular "present time" will only lead to a confusing mess of data that (1) reflect only a very few truly functional interactions, (2) is heavily dominated by "false positives" and (3) suggest only too general conclusions.

Box 14. The computer class analogy and interactome studies

We crafted this analogy in order to illustrate to an interested reader the foremost difficulty in the interactome-wide studies. In brief, this is the problem of exceedingly high rates of both the false positives (non-existing functional interactions) and the false-negatives (neglected functional interactions).

Imagine a computer class of 20..30 computers. At each computer, we have 1..2 users. Each user has spends a certain amount of time at a computer. Users are involved in a wide range of activities: some check e-mail, some are playing video games, others write letters, and yet some others, perhaps, read online books. Clearly, computers are not so far from each other, are connected by cables and we can detect a number of interactions between computers and users of the computers.

Then, imagine that there is an extraterrestrial being, the fabulous "Martian", landed near this particular place (probably, because of detecting many complex biosigns with tricoder). He/she/it is able to make a study at a distance and to observe the computers, the users and their interactions using an advanced extraterrestrial technology. First thing he will notice is that there are, definitely, interactions between almost all computers (ethernet packets running). Then, sometimes, there occur interactions between the users of the adjacent computers. Sometimes, computer entities change their user entities (or *vice versa*, depending on the point of view). After logging all of the sensor data into the main computer of the ship and using statistical analyses, our Martian might initially infer that there are complex interactions indicative of a complex, many-sided operation of a singular living entity.

From a human point of view, there are several fundamental flaws in this alien thinking. First is that computer class is a far cry from being a *singular* living entity. In fact, each user pays to use a computer for his/her own purposes. These purposes often have nothing to do with the purposes of the others present (unless, of course, the entire computer class happens, at the moment, to be filled with fans of just one multi-user game such as *Star Wars RPG*). Second, the interactions between the user-entities are, most of the time, quite superficial and do not bear much of the functional significance. Say, a user might ask a pencil from a neighbor, or, perhaps, to enquire the location of the closest café in the area. Detecting these interactions and interpreting them as "functional" neglects, apparently, the foremost functions of the users involved (namely, what in the world each of them is doing here?). Third, the presence of the ethernet/Internet entity, though it might seem to be a great discovery (for a Martian), bespeaks only of the most general function of the computer class- that it is a site for retrieval and exchange of information. Fixing attention on the particular "great discovery" will only lead to total neglect of the specific functions and interactions of the particular user-entities.

This analogy can be detailed and developed in a number of parallel directions. However deficient it might be, the analogy still conveys the central idea author holds to be important for the present discussion: the danger that a high-throughput study of interactome, even if performed with advanced post-genomic techniques, might result, *lit et verb*, in nothing. That is, it'll result merely in detection of many non-significant interactions while omitting many interactions of the foremost functional significance. In fact, the Case Study that follows the present section (*In silico interactomics: an inquiry)* does illustrate the idea using a particular molecular cascade as an example.

Albeit being very few at present, the results of the high-throughput experimental analyses of interactome-wide protein networks do seem to suggest conclusions similar to those we abstracted from the analogy. Thus, despite the recent completion of several pioneering high-throughput experiments aimed at the description of complete interactomes, the available maps of the protein-protein interactions are emphatically not sufficient to draw any biologically meaningful conclusions from even such a basic bioinformatic operation as interactome-interactome comparison [25]. Moreover, the current theoretical and experimental methods to map networks of protein-protein interactions not only produce an exceedingly high number of false positives but also have very low rates of concurrence [26]. At present, no experimental high-throughput study ever produced consistent data that would be generally compatible with already known data on the molecular cascades in physiology. By the way, a short review of the computational methods of the prediction of the maps of protein-protein interactions is provided in the Chapter II of the Volume V ("Biophysics") of the present books series while Volume III ("Molecular and cell biology") presents even more elaborated discussion on the subject of protein-protein interactions.

Thus, our computer class analogy (Box 14) indirectly points out potential problems with the interactome-wide studies. There is also actual failure of the current interactome studies (both theoretical and experimental) in producing consistent results of biological significance. The cause for this failure becomes apparent if we recall the basic facts of the biophysics of the protein-protein interactions. First, depending on the subunits involved in a particular protein-protein interaction, the stability of the resulting protein-protein complex can vary greatly (several orders of magnitude of the association/dissociation constants). Second, the functional networks of the protein-protein interactions have distinct temporal and spatial boundaries [27]. For example, dynamic protein-protein interactions, crucial for the signal transduction processes, can occur on a wide time scale: from milliseconds to seconds, depending on particular molecular cascade. At the same time, there are also static protein-protein interactions (such as enzyme assemblies) that can last for days and longer. Since high-throughput studies of any kind entirely ignore this most important factor, the time factor, it comes as no surprise that many of the functional interactions are lost and instead a lot of non-essential interactions are likely to be detected.

This major difficulty of loosing the essential aspects among numerous non-essential details (the "elephant-in-the-dark syndrome", Introduction), along with the inherent difficulties of the mathematical modeling of the molecular cascades (see previous section) necessitates, in particular, the search for adequate models of the molecular cascades (which are, again, *functional* networks of protein-protein interactions). Albeit the basics of chemical kinetics do not explicitly prohibit description of the protein-protein interactions, the existing kinetic models of the molecular cascades based on the rate equations are far from being adequate, not to mention any feasible biomedical implications. Accordingly, the researchers attempt to introduce new concepts that can serve basis for more adequate models. One of them is the the concept of a "biological network" based on the approach borrowed from graph theory.

As we might infer from our discussion in the previous section, metabolic pathways do not appear to be in a great need of new models. It is more than likely that given adequate expertise in the physical chemistry, bioinformatics and physiology, it will be possible to make reasonably accurate quantitative models of the metabolic pathways using classical chemical kinetics and the assistant concept of the metabolic contexts. Nevertheless, metabolic pathways might serve as objects of study appropriate for elaboration of the systemic methods that, subsequently, could be used for the analyses of the molecular cascades. After all, metabolic networks, deduced through countless person-hours of experimental studies in biochemistry are likely to be the most complete of all the known biological networks [26].

A number of network approaches for the analysis of the metabolic pathways have been formulated [20]. Primarily, these approaches aim at analysis of the general or "system" properties of the networks of interactions and then to associate, in some or other formalized manner, the general network topology with some aspect of the biological function. Hence, a variety of concepts such as "heuristically defined pathways", network-based "elementary modes"/ "extreme pathways" and many others continue to emerge as a result of applying the terms borrowed from the graph theory to the biological problems [20]. These concepts that refer to the general topological properties such as "network robustness" (the ability of the metabolic network to respond to changes in its structure) or "pathway redundancy" (a measure of the number of equivalent pathways with respect to identical inputs and outputs) could indeed be used for a systematic analysis of the existing pathways, in order to see how efficient these concepts might be in producing non-trivial and scientifically justified results. The non-triviality of results greatly depends, of course, on (1) the way in which these concepts are applied and (2) the purpose which is behind the application.

When application of the concept of biological network is driven, apparently, by sheer desire to apply it somewhere just to show that it's "applicable", the results are not likely to be very informative. For instance, a bioinformatic study reported presence of some giant component cluster present in any metabolic network [28] comparing the closeness of a particular node to this cluster can identify the most important metabolites/enzymes of a metabolic network. It sounds all very post-genomic and interesting. However, if we take into account general properties of the networks with an average of 2 connections per node (which is true of the metabolic pathways [29]), it is more than likely that even a randomly generated network will feature such a "core cluster" *etc*. In fact, the general topological properties of various networks are public knowledge available in books on popular science [22] and there is hardly any need in papers that regurgitate this same basic fact of network topologies.

When application of the network concepts is driven by an adequate research purpose and is based on the analysis of actual experimental data, then the results can be expected to be more specific and much less trivial. What is even more interesting is that these results can be in direct opposition to the conclusions and inferences from the general, purpose*less* sorts of analyses. For example, there was a study that analyzed the relationship between the position of mutations in yeast metabolic network and their lethality [30]. Using publicly available knock-out data, the authors have shown that lethality corresponds to the lack of alternative paths in the network affected by the enzyme deletion. Contrary to the trivial results of the over-generalized "bioinformatic" analyses, the researchers illustrated that it is the elements *outside* the giant component and with a low degree of connectivity that are the most crucial ones for the entire system to function.

So, applications of the network concept might have some promise in understanding the biological complexity of the metabolic and, perhaps, other biological networks. We can also expect that the most successful applications of the network models will be made in respect to individual cells or unicellular organisms. Nevertheless, a number of fundamental issues need to be considered before these methods of the network analyses could be used as legitimate methodologies for the analysis of metabolic pathways, not to mention molecular cascades.

The chief difficulty that these approaches tend to feature is the apparent confusion between thermodynamic and kinetic aspects of the biological processes. This confusion, albeit it might seem insignificant from the point of view of a biologist, still has serious repercussions for the analysis of the system behavior (which is what the post-genomic biologist is actually interested in). It should never be forgotten that *both* thermodynamics and kinetics determine the actual system behavior. We cannot discard, simply because it feels to be more convenient, either of the aspects.

What this talk in general science has to do with the network models? The current network models of the metabolic pathways are defined by means of a "stoichiometric matrix", which is a matrix with reaction stoichiometries as columns and metabolite participants in rows. Clearly, this concept is entirely thermodynamic and is directly related to the energies spent/produced in the corresponding reactions. In the framework of the network models, this matrix undergoes a normal mode analysis to define a set of metabolic pathways as elementary modes. Being a *thermodynamics-only* model, it cannot produce any data that would describe the actual *kinetics* of the process. And it is kinetic factors that define the important details of the system behavior. If we would care to recall the basic tenets of the chemical kinetics (Box 13), we cannot ignore the basic fact that *stoichiometry* of a particular reaction is in no way related to the actual *kinetic order* of the reaction. Accordingly, the claims similar to "steady-state flux distribution in the red blood cell could be described by a few key extreme pathways" (see the reference in [20]), however appealing and post-genomic they might sound to the ears of the magically enchanted listeners, do have serious tensions with the fundamentals of the entire building of the modern science. Clearly, it is highly questionable that a kinetic phenomenon (such as steady-state flux distribution) could be described purely in thermodynamic terms (such as extreme pathways deduced from the stoichiometric matrix). Such a description still can be possible, but only as a rare exception, hardly as a rule.

Neither should we forget that it is *functional* networks of interactions that are essential for the biological systems. Post-genomic endeavors such as systems biology have not provided us as yet with any intelligible treatment of the *known* functional aspects of the biological networks. At the present, we are still far from having a complete answer to the most basic question: what can we learn about behavior of biological systems by describing it as a network? It cannot be denied that purely topological studies neglect the identity of each node [26]. In other words, a metabolic network is not a set of ropes attached to certain pegs nailed on a board in a math class. Be deficient as it is, the computer class analogy of a protein-protein interaction network (Box 14) would be far much accurate also in the case of metabolic pathways, especially when compared to a loose set of ropes attached to pegs. Moreover, network topology alone is not sufficient to determine network behavior and there are pairs of networks with the same topology but entirely different logical behaviors [31]. In order to get a more tangible feel of these and other difficulties that lie ahead of the interactome-wide studies, let's consider a practical example presented in the Case Study below and the following discussion of it.

IN SILICO INTERACTOMICS: AN INQUIRY

"The functional genomics should not claim supersession of the classical way of studying molecular cascades in physiology. At best, it's only an instrument for taking a somewhat wider look at the biological phenomena. Given this, how physiological knowledge, available for decades, can be aligned with the results produced by a functional genomics study (such as interactomics, for example)? (from a discussion at a symposium)"

Questions of the kind occur more and more often among biomedical scientists. Indeed, given that functional genomics is more than a decade on the scene (and on the market as well), why there is such an extreme paucity of the studies that clearly show full compatibility of the results produced by functional genomics with *already available* physiological knowledge accumulated in previous decades? Are there severe problems with experimental devices, with data interpretation, with sample selection or with all of these? In this Case Study, we consider *'in silico'* interactomics of the proteins of the clotting cascade. This will allow us to see what problems an experimental study in interactomics inevitably faces, especially if there is information mismanagement.

Introduction

The primary purpose of this Case Study is to reconstruct an updated version of the protein-protein interaction map of the "core" of the blood clotting cascade (Figure 7-5a). A secondary purpose is to evaluate the molecular context of the blood clotting cascade *in vivo*. The present-day studies of the plasma proteome are very general [32,33] and no functional genomics studies of the blood coagulation were ever made. So, we'll have to use an *in silico* methodology. In this volume, the *in silico* component will be implemented in the form of ... (as the reader, no doubt, already guessed) ... searching a database of publication abstracts.

Now, the central moment in this Case Study is that only reliable data on the interactions of the human proteins will be used thus considerably reducing problems plaguing the experimental interactomics and also underlining the significance of the proper informational management even of the reliable data. The current experimental technologies, such as large-scale yeast-two assays are characterized by exceedingly high rates of both false positives and false negatives. These drawbacks of the high-throughput techniques will make the kind of analysis we are about to make next to impossible. Therefore, we raise the question at another level: what if we take reliable (for the most part) data on the protein-protein interactions? Such a collection of the protein-protein interactions might be expected to have much less false positives but will it allow to produce biologically significant results (in terms of reconstruction of molecular cascades)?

In other words, let's pretend for a moment (not forgetting, of course, that we are pretending) that nothing is known about the molecular mechanism of the blood clotting cascade. Given that experimental methods of functional genomics (in this case, *in silico* interactomics) produce reliable results, would it be possible to reconstruct or, at least, to ensure existence of such a phenomenon as blood clotting cascade? ...

Methods

In this case study, we didn't use databases of protein-protein interactions culled from literature (such as HPRD, DIP *etc*). First of all, these databases appear to contain too many interactions for particular proteins (for instance, almost 50 interactions for F2 in HPRD). Secondly, the information in such databases is highly unstructured from the point of view of the biological significance of each interaction. In terms of the graph theory, without any kinds of the "node weights" and "arc weights" it is hardly possible to manage successfully the informational complexity of the problem. Thus, direct searches of PubMed (as of 2006) were used to obtain all the required data. Over 300 publications related to the proteins in the Figure 7-5a were analyzed for reliability of the biochemical evidence and only those presenting a reliable evidence were selected.

Results

The results are presented in the Figure 7-5. Slide (b) represents the raw array of the reliable data on the protein-protein interactions, without any analysis of the relative significance. Watching this multi-node and multi-arced picture (that somehow brings to mind The Black Square), it is clear that the high-throughput studies made in the post-genomic vein can, generally, revolutionize our understanding of the blood clotting cascade. Alternatively, they can also greatly obscure otherwise clear physiological picture known for decades.

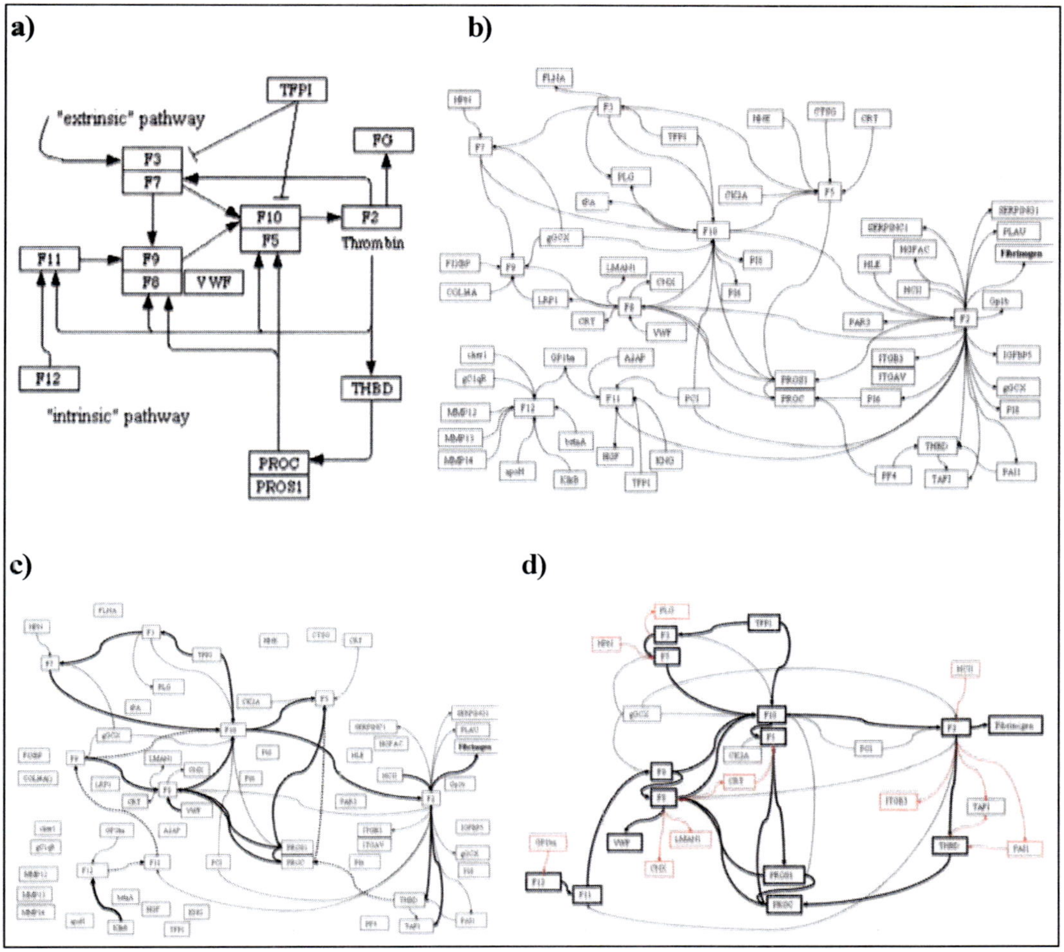

Figure 7-5. Clotting cascade: from "scratch"? a) Basic outline ("core") of the molecular circuitry of the clotting cascade (using data from KEGG; fibrinolytic and serpine branches removed); b) The unstructured array of the protein-protein interactions; c) The array after expert analysis. Thick arrows illustrate the essential links present, the thick dotted arrows the essential links that are missing; d) an expert-made reconstruction of the core of the clotting cascade illustrates an extended molecular context of the core of the clotting cascade. As in the slide (a), fibrinolytic and serpine branches were removed to make the comparisons simpler. "New" non-trivial interactions are outlined in red.

At this point, it becomes increasingly clear that an expert analysis of these "raw data" is essential to arrive at any conclusions. The raw array of data was analyzed further by careful study of each publication reporting evidence of the protein-protein interaction. The subsequent expert analysis was based, in particular, on the classical picture of the protein-protein interaction in the cascade (Figure 7-5a). Each arc in the Figure 7-5b was assigned a weight, the scale of the weights was as follows: **1.** primary physiological function; **2.** secondary physiological function; **3.** either tertiary function (minor influence) or a hypothesized physiological role; **4.** conclusions were unclear, the study is on the brink of being reliable. The Figure 7-5c represents the results of the arc-weighted analysis; in this slide only the arcs with weights **1** and **2** (that is, essential interactions and interactions of known secondary importance) were left, the rest were removed.

The data in the Figure 7-5c can illustrate to us two things. Firstly, at least half of the interactions found in an interactomics study might be of relatively low biological significance, even if they are not false positives *per se*. Secondly, quite a number of the essential protein-protein interactions are likely to be missing. In the case of the data in the Figure 7-5c, our *in silico* results neglected a number of interactions known to be essential for the consistent operation of the clotting cascade. Thus, F2 lacks interactions with F3/F7 which are present in standard scheme of the "core", F11 lacks interaction with F12 and this is the essential interaction of the intrinsic pathway, F11 lacks interactions with F9/F8, F9 lacks essential interaction with F10 *etc.* PCI is known to inhibit PROCa, urokinase, tissue plasminogen activator, F2, F10a, kallikrein and F11a but according to the figure, it only interacts with F11, F2 and F10. At the same time, there are apparently too many interactions of the F2 (thrombin, a common proteinase) and most of these interactions do not appear to have much of biological value. By the way, checking these results against HPRD, DIP and other literature-based databases of protein-protein interactions in no way alleviated the situation.

Now, we can see the difficulty: most of the reliable protein-protein interactions were identified but not all of the *essential* interactions were. According to the Figure 7-5, we can roughly estimate the false negative rate as low as 6% while the false-positive rate is about 30%. Apparently, non-zero rates of the false negatives and false negatives will be inherent to any experimental technique. It also should be recalled that the currently available techniques have rates of the false hits much higher than 6% and 30%. However, even the rate of the false negatives as low as 6% will prevent reconstructing the molecular cascade without some previously known information, while quite moderate rate of the false positives (30%) considerably obscures the biological significance of the *entire* array of interactions.

Thus, the protein-protein interaction map, obtained in our mockup of a high-throughput experiment needs to be subjected to further expert analysis. Otherwise, such a protein-protein interaction map alone (which is, once again, will be the major result of an interactomics experiment) is next to useless. The expert analysis was done by combining the data from the *in silico* interactomics with the expertise on the circuitry of the clotting cascade. The result is presented in the Figure 7-5d. As we see, despite that many of the interactions were enumerated in the Figure 7-5b, the core of the clotting cascade remained essentially the same as the well-known "classical" circuitry (Figure 7-5a). What can be considered as a "new" result is outline of the non-trivial interactions which show, in part, the molecular context of the clotting cascade. We discuss significance of these interactions further in this chapter.

Conclusions

And what would be "the moral" of this Case Study, if only we can find it? Apparently, the meaningful (for the most part) picture presented in the Figure 7-5d was constructed relying more on the expert knowledge rather than on our *in silico* imitation of the approach of experimental interactomics. The high-throughput interactome analysis *per se* had only some secondary significance. Without the pain-staking expert analysis, the interactome analysis produced such an amount of biologically questionable interactions that they obscured the meaningful picture known for decades.

Of course, this does not mean that the interactomics or other new OMIC approaches are entirely useless. It means, however, that in order to avoid this kind of unnecessary complications, elaboration of the precise logistics and rational informational management of the functional genomics studies is essential. Other points include: a) unless the experimental techniques in interactomics will have extremely low rates of false positives and false negatives, they cannot be used for the reconstruction of the molecular cascades; b) when armed with expertise in physiology, experimental interactomics can be helpful to map the molecular contexts of the known molecular cascades; c) it is essential to stratify the significance of the data collected in the post-genomic type of databases of protein-protein interactions. It does not matter whether such a database contains data from a functional genomics experiments or results of the literature searches. This stratification should be scientifically justified.

In Silico Interactomics: Some Perspectives

Blood clotting cascade is one of the best studied molecular cascades in the modern physiology. A wealth of clinical, physiological, biochemical and cellular studies of this molecular cascade were already published. However, no studies of the cascade with the use of post-genomic techniques was ever reported and the very few available post-genomic studies do not go further than listing the proteins of the cascade as belonging to some vague "plasma proteome". The Case Study entitled "*In silico interactomics: an inquiry*" illustrated some of the serious problems an interactome-wide study may run into even if the protein-protein interactions were analyzed by a reliable experimental technique. It also illustrates the comparative efficiencies of the purpose-oriented research and less purposeful database cramming.

Initially, the author planned to use HPRD database (http://www.hprd.org) to mimic the interactome-wide analysis in the Case Study. HPRD initiative is a laudable attempt of bringing in the already available information from the decades-long research into post genomic type of study. Undoubtedly, this kind of databases might be a useful tool for verification of the functional genomics experiments as, at the very least, this database contains references to interactions experimentally verified in single-protein studies. As such, the data compiled into HPRD and other similar database can be considered as representing as a model of a functional genomics (interactomics) study of the human proteome. However, to what extent these data would allow to actually reconstruct the known molecular cascades?

Unfortunately, though being a laudable attempt, HPRD and similar databases of the protein-protein interactions were compiled without much regard to primary, secondary and tertiary roles of the particular protein-protein interactions in the biological processes. In terms of the graph theory, the description of the protein-protein interactions, say, in HPRD is largely incomplete since it lacks both the node weights (relative biological significance of the proteins) and the arc weights (relative biological significance of the protein-protein interaction). Another significant drawback of large databases of the sorts is that information essential for a particular purposeful study can be missing while non-essential information will be overabundant. In relation to the data in HPRD, how reasonable would be, for instance, to include the data on interactions of common chaperones (such as interaction of hsc70 with F2) since chaperones, by definition, are involved in *highly non-specific* protein-protein interactions? Also, the HPRD entry for F2 contains over 40 different interaction partners and if the data from HPRD would be used in our Case Study, the resulting picture would come out twice as confusing in comparison to the one we presented in the Figure 7-5b. Accordingly, it was decided to make more careful analysis of the literature data instead of re-chewing the data from HPRD and other databases of the kind.

Apart from the conclusions we have formulated in the Case Study and the list of the "new" non-trivial interactions which we consider below, the results of this Case Study can also be quite illustrative as for the usefulness of the general network properties for description of the biological properties of already known molecular cascades. Currently, there are attempts to propagate the network analyses as a new word in biology and to tout them as "foundation of the systems biology". The network robustness and other general parameters are, undoubtedly, psychologically attractive math concepts that bear an air of being universally applicable. Alas, there is a profound lack of understanding, especially in the newly-grown community of the systems biology adepts that molecular cascades are ***not***, actually, some nets made of pretty identical ropes (such it is the case of, say, a fishing net). At best, a fishing net, computer class or Internet can only serve only as crude *analogies* for molecular cascades. At the same time, the general network parameters such a net robustness are much better suited for describing mechanical networks (fishing net) and/or structures similar to Internet rather than for describing the function of molecular cascades.

Let's illustrate these general considerations with just a few examples. Previously, we mentioned several concepts central in the network-based analysis. These concepts included, in particular, heuristically defined pathways, network robustness and pathway redundancy. If we read through the Case Study again and look at the accompanying diagrams, it'll be obvious that even with a largely reliable map of protein-protein interactions, the *heuristically defined pathways* will be wildly off the mark if an attempt to define them will be done by mechanically applying some set of rules without an actual human expert. For example, the *robustness of the network* presented in the Figure 7-5 will not change much if F7 be deleted. This protein, however, is essential for the activation of the entire cascade inside the biological system (blood). We can calculate *pathway redundancy* between, for example, F9 and F10 but how useful will these calculations be since the initial map (Figure 7-5b) lacks the essential protein-protein interaction F9-F10 which is essential for the operation of this entire network of protein-protein interactions? And yet another example: we previously cited a study that analyzed the relation between the position of mutations in yeast metabolic network and their lethality [30]. On the base of network analysis it was concluded that lethality corresponds to the lack of alternative paths in the perturbed network thus underlining the importance of marginal metabolic networks as opposed to best-connected. If we would overgeneralize this particular conclusion of a particular study to a principle of "systems biology" and apply it to the data in Figure 7-5b, we might erroneously conclude that F2 is the least important component of the clotting cascade. These and many other examples can illustrate that application of the network models to the molecular cascades should be taken on with extreme caution. In particular, a model of a molecular cascade that lacks even a very few key interactions might be acceptable from the point of view of the network analysis (it would have an "acceptable degree of pathway redundancy" *etc*) but it hardly be useful for any kind of practical biomedical research.

One of the positive results of an interactomics study, as the Case Study illustrated, can be elucidation of the molecular contexts of the known molecular cascades. Not all of the interactions remaining in the Figure 7-5d are non-trivial. Surface-accessible CK2A (casein kinase II) phosphorylation sites are present in the majority of proteins thus a particular interaction CK2A-F5 is not likely to be of any specific importance. A number of the clotting proteins are known to contain the Gla-domains and these Gla-domains, by definition, interact with gCCX carboxylase thus interactions of with the gCCX carboxylase are inevitable. PCI inhibitor can interact with quite a lot of proteinases and not only with those few mentioned in the Figure.

Examples of the non-trivial interactions include the interactions between the clotting and fibrinolytic branches of the coagulation cascade, the interactions across physiological systems and the interactions related to the protein secretion. Since these interactions correspond to additional feedback control loops, elucidating these interactions will be important for future qualitative and quantitative modeling of the coagulation cascade.

Interactions between the clotting and fibrinolytic branches of the coagulation cascade are related to F2, F3 and TAFI/CPB2. Interaction of PAI1 (plasminogen activator inhibitor 1) with tissue plasminogen activator is a major control point in the regulation of fibrinolysis while interaction of the PAI-1 with F2 might inhibit the clotting branch. The coagulation factor F3 (TF, tissue factor) is involved in fibrin formation, atherogenesis and angiogenesis. F3 can almost totally inhibit fibrinolysis by binding in equimolar concentrations to plasminogen [34]. Interaction between TAFI, THBD (thrombomodulin) and F2 provides an explicit molecular connection between the clotting and fibrinolytic cascades so that clotting activity down-regulates the fibrinolysis [35].

Interactions across physiological systems include interactions with integrins. As we mentioned before (Chapter II), regulation of vascular homeostasis depends upon cooperative activities between cells of the vessel wall and the clotting cascade. A direct interaction between integrin alpha5/beta3 located on endothelial/smooth muscle cells and F2 (prothrombin) may play a significant role in the maintenance of hemostasis [36]. Binding of the activated F12 to GPIb-alpha is required for the inhibition of thrombin-induced platelet aggregation so the levels of activated F12 can also control the extent of platelet aggregation in vivo [37] (providing, by the way, another indication of the important physiological role of F12 in blood coagulation, see the Case Study in the Chapter VI).

Protein secretion is another aspect of the interactions between the physiological systems. As protein levels are essential for the physiological activity of the corresponding cascades, protein secretion is essential for the proper functioning of the molecular cascades. Examples of such protein-specific interactions include interactions of F8 with CNX and LMAN1. Calnexin (CNX) and calreticulin (CRT) preferentially interact with proteins containing N-linked oligosaccharides. CNX and CRT binding is required for efficient F8 degradation and secretion while F5 does not require CNX interaction for efficient secretion [38]. An interaction between LMAN1 and F8, mediated via mannose-containing N-linked oligosaccharides in the B-domain of F8, appears to be necessary for efficient secretion of both F5 and F8 [39].

INTERACTOMICS: WHERE CAN WE GO?

From the examples we presented in this chapter, the reader might well get the impression that there would be considerably less problems with metabolic pathways in the post-genomic studies of reactome than there would be with the study of molecular cascades in the interactome studies. Indeed, apart from thousands of valuable pieces of particular information, there are, as of yet (2006), no reliable computational methods that would allow to treat the complex networks of protein-protein interactions as functionally meaningful units. Albeit classical chemical kinetics still holds a promise of being properly applied to study molecular cascades, no such proper application had ever being made. A mere block diagram of a molecular cascade complemented by an expert-written detailing text still remains the best available description of the functional peculiarities of particular molecular cascades. Moreover, the large databases of the protein-protein interactions culled from literature (such as HPRD, MIPS, DIP *etc*) do not appear to be much of a help in outlining the *functional* aspects of interactome and only obscure physiological significance of interactions with large amounts of irrelevant data. Thus, it is a legitimate question to ask, what's the alternative?

There is high degree and fascinating complexity of integration of the systems in the human body that, generally, precludes dissection into some "key parts". The researchers in the post-genomic biology do want to avoid facing the painful effects of the "where-is-the-fly" syndrome (Introduction). However, introducing general but one-sided concepts such as "biological networks" is hardly a description that reflects any significant functional aspects of this complex integration. So, the one of the most viable alternatives, as the author sees it, is the adequate use of human expertise within the framework of the purpose-oriented bioinformatics. As the Case Studies throughout this volume continue to illustrate (hopefully), a multidisciplinary expertise coupled with an efficient search of the databases of publications (PUBMED, EMBASE and others) can provide meaningful results that are both of fundamental and of practical significance. In addition, databases like "Reactome" (http://www.reactome.org/, contains data both on metabolic pathways and molecular cascades) that operate in terms of physiological cascades will potentially be of considerably more use than huge but loose collections of protein-protein interactions slammed together without any regard to physiology, either *in silico* or *in vivo*.

It is also our hope that we were able to illustrate several essential points that should be taken into account while planning and analyzing studies in interactomics. In particular, straight-forward usage of the network models of the molecular cascades should be cautioned against. To author's knowledge of the known physiological circuitry, however limited it might be, the functioning, say, of the blood coagulation cascade bears a considerable resemblance to a game of chess (we give more attention to the chess in the Volume VI "Neuroscience and AI"). There is very little resemblance, if any, between the function of this molecular cascades and some abstract mesh of ropes which is implied in the current formulations of the network-based models. Using the chess as an analogy, it would be a naïve idea to try to redefine a game of chess in such overly general terms as "networks". Moreover, it would be entirely idiotic to attempt to engineer a computer program to play chess based on some "network" model. As we remember, in the game of chess there are definite pieces, of definite value and of very specific function. These pieces are to be *definitively* used: in chess, one just can't move a piece at random without engendering considerable losses in the next few steps. As suggests the decades-long research in experimental physiology, the situation looks largely the same with the molecular cascades.

CHAPTER SUMMARY

The **interactome** of a biological system is the entire array of protein-protein interactions pertaining to that system while the **reactome** is the entire array of all the chemical reactions. We call the basic functional units of the interactomes **molecular cascades** and the basic functional units of the reactome **metabolic pathways**. Both molecular cascades and metabolic pathways feature several inherent peculiarities such as system-specific occurrence, system-specific implementations and multiple biological functions of the proteins involved. A **metabolic context** of a particular reaction or a compound is an efficient way of representing the data on metabolic pathways, especially for answering questions of biomedical importance. Moreover, the concept of metabolic contexts allows direct transition from the qualitative to the **quantitative models** of the metabolic pathways by means of the **classical chemical kinetics**. Albeit classical chemical kinetics also holds a promise for the modeling of the molecular cascades that are based mostly on the protein-protein interactions, the current results in the area are largely controversial. The current models of the **biological networks** are over-generalized and are no substitute for the classical functional descriptions of the known molecular cascades. However pain-staking it might seem, the adequate use of the human expertise within the framework of the purpose-oriented bioinformatics appears to be the only viable alternative to the network models.

REFERENCES

[1] Roitt I, Brostoff J, Mosby D. *Immunology 5th edn*, L Crowe (ed), London, 1998, pp. 99-211.

[2] Jansen H, Verhoeven A, Sijbrands E. Hepatic lipase: a pro- or anti-atherogenic protein? *J Lipid Res* 2002;43:1352-62.

[3] Kim J, Dang C. Multifaceted roles of glycolytic enzymes. *Trends Biochem Sci* 2005 Mar;30:142-50.

[4] Torshin IY. Activating oligomerization as intermediate level of signal transduction: analysis of protein-protein contacts and active sites in several glycolytic enzymes. *Front Biosci* 1999;4:557-70.

[5] Klee G. Cobalamin and folate evaluation: measurement of methylmalonic acid and homocysteine vs vitamin B(12) and folate. *Clin Chem* 2000;46:1277-83.

[6] Mikael L, Genest J, Rozen R. Elevated homocysteine reduces apolipoprotein A-I expression in hyperhomocysteinemic mice and in males with coronary artery disease. *Circ Res* 2006;98:564-71.

[7] Oberbeck R. Catecholamines: physiological immunomodulators during health and illness. *Curr Med Chem* 2006;13:1979-89.

[8] Kukreja R, Datta B, Chakravarti R. Catecholamine-induced aggravation of aortic and coronary atherosclerosis in monkeys. *Atherosclerosis* 1981;40:291-8.

[9] Stampfer M, Malinow M, Willett W, Newcomer L, Upson B, Ullmann D, Tishler P, Hennekens C. A prospective study of plasma homocyst(e)ine and risk of myocardial infarction in US physicians. *JAMA* 1992;268:877-81.

[10] Selhub J, Jacques P, Bostom A, D'Agostino R, Wilson P, Belanger A, O'Leary D, Wolf P, Rush D, Schaefer E, Rosenberg I. Relationship between plasma homocysteine, vitamin status and extracranial carotid-artery stenosis in the Framingham Study population. *J. Nutr* 1996;126:1258S-65S.

[11] Sutton-Tyrrell K, Bostom A, Selhub J, Zeigler-Johnson C. High homocysteine levels are independently related to isolated systolic hypertension in older adults. *Circulation* 1997;96:1745-9.

[12] Heux S, Morin F, Lea RA, Ovcaric M, Tajouri L, Griffiths L. The methylenetetrahydrofolate reductase gene variant (C677T) as a risk factor for essential hypertension in Caucasians. *Hypertens Res* 2004;27:663-7.

[13] Mendis S, Athauda SB, Naser M, Takahashi K. Association between hyperhomocysteinaemia and hypertension in Sri Lankans. *J Int Med Res* 1999;27:38-44.

[14] Mayer E, Jacobsen D, Robinson K. Homocysteine and coronary atherosclerosis. *J Am Coll Cardiol* 1996;27:517-27.

[15] Harker L, Slichter S, Scott C, Ross R. Homocystinemia, vascular injury and arterial thrombosis. *N Engl J Med* 1974;291:537-43.

[16] Harker L, Ross R, Slichter S, Scott C. Homocystine-induced arteriosclerosis. The role of endothelial cell injury and platelet response in its genesis. *J Clin Invest* 1976;58:731-41.

[17] Harker L, Harlan J, Ross R. Effect of sulfinpyrazone on homocysteine-induced endothelial injury and arteriosclerosis in baboons. *Circ Res* 1983;53:731-9.

[18] Zhu B. On the mechanism of homocysteine pathophysiology and pathogenesis: a unifying hypothesis. *Histol Histopathol* 2002;17:1283-91.

[19] Clarke S, Banfield K. S-adenosylmethionine-dependent methyltransferases. In: Carmel R, Jacobson DW, eds. *Homocysteine in Health and Disease*. Cambridge: Cambridge Press; 2001.

[20] Papin J, Price N, Wiback S, Fell D, Palsson B. Metabolic pathways in the post-genome era. *Trends Biochem Sci* 2003;28:250-8.

[21] Reed M, Nijhout H, Neuhouser M, Gregory J, Shane B, James S, Boynton A, Ulrich C. A mathematical model gives insights into nutritional and genetic aspects of folate-mediated one-carbon metabolism. *J Nutr* 2006;136:2653-61.

[22] Kelly K. *Out of Control: The New Biology of Machines, Social Systems and the Economic World.* Perseus Books Group NY 1995

[23] Breitlinga R, Hoeller D. Current challenges in quantitative modeling of epidermal growth factor signaling. *FEBS Lett* 2005;579:6289–6294.

[24] Stränge K. The end of "naive reductionism": rise of systems biology or renaissance of physiology? *Am J Physiol Cell Physiol* 2005;288:C968-74.

[25] Cesareni G, Ceol A, Gavrila C, Palazzi L, Persico M, Schneider M. Comparative interactomics. *FEBS Lett* 2005;579:1828–1833.

[26] Alm E, Arkin A. Biological networks. *Curr Opin Struct Biolo* 2003;13:193–202.

[27] Klemm J, Schreiber S, Crabtree G. Dimerization as a regulatory mechanism in signal transduction. *Annu Rev Immunol* 1998;16:569-92.

[28] Ma H, Zeng A. The connectivity structure, giant strong component and centrality of metabolic networks. *Bioinformatics* 2003;19:1423-30.

[29] Nielsen J, Oliver S. The next wave in metabolome analysis. *Trends Biotechnol* 2005;23:544-6.

[30] Palumbo M, Colosimo A, Giuliani A, Farina L. Functional essentiality from topology features in metabolic networks: a case study in yeast. *FEBS Lett* 2005; 579:4642-6.

[31] Guet C, Elowitz M, Hsing W, Leibler S: Combinatorial synthesis of genetic networks. *Science* 2002, 296:1466-1470.

[32] Berhane B, Zong C, Liem D, Huang A, Le S, Edmondson R, Jones R, Qiao X, Whitelegge J, Ping P, Vondriska T. Cardiovascular-related proteins identified in human plasma by the HUPO Plasma Proteome Project pilot phase. *Proteomics* 2005;5:3520-30.

[33] Misek D, Kuick R, Wang H, Galchev V, Deng B, Zhao R, Tra J et al. A wide range of protein isoforms in serum and plasma uncovered by a quantitative intact protein analysis system. *Proteomics* 2005;5:3343-52.

[34] Fan Z, Larson P, Bognacki J, Raghunath P, Tomaszewski J, Kuo A, Canziani G, Chaiken I, Cines D, Higazi A. Tissue factor regulates plasminogen binding and activation. *Blood* 1998;91:1987-98.

[35] Bajzar L, Morser J, Nesheim M. TAFI, or plasma procarboxypeptidase B, couples the coagulation and fibrinolytic cascades through the thrombin-thrombomodulin complex. *J Biol Chem* 1996;271:16603-8.

[36] Byzova T, Plow E. Activation of alphaVbeta3 on vascular cells controls recognition of prothrombin. *J Cell Biol* 1998;143:2081-92. PMID: 9864377 [PubMed - indexed for MEDLINE]

[37] Bradford H, Pixley R, Colman R. Human factor XII binding to the glycoprotein Ib-IX-V complex inhibits thrombin-induced platelet aggregation. *J Biol Chem* 2000;275:22756-63.

[38] Pipe S, Morris J, Shah J, Kaufman R. Differential interaction of coagulation factor VIII and factor V with protein chaperones calnexin and calreticulin. *J Biol Chem* 1998;273:8537-44.

[39] Cunningham M, Pipe S, Zhang B, Hauri H, Ginsburg D, Kaufman R. LMAN1 is a molecular chaperone for the secretion of coagulation factor VIII. *J Thromb Haemost* 2003;1:2360-7.

Chapter VIII

METABOLOME ANALYSES AND DRUG DESIGN

INTRODUCTION

This chapter finishes our trilogy that deals, in some or other way, with the methods of functional genomics (Chapters VI, VII and VIII). The present Chapter is relatively short since at present metabolomics, perhaps, is the least developed areas of the functional genomics (Figure 6-1). Only a few of the attempts of the actual studies of metabolome has being made and the area is largely hypothetical than experimental. That is, it cannot be denied that a biological system does have certain metabolome (Figure 1-3). However, the direct methods of the experimental study of that metabolome are largely at the age of their infancy. Accordingly, the bioinformatics-based metabolome inferences might have a prominent place in the current metabolome studies.

Metabolome is the entire array of so-called "small molecule" metabolites of the biological system (cell, tissue, organ). Apparently, the notion of metabolome does not include macromolecules (peptides, proteins, DNA/RNA, carbohydrates) and xenobiotics. The metabolome is inseparable from the reactome and vice versa (Figure 7-1) and the metabolite compounds are the vortices of the metabolic contexts (Chapter VII). From the point of view of the levels of function, the metabolome fits in somewhere in between the protein levels and the tissue metabolism (Figure 1-5). As protein levels, the small molecule metabolites such as amino acids, fatty acids, nucleotides, folate, homocysteine and many, many others represent biochemical variables (Chapters I&II). All of these peculiarities of the general function of metabolome suggest that metabolome studies can be of considerable assistance in the biomedical studies. In particular, metabolome studies of the pathological microorganisms can be of considerable use for the rational drug design. We present a methodology for the metabolome-based drug design in the final Case Study of this book (*A Case of Malaria Drugs and In Silico Metabolomics*) after providing background on the biology, medicine and functional genomics of the malaria parasite.

METABOLOMICS: A REVIEW

Metabolomics is defined as a global analysis of all metabolites in a biological sample. As we already mentioned, studies of metabolome is a new area rife with suggestions, hypotheses and, if the reader would care to make a few searches in abstract databases, there seems to be more reviews than actual experimental studies in human metabolomics (as of 2006). Despite this, the number of metabolomics-related publications in the last three years is almost four times the total number of publications on the topic in all the preceding years suggesting, if nothing else, a certain degree of interest.

Metabolomics is an extension of the existing methods that look at some specified target compounds in the tissues of interest. The most immediate predecessor of metabolomics is the metabolite profiling that first appeared in late 1950s [1,2]. Apparently, metabolomics and metabolite profiling are extensions of the clinical analyses of biochemical variables commonly known as "lab tests". As in the case of the biochemical lab tests in clinical science, the primary utility of metabolomics can also be rationalized in terms of specific "normal ranges" of a set of biochemical variables.

There are, nevertheless, at least two important differences between the lab tests as we know them and the metabolomic approach: the scope and the technological basis of the study. Firstly, the *scope* of metabolomics is considerably wider than that of a panel of lab tests. While lab tests analyze only the metabolites or other biomarkers that are *known* to be associated with particular conditions, metabolomics aims at establishing the presence and the levels of metabolites in the entire biological system (cell, tissue, organism). The data on all of the metabolites form a metabolic profile of the system. A metabolic profile can include all of the metabolites that can be registered or can be crafted in accordance with a particular purpose of biomedical research.

Secondly, metabolomic analyses differ from a panel of lab tests because in metabolomics many compounds of diverse chemical nature are analyzed using a uniform *technological basis*. Since, in perspective, metabolomics is supposed to enable the rapid and accurate analysis of metabolites, advanced physical techniques, such as high-throughput chromatography (gas or liquid), mass spectrometry and nuclear magnetic resonance are being employed as the major experimental tools. These techniques are routinely used in physics and chemistry and are known to allow detection of very diverse chemical structures.

The currently formulated ideas of the general approaches in metabolomics include metabonomics (analysis of the metabolic response to drugs), metabolic fingerprinting (establishing of an overall fingerprint of the metabolites that are produced in a system, no specific metabolites), metabolic control analysis (response of the metabolome to perturbations in the proteome) and metabolic profiling (analysis of the presence and levels of specific metabolites) [3,4]. Metabolic profiling can be restricted to metabolites involved in a particular metabolic pathway and in this case it's called "metabolite target analysis". Potentially, metabolic profiling can also be used for the discovery of new metabolic pathways.

Unlike the transcriptome profile, metabolome profile (and, to some extent, also the proteome profile) is much closer to the clinical lab tests. Indeed, the intermediate location of the metabolome in the diagram of the levels of function (Figure 1-5) does suggest the usefulness of the metabolome data in bringing together clinical science, physiology and genetics. For example, most of the common factors of cardiovascular risk we mention in the Figure 2-3 will exert some influence the metabolome. Some of these factors, such as diet, age, gender, drugs and diurnal variation are indeed known to influence the metabolic profiles [5]. Of these factors, studies involving nutrients and drugs represent are of especial interest for post-genomic medicine.

Interactions of clinical variables with **nutrients** comprise an important component of the personalized medicine. Studies of the nutrient-gene interactions are of definite interest for the post-genomic studies in medicine (Volume II). In particular, effects of the common nutrients such as fats, amino acids and carbohydrates on the metabolism can be studied by using tracers [6]. With the advent of NMR and mass spectrometry, the cumbersome radioactive tracers can be replaced by stable isotope tracers such as ^{13}C. Tracer-based metabolomics would focus on metabolite distribution and flux determination to estimate the extent of synthesis through various pathways thus providing a physiological basis for the future studies of the nutrient-gene interactions.

Studies of the nutrient-gene interactions can include not only the common nutrients but also microelements. Several examples we considered previously (Chapter I, "Disease etiology and medical diagnostics") remind us that deficiencies of microelements, nutritional or otherwise are important health factors. The same can be said about overabundance of microelements that toxicology deals with. Microelement deficiencies are often overlooked in modern clinical practice and it is expected that post-genomic studies, especially metabolomics, can be of help in solving several major problems of the microelement biology and medicine. These problems include the improvement of the diagnostics of the microelement deficiencies, quantitative determination of the requirements for microelements, biological basis for the variation in requirements and the complex interplay between genetic, epigenetic and dietary factors that determine nutritional phenotype [7]. We return to the microelement biology in another volume of the present book series (Volume IV).

It is the major assumption behind post-genomic studies related to medication **drugs** that more systematic study of the biological systems will allow to achieve a higher degree of physiological precision in the drug design and application. It is no secret that the drug design was (and currently is) a highly empirical area. Even despite the modern methods of biochemistry (Volume IV) and biophysics (Volume V) that can serve as the basis for rational drug design, the actual procedures of the drug design often involve screening of tens of thousands of compounds and this screening is often performed on entirely random basis. We employ an *in silico* approach to metabonomics in The Case Study at the end of this chapter. Metabolomics and metabonomics might show some promise in the area of toxicology and drug development. Theoretically, during both preclinical screening and mechanistic exploration, metabolic profiling offers a rapid and non-invasive method to gather toxicological information, method which is also robust and reproducible [5].

Examples of the actually made experimental studies in human metabolomics (not mentioning a thing concerning the study quality) are very few. These studies often appear to be unpublished in academic journals. For example, a certain web site (http://www.metabolon.com/index.htm) presents an intriguing example (without presenting, however, any key details) that sounds like an outstanding result. In this case, the researchers sought for biomarkers of premature birth, the leading cause of neonatal fatalities. Using a preliminary (unblinded) metabolomics study as a training set, the researchers conducted a second (blinded) metabolomics study. The prediction scheme allowed classification of the patients into the clinical groups with almost 89% accuracy. A number of metabolites were identified as diagnostic biomarkers for the preterm conditions which included certain amino acids, carbohydrates and lipids.

The analysis of the values of lab test variables is a major tool in modern clinical diagnostics. A single lab test is insufficient to draw any defendable clinical conclusions. Hence, metabolomics is a promising field for numerous clinical applications. There are several important tasks of technological and informational management that need to be solved before metabolomics becomes a valuable instrument of the clinical science. Much of the up-to-date work on metabolomics has been of technological and qualitative in nature rather than representing any attempts to assess the flux of individual metabolites. Nuclear magnetic resonance and mass spectrometry are held to be the most common techniques to be used in metabolomics. In particular, identification of specific metabolites by NMR can be problematic in complex mixtures and the development of specific NMR techniques (2D NMR) that can overcome the problem is underway. It is essential that the high-throughput application of both NMR and MS be developed to such a level of accuracy (both qualitative and quantitative) so that they can indeed be routinely applied for the metabolome-wide studies. This task, in addition to the development of the technical procedures *per se*, also requires a uniform definition of the appropriate data standards and the development of the appropriate libraries of spectra [2].

Apart from the significant technical difficulties in the development of metabolomics, the proper physiological expertise is also essential for the metabolomics studies not to become just another huge waste of resources. Unlike most of the researchers currently involved in the transcriptomics and proteomics (who, generally, tend to disregard the wealth of information that classical physiology offers), the researchers in metabolomics appear to recognize the need for the adequate expertise in physiology and medicine as well as the need for proper informational management of the research [1-4].

A Few Pathway Databases

In metabolomics, a set of the substances that exist in a particular metabolome can be built using experimental techniques (NMR, MS and chromatography). Since metabolome is inseparable from the metabolic pathways that comprise the reactome, a theoretical approximation of the metabolome constituents can also be generated using methods of bioinformatics. Accordingly, using a properly organized database of metabolic pathways for the biological system of interest, it can be possible to make a number of inferences prior to experimental studies in metabolomics. Although studies in metabolomics will require various types of databases [3], the databases that contain the data on the metabolic pathways/reactomes are of primary importance for metabolomics, both theoretical and experimental. In this section, we mention the databases that can actually be used during feasibility assessment, design or elaboration of various biomedical applications: KEGG [8], MetaCyc/BioCyc/HumanCyc [9] and the Reactome [10].

KEGG stands for "Kyoto Encyclopedia of Genes and Genomes" [8]. The KEGG resource (http://www.genome.jp/kegg/) includes, in particular, data on individual genes (KEGG GENES), small molecule metabolites (KEGG COMPOUND) and metabolic diagrams or protein-protein interaction networks (KEGG PATHWAY). The KEGG GENES component is a collection of gene for the complete genomes (31 eukaryotes, 235 bacteria, fall of 2005). There are over one million gene entries listed in KEGG GENES. Other KEGG components include ENZYME (enzyme nomenclature), REACTION (reaction equations) and DRUG (drug information). The KEGG PATHWAY database is a collection of manually drawn pathway maps for metabolism, genetic information processing and some molecular cascades. Database appears to be based on an extensive survey of published literature. Although KEGG is not one of the first databases of the metabolic pathways, it definitely contains a lot of useful reference material. In this book, we have used the KEGG diagrams on a number of occasions (for instance, in Chapters I and VII).

Unlike KEGG, in databases associated with the **MetaCyc** project [9] (http://MetaCyc.org/) we find not so much metabolic pathways but rather the entities we call "metabolic tracks" (see Volume IV). In brief, a metabolic track is a route along which a certain substance is being processed to obtain some other substance as the outcome. Like KEGG, MetaCyc serves as a metabolic encyclopedia, containing a collection of non-redundant pathways central to small-molecule metabolism and which were reported in experimental literature. Most of the pathways in MetaCyc occur in microorganisms and plants, although animal pathways are also represented. There are some differences between the KEGG and MetaCyc. In particular, MetaCyc contains some additional comments that describe individual metabolic pathways and enzymes while KEGG has much fewer comments of the kind but rather additional references to the relevant databases. MetaCyc resource often cites the primary literature sources from which pathway and enzyme data were obtained while the references to the particular sources for the KEGG data appear to be less clear. The KEGG database is more oriented to be used by an expert since it does not actually contain predictions of particular pathways in particular organisms but simply colors a generalized KEGG map for a pathway to indicate which enzymes within a map are present within the corresponding genome. Limitations of the MetaCyc approach (which are true, by the way, in the case of any other pathway database) include incompleteness (some known or unknown metabolic pathways are likely to be missing), false-positive and false-negative pathway predictions as well as the absence of information about the cell compartments, cell type and tissue type in which a particular pathway occurs [9]. Accordingly, presence of a pathway could be inferred incorrectly because the enzymes that make up a pathway are never expressed in a common location [9,11,12].

In conjunction with additional software, MetaCyc can be used to predict metabolic networks using a list of annotated sequences from a completely sequenced genome. The **BioCyc** collection features pathway predictions for 160 eukaryotic and prokaryotic species with fully sequenced genomes [11]. In case of bacterial genomes, operon predictions are also available. **HumanCyc** [12] is a part of BioCyc which represents a prediction of the human metabolic map (http://HumanCyc.org/). This prediction includes 2709 human enzymes involved in 896 reactions; about 600 of the enzymes were assigned to 135 predicted metabolic pathways. The BioCyc databases are one of the most suitable for the purposes of comparative research since 1. they were uniformly generated using the same computational methodology and 2. they were generated by human experts and not by merely running a program. As such, BioCyc provides a uniform base for the reactome/interactome metabolome comparisons among various genomes. For example, using Metacyc it was possible to identify probable omissions in the human genome annotation in the form of essential enzymes missing within the predicted pathways [12]. A comparative analysis of the predictions of metabolic pathways for human, E.coli and A. thaliana has allowed identification of 35 pathways potentially shared among all three organisms [11,12]. Our metabolome comparisons in *The Case of Malaria Drugs and In Silico Metabolomics* we present later in this chapter were based on the HumanCyc and PFAcyc components of the BioCyc.

As we already mentioned, a block diagram of the molecular cascade complemented by a detailing text still remains the best theoretical description of the functional peculiarities of particular molecular cascades. Unlike high-throughput but low-intelligence techniques, this approach, at least, does combine an adequate use of human expertise with the computational techniques. **Reactome**, a curated database of biological pathways (http://www.reactome.org/) is an example of a rationally built database of the sort [10]. Unlike the two previously mentioned database complexes, the KEGG and the MetaCyc, the basic unit of the Reactome database is a reaction. The reactions are then grouped into causal chains to form pathways. Again, the reaction unit in Reactome is not restricted to purely metabolic reactions but is broadly defined as "any event that converts inputs to outputs, where inputs and outputs are physical entities such as small molecules, proteins, lipids or nucleotides, or complexes of these". This definition, apparently, includes intermolecular binding, dissociation, complex formation, translocation, polymerization and even conformational changes. Thus, the notions of the "reactome" and "interactome" are largely diffused as there are no distinct differences between the entities we call "metabolic pathways" and "molecular cascades" (Chapter VII). In Reactome database, both are referred to by adopting a more generic term "biological pathways". The current version of the database (as of 2006) covers "biological pathways" involved in various cellular processes such as cell cycle, DNA repair and replication, gene expression cascades, mRNA processing, metabolic pathways involving sugars, ethanol, amino acids, nucleotides and lipids and others. Albeit the database is compiled with the help of the experts in the field, the automated interferences based on the expert-built diagrams and the data from newly sequenced genomes, also presented in the Reactome project, should be taken with much caution. Indeed, one can hardly imagine presence of the blood coagulation cascade in unicellular organisms, despite whatever automated predictions made on the base of genomes..

In sum, although incompleteness, false-positive pathway predictions and absence of the information about the biological location (compartments, cell types, tissues) can be considered as common limitations of any of the currently available pathway databases, these bioinformatic resources often prove to be of help in actual research projects, both of fundamental and practical significance. As practical experience suggests, data from various pathway databases need to be analyzed together along with the appropriate articles from

PubMed. That's the best strategy to avoid being trapped by database errors and by other unpleasant surprises that do tend to occur when dealing with large databases.

Malaria: The Current Situation

"...to get a chance of moving in safety from the stockade to the two-pointed hill, there to be clear of malaria and keep a guard upon the money..."

R.L. Stevenson, Treasure Island.

How the methods of functional genomics and, in particular, of metabolomics can be used for practical biomedical tasks? We use malaria as an example to illustrate an application of a metabolome-wide analytical strategy in *The Case of Malaria Drugs and In Silico Metabolomics*. The rest of this chapter prior to the Case Study provides some background on the medicine, biology and functional genomics of the malarial parasite.

Malaria is caused by *Plasmodia* protozoans *P.falciparum, P.vivax, P. ovale and P.malariae*. Infection with any of these four types of protozoans accounts for the most of the 300-500 million of malaria infections *per* year. Infection with *P.falciparum*, in particular, carries a substantial risk of death. These protozoans are transmitted through the bite of an infected *Anopheles* mosquito. Malaria is endemic in parts of Asia, Africa, Central and South America and Oceania and is characterized by extreme exhaustion associated with paroxysms of high fever, sweating, shaking chills, and anemia (http://www.who.int/topics/malaria/en/).

According to WHO, the current situation with the malaria management is far from being optimal. In particular, the growing resistance of the *Plasmodia* parasites to the currently common drugs is a major concern for public health (http://www.who.int/drugresistance /malaria/en/). For example, no country in sub-Saharan Africa had the projected "substantial decline" in the disease largely because of the advance of chloroquine-resistant malaria. Chloroquine is an inexpensive, safe and effective drug that was effective during the preceeding decades but gradually looses its effect on the resistant strains [13]. Main reason for the increase of malaria-related mortality is the continued deployment of ineffective antimalarial drugs in the face of increasing resistance [14]. Taking into account that any new approved drug usually gets developed in 6..10 years and at costs approaching 1 billion dollars [15,16], we can appreciate both sides of the sentiment expressed in the quotation we used at the beginning of this section: namely, not only "to be clear of malaria" but also "to keep a guard upon the money"...

Although the growing resistance to antimalarial drugs seems to be the major problem in tackling the disease worldwide, it is far from being the only one. There are considerable problems with the actual availability of the currently known antimalarial drugs, especially in the impoverished countries of the Asia and Africa. Apart from the problem of the fake drugs [17], the cost is often the major barrier preventing deployment of already developed new drugs, such as a promising class of the antimalarial drugs known as "artemisins" (the drugs based upon the antimalarial properties of compounds extracted from the plant *Artemisia annua,* a plant oft found in the medical treatises of the past centuries). The cost is also the major barrier for the effective usage of the drug combinations [13]. In fact, the essential value of using artemisins in combination therapies appears to be so great that WHO even requested leading pharmaceutical companies to cease production of artemisinin-only malaria treatments [18].

Paradoxically, the availability of the anti-malarial drugs is far from being limited to the poorest countries and the problem also appears to be of some concern, for example, for US [19]. While the cost is the major barrier for the poorest countries of the world, there are also considerable problems with actual using of the available drugs even in USA: neither intravenous quinine nor oral, rectal, or intravenous artemisinins have been approved by FDA nor are available in this country.

Malaria: Post-Genomic Studies

What potential do the post-genomic technologies have in respect to the malaria problem? There are, of course, numerous expectations and there is a realistic *status quo* in the area. It is expected that post-genomic studies of the malarial parasite would allow the researchers to establish, with great certainty, the weaknesses in the parasite's defenses. Then, exploiting these weaknesses, it might somehow become possible to produce more efficient drugs to which, potentially, resistance is less likely to be developed. In addition, the resistance mechanisms to the currently available drugs are far from being sketched genome-wide, let alone being fully understood [17]. Molecular genetic markers of *P. falciparum* resistance are available only for a few of the currently known drugs (see the Case Study in the Volume III) so it is also assumed that post-genomic studies can help in these major tasks.

Nevertheless, since 2002, when an international consortium presented sequence of the *P.falciparum genome,* no discernable progress has being made in these endeavors [20]. Despite the conventional cheery statements of the kind "the sequencing of the genome of *P.falciparum* promises to revolutionize the way in which malaria research will be carried out", we do need to recall that the verb "will" in English does not specify any distinct time frames: it may be decades or it may be a century. Indeed, among the 5,300 of the putative genes predicted for Plasmodium falciparum genome, at least 50% encode "hypothetical, unclassified, unknown" proteins. Given this, the standard approach of the forward/reverse genetic manipulation to elucidate the function of the hypothetical genes will be fraught with considerable difficulties and will be neither cost- nor time-efficient. Thus, a number of genomic, transcriptomic and proteomic studies of the *Plasmodia* had been undertaken.

It appears that both **genomic and transcriptomic** studies of *Plasmodia* are, for the most part, general analyses that produce rather general an information. By this I mean that this information is not practically applicable in any observable amount of time and is quite far from the actual clinical medicine. Sometimes, it even represents apparently non-falsifiable statements rather than any substantiated conclusions. Comparative genomic analysis revealed that approximately 60% of genes are shared across all *Plasmodia* [21]. The major differences in gene content among *Plasmodium* species occur between genes likely to be involved in interaction with the host immune system. *P.falciparum* contains other gene families that would encode proteins potentially involved in antigenic variation and evasion of immune responses [22]. Comparative analysis of the collections of experimental data coming from genome annotation, cDNA sequencing and gene expression studies produced long lists of genes that are differently expressed during the intra-erythrocytic stages of the parasite [21].

Since publication of the *P.falciparum* genome, several **transcriptome studies** of various life-cycle stages have been completed [23-25]. At present, it's hard to judge the significance of these and other studies dealing with the *Plasmodia* transcriptomics. There is quite a low consensus among the data obtained in the high-throughput transcriptomics studies of *Plasmodia*. The most disconcerting point about these studies is that there appears to be little correspondence between the results they present and the results from molecular biology, microbiology and physiology obtained in the previous decades of *Plasmodia* research.

With the exception of a few of the genes already known to be important for the *Plasmodia* function for at least for a decade, the current transcriptomics studies did not seem to produce the promised wealth of substantiated and practically useful information. At best, these studies confirmed a few of the bits and pieces of information which are already known from transfection and knock-out studies of *Plasmodia*. Such is the case of, for example, gene PfMDR1 implicated in the mechanism of chloroquine action/resistance [26,27], dihydrofolate reductase–thymidine synthetase, knob-associated histidine-rich protein (KAHRP), located within electron-dense knob structures on the surface of infected erythrocytes. However, at least several more dozen genes became known during the decades of research (see, for instance, the table in [20]) which were rarely found in the course of the transcriptomic studies.

At present, **proteomics** studies of the *Plasmodia* appear to produce more coherent and much wider picture that also tends to agree with the currently known data on the biology of the malarial parasites. Perhaps, the best available study of the kind was the paper entitled "A proteomic view of the *P.Falciparum* life cycle" [28]. Not only does it feature a proper informational management of the entire array of the experimental data, but also it includes accurate comparisons to the available physiological data. In this study, the researchers characterized proteomes of the four selected stages of the Paslmodium Falciparum life cycle (sporozoites, merozoites, trophozoites and gametocytes) using multidimensional protein identification. Totally, over 2,400 proteins were analyzed. It was found that most of these proteins are stage-specific since only 152 of the proteins (6%) were common to all four stages (these common proteins were ribosomal proteins, transcription factors, histones and cytoskeletal proteins). We summarized the data in the Figure 8-1. This summary, apart from its general educational value, will also serve the purposes of the present book series (since we are going to return to the subject of malarial cell biology in the Volume III). The readers interested in additional details will, undoubtedly, peruse the original publications [28-30].

What is of especial interest for post-genomic studies is that a carefully made proteome study also allows making inferences about the transcriptome and genome structure. For example, the carefully performed proteome analyses made in [28] allowed compilation of the chromosomal clusters of the co-expressed proteins. Again, this was done without involving mRNA arrays using solely proteomics data and then bioinformatic techniques to map the proteins to the genes and the chromosomes. Over 60 of the gene clusters were identified and almost a half of them contained genes encoding functionally-related proteins. For example, there were ten sporozoite-specific clusters, two clusters on chromosomes 2 and 11 highly specific to the trophozoite stage; chromosome 14 contained a cluster of four aspartic proteases co-expressed in all of the blood stages *etc* [28].

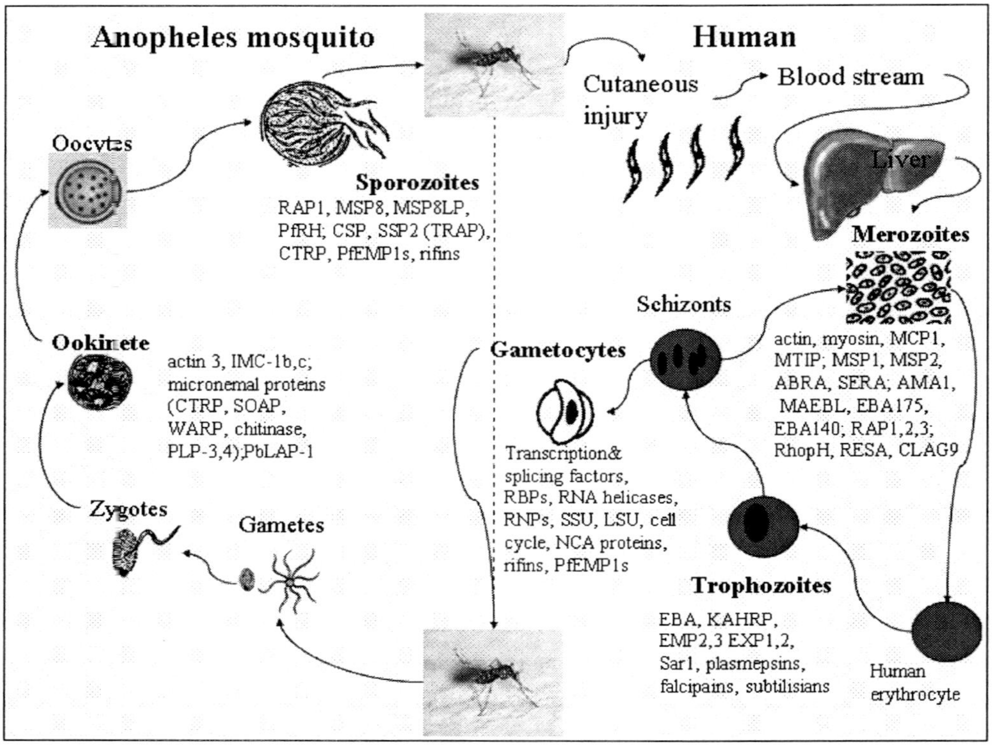

Figure 8-1. Life cycle of *Plasmodia* and the data from proteomics compatible with known Plasmodia physiology (see references in the text). The life cycle of Plasmodium species is quite complex: the parasite survives in both invertebrate and vertebrate host environments, both within cells and outside of the cells, can invade multiple cell types and evades immune responses of both human and insect hosts. Accordingly, the would-be interventional strategies based on these data will be most effective when oriented at specific life stages of the parasite. This map represents only a brief summary of the comparative proteomics throughout the cycle and, as such, in neither complete nor entirely accurate. Some acronyms in the figure might refer to particular proteins but also protein complexes, not only particular genes but also to gene groups, no differences between stage-specific and stage-abundant proteins was made *etc*.

A post-genomic study as good as the one we just cited is rather an exception than a rule. Most of the post-genomic studies, including those dealing with *Plasmodia*, merely lump together various physiological stages, processes and concepts using some arbitrary classifications and almost entirely neglect the accumulated array of the physiological data abundantly available in the specialist literature. This state of affairs once again indicates that not so much post-genomic technique *per se* is important to obtain data of fundamental and practical consequence, but rather efficient and thoughtful application of this or that post-genomic technique. The crucial components in the efficiency of the application are purpose-oriented approach, proper informational management and sufficient expertise in the physiology of interest.

It is essential to integrate both the new data from proteomics and the facts established during the decades of the physiological research of the malarial parasite [30]. The rapid development of a wide range of experimental methods in post-genomic studies will be of biological significance only if application of these methods will be kept in pace with the adequate development of the points we just mentioned. The more a post-genomic study is integrated into the existing framework of physiological knowledge, the more productive and practically applicable the results become. Sometimes, bioinformatics can be of considerable help, as the Case Study below and the following discussion might indicate.

The Case of Malaria Drugs and In Silico Metabolomics

"Almost all tropical diseases, malaria included, are largely neglected by modern medicine. Elaboration of novel drugs is very expensive and time-consuming. Many drugs have extremely adverse side effects. Could post-genomic bioinformatics be of help in this endeavor?"

Introduction

The waning efficacy of the common anti-malarial drugs points out both the need to elaborate efficient drugs and to elaborate them at a reasonable cost. A post-genomic perspective can allow an efficient elaboration of new and efficient drugs provided, of course, an efficient research strategy. In this Case Study, we'll use metabolome reconstructions and apply a technique we call differential metabolomic analysis or 'comparative metabolomics' [31]. To do so, we'll hold to the idea that both the problems of finding the drug targets and of dealing with side effects can be dealt at once (in a post-genomic manner, sort of thing): through the comparative metabolome analysis. In other words, compounds that are structurally dissimilar to those occurring in the human metabolism but considerably structurally closer to the compounds involved in PFA (*Plasmodium falciparum)* metabolism are likely to be the good starting points (or, at least, structural templates) for the new plethora of anti-malarial drugs. At the same time, they might have the least interference with the human metabolism.

Methods

For this kind of analysis, two information items are essential: the *Plasmodium* metabolome (as a list of substances) and the human metabolome. There are a number of variants of the reconstructions of both. We used here the BioCyc reconstructions the human and the *P.Falciparum* metabolomes. Briefly, BioCyc was compiled using genome data from Ensembl, Gene/LocusLink and GenBank; the Path-o-Logic software was used to generate the database and to predict metabolic pathways from functional information contained in the genomes' annotation. More detailed information on how BioCyc databases were generated is available at the BioCyc web-site. The two lists of compounds were directly obtained at the BioCyc site (www.biocyc.org), a list of 649 compounds for *P.Falciparum* and of the 950 compounds for the human metabolome. As these lists of compounds contain not only metabolites but also a number of xenobiotics, they rather should be called as "metabolome approximations". The subsequent analyses in the framework of the present Case Study included three stages: initial comparisons, analysis of the structural dissimilarities and the impact on the metabolism of *Plasmodium*.

Results

At the stage of **initial comparisons**, a raw list of 157 compounds unique to *P.Falciparum* was obtained by a mere string-to-string comparison of the two lists of compounds. Sieving out common nucleotide derivatives, simple sugars and other relatively simple metabolites such as common aliphatic acids (butyric etc), coenzyme conjugates, ethers with inorganic acids (easily hydrolysable phosphates, for the most part) gives a list of 81 compounds unique to *P.Falciparum* metabolism.

At the second stage, the **analysis of the structural dissimilarities** of the compounds unique for the *P.Falciparum* was performed in relation to the compounds in the human metabolome. At this stage, a recursive analysis of the SMILES strings that represent chemical structure was performed. After exclusion of the compounds that are over 85% structurally similar to a compound in the human metabolome (in terms of the overlap of the molecular backbones), a list of about 40 compounds unique for *P.Falciparum* metabolism was obtained. These compounds were of 250-700 Da molecular weight and included quinone derivatives, nucleosides and nucleoside-analogs, immunicillins and a number of other substances.

The third stage, the **analysis of the impact on the *P.Falciparum* metabolism**, was performed by clustering the remaining 40 compounds according to their chemical formulas and then retracing each substance to the metabolic pathway(s). The results were quite interesting. Thus, 10 out of the 40 PFA-unique compounds were in the metabolic pathway of the ubiquinone biosynthesis. This pathway, as it is well known, is a current target for the anti-malarial drug design. Further, as we already mentioned, the BioCyc "metabolomes" are not entirely metabolomes but rather approximations. Since AI (artificial intelligence) methods were used to produce both HumCyc and PFAcyc, some of the known antibiotics targeting *P.Falciparum* turned to be associated with the model of PFA metabolome and were thus identified in the final list of the compounds. These included pyrimethamine (a folic acid antagonist), sinefungin (arrests PFA growth), psicofuranine (interrupts PFA growth), coformycin (a potential antimalarial drug). About 5 out of the PFA-unique compounds were certain nucleosides/nucleoside analogs. The nucleosides and nucleoside analogs as potential anti-malarial drugs are quite plausible an idea since *P.Falciparum* is unable to synthesize purine bases and relies upon purine salvage and purine recycling. Other known antibiotics included cerulenin (antifungal antibiotic that also kills *P.Falciparum*), triclosan (common bactericide and pesticide that also kills *P.Falciparum*) and thiolactomycin (a potential antimalarial and antitrypanosomal drug).

Other compounds included several immunicillins, of which is known next to nothing (except that immunicillin-H has a clinical potential in treating autoimmune disorders and tissue rejection), specific derivatives of S-adenosylhomocysteine (which is also involved in the ubiquinone biosynthesis), and several pyrimidine and imidazole derivatives.

Of course, it would be naïve to expect any immediately applicable results from this kind of bioinformatic study. Given that genomes are partially annotated and the current metabolic pathways and metabolomes are even more incomplete, the chances of finding any significant hits are slim indeed. However, this analysis shows us how it is possible to narrow the search for the drugs in order to concentrate on the substances that can really be significant for the purposes of the rational drug design. Also, this Case Study aimed at describing *the ways* of using available information on metabolomes. Nevertheless, although this Case Study operated with a third-derivative sort of information (metabolome model that is based on MetaCyc that uses the proteome model that was based on the gene predictions), it provided an immediate link to the known drugs (which, for the most part, are analogs of the known *P.Falciparum* metabolites) as well as suggested a few more ideas for the design of the new drugs. Accordingly, the results presented in this Case Study still might be helpful for the future systematic studies based on more complete genome annotations and on more precise maps of the metabolic pathways, apart from the experimental studies dealing with the rational drug design.

MALARIA: INVOLVING BIOINFORMATICS

Apart of compilation of various databases, bioinformatics can be helpful for a number of other branches of the malarial research. It can help with the purpose-oriented extraction, analysis and synthesis of the published data on the parasite physiology, in a way we mentioned in the previous section. Bioinformatics can help with the collection and systematic review of the data on the genetic determinants of malaria susceptibility (see the corresponding Case Study in the Volume II of the present book series). And, as the *Case of Malaria Drugs and In Silico Metabolomics* illustrates, we can also apply bioinformatic techniques for metabolome-wide analyses that can assist the rational design of the novel anti-malarial drugs.

We intentionally present this Case Study and its discussion *after* we provided a summary of experimental post-genomic research. This was done in order to illustrate how a study in metabolomics, even an *in silico* one, can provide considerably more direct approach to the problems of drug design than other kinds of the post-genomic studies. While, for example, proteomic studies of *Plasmodia* (Figure 8-1) produce physiologically plausible picture which, potentially, can be of practical significance too, the approach of differential metabolomics we employed in the Case Study has considerably more direct repercussions for the rational drug design.

For example, the modern drug design often includes high-throughput screening of the compounds libraries that include hundreds of thousands of various substances. Although this blunderbuss sort of approach might be assumed to be very effective, it is clear that by gaining on the speed and high-throughput aspects of the study, we will inevitably loose on the accuracy of the entire study. In other words, the truly effective substances that can become efficient drugs can be lost in a high-sweep study of the sorts, especially if a high-throughput study is not based on a physiologically sane logistics. We give just one example to illustrate the extent of this loss which should be clear *a priori*.

As it is known from the microbiology, there are substances that kill the microbes and other parasites through different biological mechanisms and at different rates. Some substances would kill almost instantly the microbes but will be quickly eliminated from the blood stream of the patient or will have considerable deleterious effects on the human physiology. Some drugs kill a wide range of microbes and others can target specific microbes. And, for instance, there are drugs that kill a specific microorganism at considerably slower rate than the other drugs but, at the same time, these drugs allow considerably more complete elimination of the microorganism from the human tissues.

Now, imagine we are to make a high-throughput scan of about 100,000 compounds for the purposes of malaria research. Certainly, it will be largely the same amount of time allotted for analysis of each compound. This amount of time, often, will be quite restricted and arbitrary chosen. Therefore, the substances that kill the parasite at a considerably slower rate but allow more complete elimination of the parasite are likely to be discarded already at the initial stages of the high-throughput sweep.

At the same time, if we use some kind of biological rational (such as the one outlined in the *Case of Malaria Drugs and In Silico Metabolomics)*, we will be less in a hurry and will be more free to plan the set of the substances to be analyzed. This deliberate planning will be made in accordance with the existing expertise in microbiology/physiology and not in accordance with arbitrary whims and hurries.

Returning again to the basic post-genomic tetrad (Figure 1-3) and giving it some further thought in accordance with the current purpose (which is the drug design against a microorganism), we can observe that there are at least four major groups of bioinformatic methods that can have implications for the drug design. Clearly, there are certain difficulties with all of these methods since any genome is, usually, far from being fully annotated. In the first group of the methods, SNPs of the genome can be studied in relation to the known data on the drug resistance. In the second, *in silico* and *in vitro* transcriptomics can be employed to analyze gene regulation is response to certain conditions and substances. In the case of the third group, we mentioned the value of the proteomic studies in the previous section. As for the fourth group, reconstructions of the metabolic pathways of *Plasmodia* can, at least in theory, be directly used for the purposes of drug design. In the above Case Study, we used the last approach.

The inherent difficulty of any project of the drug design involves taking into account the side effects that the potential drug might exert on the human organism. All of the current anti-malarial drugs are known to produce quite many adverse effects that range from dizziness and palpitations to death from overdose [17]. Using an expertise in various fields of biological research, it can always be proven that each of these adverse effects happens because the substance used as a drug interferes with those parts of human metabolism where it, literally, has no business to interfere. Giving further thought to this simple idea can be productive. Thus, biochemical expertise suggests that most of the drugs, effectively, mimic the existing metabolites rather than interacting at altogether different allosteric sites in proteins. Accordingly, taking into account the integrative view inherent for the post-genomic biology, we are advised that both the problems of finding the drug targets and of dealing with side effects can be dealt with a single shot: by comparing the metabolomes of the *Plasmodia* and of the human host and coming to the notions of "metabolome overlaps" and "metabolome gaps" [31].

In other words, the main idea of a post-genomic method of the drug design we presented in *The Case of Malaria Drugs and In Silico Metabolomics* is that compounds that are structurally dissimilar to those occurring in the human metabolism but considerably structurally closer to the compounds involved in *Plasmodium* metabolism are likely to be good starting points ("structural templates") for the new anti-malarial drugs. Apparently, this Case Study is the initial stage of a more elaborated scheme of the rational drug design that includes six main stages: 1. finding *P.Falciparum* metabolites most dissimilar to the human metabolites; 2. database searches of the known small molecules/drugs to find structurally similar drugs or commercially available compounds; 3. retracing the compounds to the metabolic pathways and the relevant proteins with subsequent molecular modeling and protein structure analysis; 4. large-scale *in vitro* experiments using functional genomics; 5. *in vivo* experiments including animal studies and 6. clinical trials.

In this Case Study that imitates the metabolome-wide experiments we deliberately avoided using pathway analysis at the first stages of analysis. Instead, we just compared the lists of the metabolites from both organisms, analyzed chemical structure of the substances and only then looked at the corresponding pathways. Of course, the general idea (to identify chemical structures that are structurally unique to *Plasmodium* species and then to make a drug by mimicking them) is, at present, no more than a research scenario. One cannot expect to obtain ready-made drugs in one sweep using an imaginary research scenario.

However, this kind of analysis produced ideas concerning rational choice of the starting points for the further procedures of the rational drug design. Some of the known anti-malarial drugs were identified, just in this one sweep. The common groups of the anti-malarial drugs include anti-folates, anti-mitochondrial drugs and drugs targeting redox reactions. The malaria parasite produces folates from para-aminobenzoic acid and structural analogues of these substances are used as drugs known as **anti-folates**. Several substances related to folate synthesis were in the final list in our Case Study: pyrimethamine (a known anti-malarial drug), 6-hydroxymethyl-dihydropterin and dihydroneopterin. Many of the substances on the final list were related to the ubiquinone pathways(s) and thus are directly relevant to **anti-mitochondrial drugs** and **drugs targeting redox reactions** of *Plasmodium*. For example, an anti-mitochondrial drug atovaquone is a naphthoquinone which inhibits the electron transport chains. The atovaquone molecule (Figure 8-2) bears some degree structural similarity to the ubiquinone molecule ("coenzyme Q"). By the way, quinine is the key anti-malarial drug that also, apparently, does not induce resistance in *P.Falciparum*. Mechanism of the quinine action is largely unknown and it was surmised that quinine and the related drugs interfere with the hemoglobin metabolism in *P.Falciparum*. As it is known, hemoglobin is the central component of the cytochromes that are involved in the redox reaction of the ubiquione-related pathways.

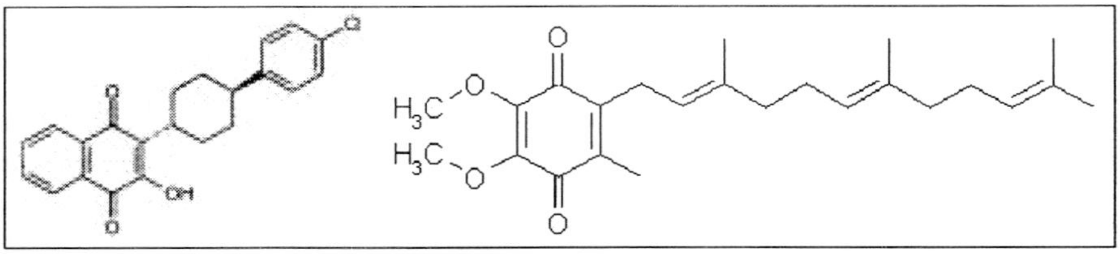

Figure 8-2. Malarial drug atovaquone (left) and ubiquinone molecule (right).

The conceptual approach we used in the *Case of Malaria Drugs and In Silico Metabolomics* can also allow us to interpret the side effects of the drugs in terms of the metabolome overlap (that is, the structurally similar compounds in the metabolomes). The concept of the metabolome overlap implies that the drug, in addition to targeting the pathological microorganism, also mimics one or several of the compounds involved in the host metabolism thus producing side effects. As the compounds selected to be the initial starting points were the most dissimilar to those in the human metabolome (in terms of the differential metabolomics, belong to a "metabolome gap"), it might be expected that they would have considerably less side-effects than a random compound from a library of hundreds of thousands of compounds usually used in screening for new drugs. Of course, the approach is not, generally, free from drawbacks and one of them is that it totally neglects biocompatibility of the potential drugs/drug templates with the bacterial symbionts of humans: some of the substances in the final list of the Case Study are very powerful antibiotics which can have strong deleterious effects on the beneficial bacteria living in the human body though not on the human physiology *per se*. Nevertheless, although this approach might be simplistic, it's also fits well with the post-genomic paradigm in biomedical research.

CHAPTER SUMMARY

Metabolome is the entire array of the small-molecule ligands of the biological system and **metabolomics** deals with a global analysis of all metabolites in a biological sample. Metabolomics is one of the least developed areas of the functional genomics which, nevertheless, has the greatest potential for the practical applications which include **metabonomics, metabolic fingerprinting, metabolic control analysis and metabolic profiling**. As metabolome is inseparable from the metabolic pathways, theoretical approximations of the metabolome can be made through the methods of **bioinformatics**. After providing some background on physiology, medicine and functional genomics studies of the **malarial parasite**, we attempted to illustrate how effective combination of metabolomic and bioinformatic approaches might be for the purposes of the **rational drug design**.

REFERENCES

[1] Weckwerth W, Morgenthal K. Metabolomics: from pattern recognition to biological interpretation. *Drug Discov Today* 2005;10:1551-1558.

[2] Rochfort S. Metabolomics reviewed: a new "omics" platform technology for systems biology and implications for natural products research. *J Nat Prod* 2005;68:1813-20.

[3] Goodacre R, Vaidyanathan S, Dunn W, Harrigan G, Kell D. Metabolomics by numbers: acquiring and understanding global metabolite data. *Trends Biotechnol* 2004;22:245-52.

[4] Nielsen J, Oliver S. The next wave in metabolome analysis. *Trends Biotechnol* 2005;23:544-6.

[5] Keun H. Metabonomic modeling of drug toxicity. *Pharmacol Ther* 2006;109:92-106.

[6] Lee W, Go V. Nutrient-gene interaction: tracer-based metabolomics. *J Nutr* 2005;135:3027S-3032S.

[7] Young VR. Trace element biology: the knowledge base and its application for the nutrition of individuals and populations. *J Nutr* 2003;133:1581S-7S.

[8] Kanehisa M, Goto S, Hattori M, Aoki-Kinoshita K, Itoh M, Kawashima S, Katayama T, Araki M, Hirakawa M. From genomics to chemical genomics: new developments in KEGG. *Nucleic Acids Res* 2006;34:D354-7.

[9] Caspi R, Foerster H, Fulcher CA, Hopkinson R, Ingraham J, Kaipa P, Krummenacker M, Paley S, Pick J, Rhee SY, Tissier C, Zhang P, Karp PD. MetaCyc: a multiorganism database of metabolic pathways and enzymes. *Nucleic Acids Res* 2006 ;34:D511-6.

[10] Joshi-Tope G, Gillespie M, Vastrik I, D'Eustachio P, Schmidt E, de Bono B, Jassal B, Gopinath GR, Wu G, Matthews L, Lewis S, Birney E, Stein L. Reactome: a knowledgebase of biological pathways. *Nucleic Acids Res* 2005;33:D428-32.

[11] Karp P, Ouzounis C, Moore-Kochlacs C, Goldovsky L, Kaipa P, Ahren D, Tsoka S, Darzentas N, Kunin V, Lopez-Bigas N. Expansion of the BioCyc collection of pathway/genome databases to 160 genomes. *Nucleic Acids Res* 2005;33:6083-9.

[12] Romero P, Wagg J, Green ML, Kaiser D, Krummenacker M, Karp PD. Computational prediction of human metabolic pathways from the complete human genome. *Genome Biol* 2005;6:R2

[13] Arrow K, Gelband H, Jamison D. Making antimalarial agents available in Africa. N Engl *J Med* 2005;353 333-5.

[14] White N. Malaria--time to act. *N Engl J Med* 2006;355:1956-7.

[15] DiMasi J, Hansen R, Grabowski H. The price of innovation: new estimates of drug development costs. *J Health Econ* 2005;22:151-185.

[16] Goozner M. The $800 million pill: the truth behind the cost of new drugs. University of California Press, Berkeley, 2004.

[17] Baird J. Effectiveness of antimalarial drugs. *N Engl J Med* 2005;352:1565-77.

[18] 18.R. Khamsi. Drug-makers to withdraw single-drug malaria treatment. New Scientist, 12 May 2006 (web reference: http://www.newscientist.com/article/dn9153-drugmakers-to-withdraw-singledrug-malaria-treatment.html)

[19] Magill A, Panosian C. Making antimalarial agents available in the United States. *N Engl J Med* 2005;353:335-7.

[20] Horrocks P, Bowman S, Kyes S, Waters A, Craig A. Entering the post-genomic era of malaria research. *Bull World Health Organ* 2000;78:1424-37.

[21] Di Girolamo F, Raggi C, Bultrini E, Lanfrancotti A, Silvestrini F, Sargiacomo M, Birago C, Pizzi E, Alano P, Ponzi M. Functional genomics, new tools in malaria research. *Ann Ist Super Sanita* 2005;41:469-77.

[22] Hall N, Carlton J. Comparative genomics of malaria parasites. *Curr Opin Genet Dev* 2005;15:609-13.

[23] Le Roch K, Zhou Y, Blair P, Grainger M, Moch J, Haynes J,de la Vega P, Holder A, Batalov S et al. Discovery of gene function by expression profiling of the malaria parasite life cycle. *Science* 2003; 301:1503-1508.

[24] Le Roch K, Johnson J, Florens L, Zhou Y, Santrosyan A, Grainger M, Yan S et al. Global analysis of transcript and protein levels across the Plasmodium falciparum life cycle. *Genome Res* 2004;14:2308-2318.

[25] Bozdech Z, Zhu J, Joachimiak M, Cohen F, Pulliam B, DeRisi J. Expression profiling of the schizont and trophozoite stages of Plasmodium falciparum with a long-oligonucleotide microarray. *Genome Biol* 2003;4:R9.

[26] Gunasekera A, Patankar S, Schug J, Eisen G, Wirth D. Drug-induced alterations in gene expression of the asexual blood forms of Plasmodium falciparum. *Mol Microbiol* 2003;50:1229-39.

[27] Myrick A, Munasinghe A, Patankar S, Wirth D. Mapping of the Plasmodium falciparum multidrug resistance gene 5'-upstream region, and evidence of induction of transcript levels by antimalarial drugs in chloroquine sensitive parasites. *Mol Microbiol* 2003;49:671-83.

[28] Florens L, Washburn M, Raine J, Anthony R, Grainger M, Haynes J, Moch J, Muster N, Sacci J, Tabb D, Witney A, Wolters D, Wu Y, Gardner M, Holder A, Sinden R, Yates J, Carucci D. A proteomic view of the Plasmodium falciparum life cycle. *Nature* 2002;419:520-6.

[29] Lasonder E, Ishihama Y, Andersen J, Vermunt A, Pain A, Sauerwein R, Eling W, Hall N, Waters A, Stunnenberg H, Mann M. Analysis of the Plasmodium falciparum proteome by high-accuracy mass spectrometry. *Nature* 2002;419:537-42.

[30] Sinden R. A proteomic analysis of malaria biology: integration of old literature and new technologies. *Int J Parasitol* 2004;34:1441-50.

[31] Torshin IY. New anti-malarial drugs: post-genomic bioinformatics. 2nd Modern Drug Discovery and Development Summit, Dec 4-6, 2006, Philadelphia, PA, USA.

CONCLUSION

However fast you run,
you can't run away from your feet.
(A folk wisdom)

With the advent of the post-genomic era, biomedical research and clinical practice has witnessed an explosion in various goals and expectations. But, as numerous examples in the present volume seem to suggest, this explosion in goals and expectations does not seem to be balanced with the tangible results of biomedical and clinical significance. Moreover, the situation gets greatly aggravated when the fast-running, high-throughout technologies are applied in a low-intelligence, mechanical mode and with little regard to already acquired data on the human physiology.

As we already said, the purpose of this volume is not producing various criticisms but rather to be practical and descriptive. Being descriptive or practical is not the same as being critical. Nevertheless, anyone who writes anything that conflicts with current beliefs and opinions of a certain social establishment always risks being misunderstood, vilified and attacked. The scientific community is far from being free from this sort of primitivism so the sceptics might well accuse the author in a wide range of demeanors from "picking data", "insufficient descriptions of methods", "being too didactic" down to "being arrogant", "data fabrication" *etc.* This is quite understandable a phenomenon since people tend to hold the privilege to attack whatever and whomever they dislike, without any logic at all. It's apparently because we haven't yet reached the stage of our human civilization at which the causes underlying the likes and dislikes are clearly observed and taken into account as sorts of, shall we say, "risk factors".

Hence, there is the need for the objective assessment of any phenomenon in biology and medicine. Hence, certain theoretic methods and certain intellectual attitudes are required for this assessment to be made. In the present volume, we provided not so much an assessment of the current state of the post-genomic research. Rather, to the extent of the author's limited abilities, we presented a set of theoretical and computational tools that can be used in post-genomic research for numerous purposes (Chapters II-VIII). Quite obvious, the examples of the data structures and general methods presented in this book are not supposed to be taken as some once-and-for-all rigid classifications. The readers are encouraged to adopt an open-minded posture and, when interested, to do a little research themselves, be it even just a few searches in PubMed or any other abstract database.

This volume might also seem to be quite unusual for a reader thinking of bioinformatics only as a narrow field that deals mostly with writing nice-looking computer interfaces, using DBASE, ORACLE *etc* to cram gigabytes of data into impressively huge databases. This book concentrates not so much on the computer programs and some pretty abstract math modeling of who-knows-what, though algorithms do feature in this volume. Instead, the focus is rather on the most efficient logical strategies to find, to extract and to apply what is *already known* but is scattered among many millions of abstracts in PubMed and other publication databases. That is why almost any of the "Case Studies" the respected reader finds in the present volume aims at the efficient usage of already available information and methods: in particular, databases and algorithms.

Indeed, common sense suggests that it is not enough to compile a database. It is not less important to know whether it can be practically used for a particular research purpose and if it is, to learn how to use it most effectively. The purpose-oriented approach to bioinformatics often happens to be more of a help to actual clinical research than compilation of databases of enormous size containing numerous factual errors. Also, bioinformatics is far from being restricted to the matter of computer programs. The research tasks of bioinformatician, especially those considered in the present volume, imply, in particular, usage of a universal human skill that has to be developed to a certain level of efficiency: the skill of rapid and efficient reading accompanied by clear understanding of what was read and with what purpose in mind. Given the current amounts of information, developing this and other human skills is essential for the genuine researchers to survive in the overflow of biomedical information (Chapter III).

Another purpose of the present volume is to make clear that post-genomic biology should not even attempt to disregard the preceding decades of the medical knowledge and experience. Unfortunately, this "revolutionary" (or "terrorist", using more recent political terminology) disregard to the decades of physiological research is not uncommon. Without preserving the historical continuity of the scientific research, expressed by the principle "humans proceed from the known to the unknown", it would be foolish to expect that any sort of a post-genomic "revolution" will lead to anything but financial "bloodshed". The actual *evolution* of scientific thinking, from primitive conditioned reactions to more advanced modes is, perhaps, the only realistic way for the post-genomic biology. The continuity of scientific knowledge, which is the essence of post-genomic biology, can never be achieved by simply disregarding the numerous facts of human physiology. As we attempted to illustrate in a number of Case Studies, an important role of the bioinformatician is to enable the other researchers to look at the entire logical structure of the study at all the levels of function involved and to watch out for logical flaws and lack of consistency across the levels of function (Chapters I,III&V).

At the very least, methods of post-genomic bioinformatics allow to generate working hypotheses of a type that help to put thoughts and known facts into order. Take, for instance, our analysis of the link between the estrogens and venous thrombosis (Chapter VI). Whatever the physiological mechanisms of the actual link are, the methods of bioinformatics, be they the "*in silico* transcription profiling" in Chapter VI or genome-wide sequence analysis

(Volume II) do allow to generate physiologically plausible, experimentally falsifiable and logically substantiated hypotheses. Provided that the researcher truly aims to be objective, the bioinformatic methods will be allow formulation of viable working hypotheses for a wide range of physiological phenomena, from molecular mechanisms of disease (Chapters II, VIII) to the rational drug design (Chapter VIII).

What is often the real problem, which becomes more and more apparent in the post-genomic era of biological research, is that for quite a significant number of biological researchers it's often very hard to be objective or even rational in doing their research. An acquaintance of mine, a professor in medicine, once said that "the research models like 'high-throughput genotyping' can easily be twisted into huge waste of resources and no medical value whatsoever. By the way, a persistent waste of resources with no results coming is a peculiar characteristic of any delusional state".

What could be the source of this sort of delusion in many of the contemporary attempts to develop post-genomic technologies as biomedical instrument? Again, relying on the PubMed searches, it seems that the researchers in social psychology have explicitly confirmed one of the key reasons why people fall into what might be called as "costly entrapment in losing endeavors" [1]. Say, a gambler plunges deeper into debt when, from a rational point of view, the huge losses should stop him from doing so. Or, more and more money spent on so-called "post-genomic research", while these expenditures result in hardly any actual clinical applications. Moreover, the actual design of the new drugs seems to suffer but the waste still continues. These and other scenarios of the kind have something in common: an individual or a group of individuals entangled in a very costly, a very protracted and quite losing a venture. As the researchers concluded from four related studies [1], pursuing favorable views of the self could be costly to decision makers' financial well-being and produce self-defeating behaviors. Their findings, based on a population study of monetary choices, might be unsurprising to many observers of human nature: it comes down to the most primitive instincts of the ego that go out of control. What is most interesting about this psychological study is that the individuals entrapped in these sorts of scenarios lost, eventually, both money and face (see [1] and a related news item at http://www.world-science.net/exclusives/060613_ego.htm). Clearly, this kind of social analysis has direct implications for the costly post-genomic research.

It can produce an interesting sensation to compare a situation such as this with a hypothetical situation in an underdeveloped country where an advanced Western technology have penetrated but, through lack of accurate rational application, became adopted by the natives in a superficial manner. Some of the natives will use the technology as a totem (and will loose, apparently, its actual value). Some others will oppose the technology saying that it contradicts to all that their ancestors believed (and will loose all the benefits of using the technology)...

And, yet again, being descriptive is not the same as being critical or derisive. It's exceedingly important for the public health that the actual value of the post-genomic techniques should be understood by the wide audience of the doctors and medical practitioners. It cannot be allowed that some modern quacks, under the guise of developing novel technologies, would extort money from numerous governmental and private institutions and will not give in return anything of medical significance. Indeed, multiple examples, cited anonymously and explicitly throughout the present volume do suggest that, alas, similar situations tend to occur quite often in almost any area of post-genomic research.

All of our Western science rests on the assumption that the Universe is orderly and that this order can be discovered, described and understood. Discovery and understanding of this order produces scientific concepts, hypotheses and then theories. At the same time, the opposite process of projecting certain biased concepts on the objective physical reality is another common feature of any delusional state. Therefore, it's no use to worship any concept or methodology from science expecting that mere reiteration of it would magically produce the craved results. Symbols and techniques employed in science are not idols that magically produce grant money from funding agencies. So, being obsessed, for instance, with the idea of "systems biology" won't do any good for the process of actual scientific discovery.

As it often happens in actual science, the true advancements and real discoveries are often made 'almost incidentally'. That is, it is not uncommon to read in works of one or another of the great scientists such as Newton, Einstein, Virchow, Pavlov statements similar to "we were trying to learn about something completely different". However, in order for this 'incident' to occur, the mental state of the observer should be clear from the prejudice and hatred towards anyone who points out the obvious drawbacks of the adopted research strategies. It also might be helpful to remember about the peculiar egoistic admixtures that largely inhibit clear scientific thinking. This list is as old as the Western science itself (since it's attributed to Roger Bacon) and includes the following interesting items: 1. too much reliance upon a mere authority; 2. enslavement by custom; 3. unquestioning acceptance of popular beliefs and current opinions; and 4. pretensions to knowledge. Each of us, scientists, can be deficient in respect to some or other of these principles...

While medicine is plagued by considerably more pressing problems such as unscientific practices [2,3], fallacious research [4] and outright medical errors [5], enormous prevalence of unprofessional behavior among medical students (see, for instance, the letters at http://content.nejm.org/cgi/content/full/354/17/1851-a), considerable problems with availability of already known drugs [6,7] *etc etc*, this does not mean that the perspective trends in the biomedical research should be entirely abandoned in favor of dealing with those more pressing concerns. On the contrary, instead of becoming even more rigid and ultra-narrowly specialized under the pressure of the problems and of the increasing flow of information, biomedical researchers, with the help of bioinformatics methodologies, have a chance to transform in the direction of a greater flexibility of scientific thinking. And post-genomic bioinformatics is one of such promising trends that can help the medical science to be more exact, more effective and, paradoxically it may seem, more humane.

References

[1] Zhang L, Baumeister RF. Your money or your self-esteem: threatened egotism promotes costly entrapment in losing endeavors. *Pers Soc Psychol Bull* 2006;32:881-93.

[2] 2.E. Young. Faking it. New Scientist, 12 Dec 2000 http://www.newscientist.com /article/dn253.html)

[3] Smith R. Unscientific practice flourishes in science. *BMJ* 1998;316:1036.

[4] Ioannidis JP. Why most published research findings are false. *PLoS Med* 2005;2(8):e124.

[5] Committee on Quality of Health Care in America, L Kohn, J Corrigan, M Donaldson (eds). To Err Is Human: Building a Safer Health System. Institute of Medicine (2000), http://books.nap.edu/books/030906681/html/index.html

[6] Arrow K, Gelband H, Jamison D. Making antimalarial agents available in Africa. *N Engl J Med* 2005;353:333-5.

[7] Magill A, Panosian C. Making antimalarial agents available in the United States. *N Engl J Med* 2005;353:335-7.

APPENDIX I. CVD GENESCAPE DIAGRAM: THE BRIEF DESCRIPTIONS OF THE GENES

This Appendix presents brief descriptions of the genes mentioned in the *GeneScape* diagram (Figure 2-5). The genes are listed alphabetically. Entry for each gene is supplemented by the OMIM ID code and a tag that refers to the general physiological function of the gene. In *CVD GeneScape* we used a three-partite classification (cardiovascular system, coagulation cascade, bloodstream/vessel wall). Here, the genes were grouped into the eight major categories: atherosclerosis (lipoproteins), atherosclerosis (inflammation), adipose tissue, vascular constitution and remodeling, vasoconstriction and vasodilatation, renin-angiotensin system, electrolyte balance, blood coagulation cascade. This grouping is relative since many of the genes should be mentioned in at least two of the eight categories above. For example, renin-angiotensin system can influence vascular fibrinolytic balance: angiotensin II peptide increases production/secretion of PAI1. MTHFR is a highly pleiotropic gene. ACE enzyme can reduce production of tissue plasminogen activator and synthesis of the endothelial nitric oxide *etc.*

The three largest groups of of the eight appear to be blood coagulation cascade (both clotting and fibrinolysis), atherosclerosis (lipoprotein involvement) and atherosclerosis (inflammation-related). Since these groups appear to be the major functional groups in the *CVD GeneScape* diagram, below we provide short general descriptions for these groups of genes.

Blood coagulation cascade is involved in hemostasis: the process of blood clotting and then of the subsequent clot dissolution which accompanies the tissue repair. Hemostasis can be divided into four major stages: 1. vascular constriction to limits the blood flow to the injury; 2. platelet activation by thrombin and formation of the initial platelet plug at the site of injury; 3. cross-linking of the initial platelet plug by the fibrin (clotting) and 4. dissolution of the finrin clot to restore the normal blood flow (fibrinolysis).

The formation of the fibrin clot is said to occur by two pathways: the extrinsic and the intrinsic pathway. The *extrinsic pathway* is initiated upon vascular injury which leads to exposure of tissue factor (F3). Following the injury of the blood vessel, endothelium, tissue factor is released into blood stream forming TF-F7a complex, which activates F9 and F10. In turn, the F10a and its co-factor F5a form the prothrombinase complex which, in turn, activates thrombin, the central component of the coagulation cascade.

The *intrinsic pathway* is a reaction to abnormal vessel wall in the absence of considerable tissue injury and is activated when blood is exposed to negatively charged surfaces. It involves conversion of prekallikrein to kallikrein and activation of the coagulation factor F12. F12 activates F11 which, in turn, activates F9, F8 and F10. The intrinsic pathway is thought to have a "minor role" in hemostasis, a prejudice based largely on a single argument torn out of the context of human physiology which goes like "patients with severe F12 deficiencies do not tend to have severe bleeding disorders". Clearly, there is something missing in a statement like this. Namely, it's the phrase equivalent to "...under the test conditions investigated". In other words, abnormal physiological states (hyperlipidemia, inflammation *etc*) can lead to activation of coagulation via the intrinsic clotting cascade and result in thrombotic tendencies (see a Case Study in the Chapter VI).

As recent research shows, the concepts of "extrinsic" and "intrinsic" pathways in blood clotting are rather useful approximations. The fact is that these two pathways appear to be highly interconnected. For example, F3-F7a dimer does not only activate F10 but also F9 of the intrinsic pathway. Severe bleeding associated with the F8 and F9 deficiencies also suggests that "extrinsic pathway" alone is not sufficient for normal hemostasis.

Finally, activation of prothrombin to thrombin results in subsequent conversion of the fibrinogen to fibrin. Degradation of fibrin clots is the function of plasmin. Plasmin is converted from plasminogen by tissue plasminogen activator and urokinase.

Atherosclerosis: lipoproteins. The ten principal apolipoproteins (A1, A2, A4, B48, B100, C1, C2, C3, D and E) are synthesized and secreted by the liver and intestine. Lipids obtained from the food stuffs emerge from the small intestine as the 200-500nm particles known as "chylomicrons". The 99% of the volume of these fat particles consist of triglycerides while apoB-48 forms the amphipatic shell around the fat particle so that to form the external hydrophilic surface. Other apolipoproteins comprising the chylomicrons are APOC3, APOE and APOC2. The chylomicrons are transported in the blood stream and then to all the tissues that require fat.

When the consumed foodstuffs contain more fats than is required by the organism, the excess of triglycerides and cholesterol is exported by the liver into blood in the form of 80-100nm VLDL (very low density lipoprotein) particles stabilized mostly by ApoB100 isoform and ApoE (also APOC3 and APOC2). VLDL particles are transported to adipose where the triglycerides are released from VLDL with the help of the LPL enzyme. The release of the triglycerides from VLDL to adipocytes leads to transformation of VLDL particles into 20-30nm LDL particles. Each LDL (low density lipoprotein) particle is enclosed into a shell of phospholipids and is stabilized by one molecule of ApoB100. The LDL particles contain large percentage of cholesterol which they carry to the peripheral tissues that require cholesterol intake. The LDL receptors of the appropriate cells of the peripheral tissues recognize the LDL particles through interaction with the APOB100.

In parallel, the liver synthesizes the small 8-20nm HDL (high density lipoprotein) particles that contain an entire spectrum of apolipoproteins: APOA1 that activates the LCAT enzyme (the lecithin cholesterol acyl transferase which catalyzes formation of cholesterols from lecithin); APOC2 that activates LPL; APOC3 that inhibits LPL; APOD as well as APOE that activates the clearance of VLDL particles from the blood stream. The HDL particles serve as scavengers of the cholesterol from the bloodstream which they carry back to the liver. Then the HDL particles enriched with the cholesterols are taken up by liver and degraded, the excess cholesterol is converted into bile salts or, again, repackaged into VLDL.

If the normal functioning of this fat processing system is thrown out of balance either by excess fats in general, excess of cholesterol in particular or because of some genetic defects, a number of pathogenic states can issue. Many genes mentioned in the *CVD GeneScape* diagram mediate the packaging of the lipoproteins and the above-mentioned transformations so the genetic polymorphisms of these genes can affect the fat processing in various ways. Despite that there are many genes in the molecular cascades involved in fat metabolism, polymorphisms of only few of them might serve as common genetic markers. The main physiological mechanism by which the defects in the fat metabolism influence the cardiovascular pathology is through the *atherosclerosis*- the pathological process of deposition of cholesterols and other lipids as plaques on the vessel walls. The formation of the atherosclerotic plaques depends on the levels of the lipids and lipid particles in the blood stream as well as on the interactions of the lipid particles with the vessel walls.

Atherosclerosis: inflammation-related. The process of deposition of the cholesterol in the lipoprotein particles on the vessel walls is far from being the only process that contributes to the formation of the atherosclerotic lesions in the blood vessels. If, for instance, the wrong diet is not addressed in the treatment course, then the cholesterol deposits (which, initially, are just streaks of fat on the vessel walls) become aggravated by adsorbing the fibrous deposits which cap the initial fat streaks. Further progression includes calcification and ulceration followed by hemorrhage. Hemorrhage, in turn, activates thrombotic processes until the lumen of the vessel, garbled with all this junk, puts considerable restrictions on the blood flow through this vessel.

Apart from this scheme that describes the progression of the atherosclerotic process from the fat streaks to thrombosis, there are many other factors that determine predisposition to atherosclerotic plaques and their parameters. These factors are, primarily, related to inflammation processes. The causes of inflammation, direct or indirect, include bacteria, genetic predispositions, improper diet or destructive life style. There are several distinct mechanisms related to inflammation (such as oxidation of phospholipids, peroxide production, cytokine chemotaxis, leukocyte adhesion *etc*) and a number of genes variously expressed in different types of cells are involved in the process of inflammation and in the responses to inflammation.

For example, the oxidation hypothesis of atherogenesis focuses on proinflammatory phospholipids that result from oxidation of LDL phospholipids containing arachidonic acid. These oxidized phospholipids, after being produced in the lipoxygenase and myeloperoxidase pathways are recognized by the immune system and induce monocytes. Therefore, the gene products involved in the oxidation of lipoproteins or the gene products involved in the protection from oxidation are important for this particular physiological mechanism related to atherosclerosis. It has also been demonstrated that the inflammatory response elicited by these oxidized phospholipids is mediated, in particular, by induction of monocyte chemoattractant protein (MCP)-1, interleukin-8 and macrophage colony stimulating factor. Accordingly, the polymorphisms in all of these genes are potential atherosclerotic markers.

THE GENE DESCRIPTIONS

ABCA1 (OMIM 600046) **Atherosclerosis (lipoproteins).** ATP-binding cassette transporter 1 is a transmembrane channel that functions as a cholesterol efflux pump and regulates the outward flow during removal of the cellular lipids. Overexpression of ABCA1 results in an antiatherogenic profile. Polymorphisms in ABCA1 were associated with severity of atherosclerosis and CVD as well as with triglyceride levels.

ACE (OMIM 106180) **Renin-angiotensin system.** Angiotensin-converting enzyme hydrolyzes angiotensin I into angiotensin II. The protein was localized to a number of tissues including kidney and heart. The common I/D (insertion/deletion) polymorphism was reported to correlate with plasma ACE; DD homozygote being associated higher ACE level. In many studies, DD variant was associated with MI, stroke, tendency to restenosis and salt-sensitivity in hypertension.

ADD1 (OMIM 102680) **Electrolyte balance.** Adducin-alpha promotes the assembly of the spectrin-actin network in the membrane cytoskeleton. The gene is expressed in most of the tissues; in particular, the protein regulates signal transduction in the renal tubules. G460W polymorphism of ADD1 was associated with the salt-sensitive hypertension.

ADRB1 (OMIM 109630) **Vasoconstriction and vasodilatation.** Adrenergic receptors alpha1, alpha2, beta1 and beta2 mediate physiological effects of the epinephrine and the norepinephrine. Variants in ADRB1 were associated with acute MI and cardiovascular survival.

ADRB2 (OMIM 109690) **Vasoconstriction and vasodilatation.** Variants Argl6Gly, Gln27Glu of ADRB2 were associated with predisposition to asthma and hypertension. The polymorphisms can also adjust response to physical exercise.

ADRB3 (OMIM 109691) **Adipose tissue.** Adrenergic receptor-beta 3 is located mainly in adipose tissue and is important for lipolysis and thermogenesis regulated by the autonomic nervous system. Polymorphism W64R of ADRB3 was associated with higher plasma triglyceride, hypertension and CHD.

AGT (OMIM 106150) **Renin-angiotensin system.** Angiotensinogen is precursor of angiotensin II- the key effector of the renin-angiotensin system that plays central role in the regulation of blood pressure and electrolyte balance. Angiotensin II is a potent vasopressor hormone which stimulates aldosterone secretion. Angiotensin can also induce inflammatory response in the vascular wall. Higher AGT level was associated with arterial hypertension, carotid intimal-medial wall thickening, aortic stiffness and, in general, with CHD; common AGT polymorphisms (in particular, T174M and M235T) were associated with increased AGT level and hypertension.

AGTR1 (OMIM 106165) **Renin-angiotensin system.** Angiotensin receptor 1 is the major angiotensin receptor that occurs, in particular, in liver and kidney cells. The receptor mediates the cardiovascular effects of angiotensin. A common polymorphism A1166C in AGTR1 was associated with CHD in general, MI and hypertension in particular.

ANXA5 (OMIM 131230) **Blood coagulation.** Annexins are implicated in membrane-related events during exocytosis and endocytosis. Annexin 5 has a potential role in cellular signal transduction, inflammation, growth and differentiation. Annexin 5 (placental protein 4, endonexin II) is also an anticoagulant protein that acts as an indirect inhibitor of the tissue factor (F3) protein-protein complex. A polymorphism in annexin 5 was associated with a reduced risk of MI.

APOA1 (OMIM 107680) **Atherosclerosis (lipoproteins).** Apolipoprotein A1 is the major HDL apolipoprotein relatively abundant in plasma. Common polymorphisms -76 G>A and +83C>T in the promoter region were associated with decreased promoter activity, lower HDL and lower APOAI levels.

APOB (OMIM 107730) **Atherosclerosis (lipoproteins).** Apolipoprotein B is the main protein fraction of LDL (ApoB100, hepatic form). The common polymorphisms SpIns/Del and XbaI were associated with CHD and MI.

APOC3 (OMIM 107720) **Atherosclerosis (lipoproteins).** Apolipoprotein C3 is a major component of the VLDL and HDL particles. The promoter polymorphisms -455T/C and -482C/T were associated with increased plasma triglycerides and APOC3 as well as with low HDL levels.

APOE (OMIM 107741) **Atherosclerosis (lipoproteins).** Apolipoprotein E is a component of VLDL and HDL particles. The common polymorphism E2 (R158C) was associated with lower plasma cholesterol levels while another common polymorphism E4 (C112R) was associated with higher plasma cholesterol levels, increased risk for atherosclerosis and MI.

CBS1 (OMIM 236200) **Atherosclerosis (inflammation).** Cystathionine-beta synthase catalyzes conversion of L-serine and L-homocysteine to cystathionine. The enzyme is most abundantly expressed in cytoplasm of liver and pancreatic cells. Hyperhomocysteinemia is an established risk factor for atherosclerosis and polymorphism 833T/C in CBS1 was associated with venous thrombosis, MI and atherosclerosis.

CD14 (OMIM 158120) **Atherosclerosis (inflammation).** Monocyte differentiation antigen is a myeloid leucine-rich glycoprotein expressed in monocytes, macrophages and granulocytes. CD14 functions as a mediator for the activation of macrophages by endotoxins from the envelope of gram-negative bacteria. Common polymorphisms -260C/T and -159T/C were associated with MI.

CD36 (OMIM 173510) **Blood coagulation.** Thrombospondin receptor is the 4th major glycoprotein of the platelet surface. CD36 binds to collagen, thrombospondin, anionic phospholipids and oxidized LDL. A promoter polymorphism in CD36 is, on one hand, associated with an atherogenic lipid profile and increased cardiovascular risk while, on the other hand, with protection from cerebral malaria.

CETP (OMIM 118470) **Atherosclerosis (lipoproteins).** Cholesterol ester transfer protein is involved in the transport of insoluble cholesteryl esters between different types of lipoprotein particles. Polymorphisms in CETP were associated with susceptibility to atherosclerosis and with the extent of CHD.

CRP1 (OMIM 123260) **Atherosclerosis (inflammation).** C-reactive protein is a major acute-phase plasma protein. CRP is a marker of low-grade inflammation that may have a role in the pathogenesis of atherosclerotic lesions. CRP plasma levels can predict recurrent ischemia and MI; these diseases were also associated with the common polymorphisms in the CRP1 gene.

CSF1 (OMIM 120420) **Atherosclerosis (inflammation).** Macrophage colony-stimulating factor controls the production, differentiation and function of macrophages. CSF1 protein is likely to be involved in the formation of the initial fatty streaks as well as in the progression of atherosclerosis to the fibrous lesions. Plasma levels of CSF1 correlate with atherosclerosis.

CSF2 (OMIM 138960) **Atherosclerosis (inflammation).** Granulocyte-macrophage colony stimulating factor is a cytokine that controls the production and differentiation of granulocytes and macrophages. A common polymorphism I117T of CSF2 was associated with severe coronary atherosclerosis and MI.

CYBA (OMIM 608508) **Atherosclerosis (inflammation).** Cytochrome B-alpha, also known as "P22phox" and "NADH/NADPH phagocyte oxidase", has a critical role in the generation of superoxide anions in the microbicidal oxidase system of macrophages and other phagocytes. The superoxide production is implicated in the pathogenesis of both hypertension and atherosclerosis; the CYBA gene polymorphisms were associated with CHD and hypertension.

CYP11B2 (OMIM 124080) **Electrolyte balance.** Cytochrome 11B2 synthesizes aldosterone and 18-oxocortisol. Aldosterone diminishes secretion of sodium ions and stimulates the secretion of potassium ions through the kidneys. Aldosterone is synthesized from cholesterol in the adrenal gland in response to an increase in either angiotensin II or plasma potassium levels. Common polymorphisms in CYP11B2 were associated with increased aldosterone levels, hypertension and MI.

CYP7A1 (OMIM 118455) **Atherosclerosis (lipoproteins).** Cholesterol 7-alpha-hydroxylase catalyzes the first step in bile acid synthesis by converting cholesterol. Common polymorphism -204 A>C was associated with variations in LDL and triglyceride levels.

ECE1 (OMIM 600423) **Vasoconstriction and vasodilatation.** Endothelin converting enzyme 1 cleaves endothelin-1 to biologically active peptide(s). Altered gene expression can influence arterial stiffness and blood pressure. Promoter polymorphisms in ECE1 were associated with hypertension.

EDN1 (OMIM 131240) **Vasoconstriction and vasodilatation.** Endothelin 1 is a vasoconstrictor peptide produced by vascular endothelium. A common polymorphism K198N was associated with an increase of blood pressure.

EDNRA (OMIM 131243) **Vasoconstriction and vasodilatation.** Endothelin receptor type A. Endothelin-1 exerts its effects through two receptors, A and B. EDNRA 1363 C>T polymorphism was associated with alterations in the blood pressure.

F2 (OMIM 176930) **Blood coagulation.** Prothrombin is the key protein of the clotting cascade: the thrombin precursor. Prothrombin also plays a role in maintaining vascular integrity during development and postnatal life. Common polymorphism 20210 G/A in 3'UTR was associated with elevated plasma prothrombin and is considered to be the second most significant genetic risk factor for venous thrombosis in Caucasians.

F3 (OMIM 134390) **Blood coagulation.** Coagulation factor 3 (tissue factor, TF) is a receptor for FVII which is the primary triggering agent of the extrinsic branch of the clotting cascade. Common polymorphisms in F3 were associated with reduced plasma TF levels, lower risk of venous/coronary thrombosis.

F5 (OMIM 227400) **Blood coagulation.** Factor V is the essential cofactor for the factor Xa-catalyzed activation of prothrombin. The common polymorphism known as "Leiden" variant (R506Q) is considered to be one of the most significant genetic risk factors for venous thrombosis in Caucasians.

F7 (OMIM 227500) **Blood coagulation.** Factor VII is a vitamin K-dependent clotting factor. Converted into an active form by proteolysis, F7 binds to F3 and further activates the clotting cascade through the extrinsic pathway. Elevated FVII activity has been associated with increased risk of coronary thrombosis. A common polymorphism R353Q of FVII is associated with somewhat lower plasma levels of FVII and might correspond to a lower risk for DVT, MI and stroke.

F10 (OMIM 227600) **Blood coagulation.** Coagulation factor X is vitamin K-dependent. It is activated by factor IXa (intrisic pathway) or by factor VIIa (extrinsic pathway). FXa converts prothrombin to thrombin. High levels of FX were associated with increased risk of thrombosis. No common and, at the same time, functionally significant polymorphisms in F10 were established as yet.

F12 (OMIM 234000) **Blood coagulation.** Upon proteolytic activation, coagulation factor XII activates coagulation factors VII and XI. F12 also participates in fibrinolysis as well as generation of bradykinin and angiotensin. Common polymorphism 46C>T was associated with MI, CHD, venous thrombosis and stroke.

F13A1 (OMIM 134570) **Blood coagulation.** Factor XIII (fibrin stabilizing factor) is precursor of the fibrinoligase that cross-links the fibrin molecules thus stabilizing the fibrin clot. The biologically active form is comprised of the subunits of two types: alpha and beta. The common polymorphism V34L in the alpha-subunit was associated with reduced risk of venous thrombosis, MI and stroke.

FGA (OMIM 134820) **Blood coagulation.** Fibrinogen is the central component of the coagulation cascade and is also an acute phase response protein. This plasma glycoprotein is comprised of the three types of subunits: alpha, beta and gamma. Polymorphisms in the alpha-subunit gene FGA were associated with higher incidence of MI.

FGB (OMIM 134830) **Blood coagulation.** Higher plasma levels of fibrinogen-beta indicate heightened CVD risk. Synthesis of fibrinogen-beta is the rate-limiting step in fibrinogen production. Common polymorphism -455 G>A was associated with adjustments in fibrinogen level and with an increase in CVD risk.

GCCR (OMIM 138040) **Adipose tissue.** Glucocortid receptor acts as a transcription factor and binds to glucocorticoid response elements (GRE) of many genes. The protein can also act as a modulator of other transcription factors. The GCCR gene is expressed in many tissues and the levels of the GCCR gene expression affect cellular proliferation and differentiation in those tissues. Common polymorphisms in GCCR were associated with increased plasma cholesterol, obesity and CHD.

GJA4 (OMIM 121012) **Atherosclerosis (inflammation).** Gap junction protein alpha-4 ("connexion 37" or CX37) mediates interactions between endothelial and smooth muscle layers. The protein levels are higher during early atherosclerosis; a common polymorphism P319S was associated with higher incidence of atherosclerosis.

GNAS1 (OMIM 139320) **Vasoconstriction and vasodilatation.** Guanine nucleotide-binding protein, alpha-stimulating activity polypeptide 1 is essential for activation of adenylyl cyclase in cardiac and vascular smooth muscle. A common polymorphism at codon 131 is potentially associated with hypertension and might explain some differences in responses to beta-blocker treatments.

GNB3 (OMIM 139130) **Vasoconstriction and vasodilatation.** G-protein beta polypeptide 3 is the beta-subunit of heterotrimeric guanine nucleotide-binding proteins (G proteins) which transmit signals from the receptors down to subsequent effector proteins. Polymorphisms in GNB3 were associated with vasoconstriction, lower renin level and arterial hypertension.

GP1BA (OMIM 138720) **Blood coagulation.** Platelet glycoprotein Ib is a major platelet receptor for von Willebrand factor mediating platelet aggregation and adhesion. It is composed of 4 subunits: GPIba, GPIbb, GPIX and GPV. Common polymorphisms T145M and -5 T>C in the gene GP1BA (alpha-subunit) were associated with CVD.

HIF1A (OMIM 603348) **Vascular constitution and remodeling.** Hypoxia-inducible factor-1 (HIF1) is a transcription factor that plays an essential role in cellular and systemic homeostatic responses to hypoxia, including regulation of genes involved in energy metabolism, angiogenesis and apoptosis. An impairment of the VEGF production is likely to be linked with decreased HIF1 activity in response to hypoxia. A polymorphism in the HIF1 subunit A was associated with oxygen consumption and course of CHD.

HSD11B2 (OMIM 218030) **Electrolyte balance.** Cortisol is a corticosteroid hormone involved in the response to stress; it increases blood pressure and blood sugar level. The enzyme 11-hydroxysteroid dehydrogenase encoded by HSD11B2 converts cortisol to inactive cortisone thus modulating intracellular glucocorticoid levels and preventing interactions between mineralocorticoid (aldosterone) receptor and glucocorticoids. The protein is most abundant in kidney, pancreas and prostate. Polymorphisms in the HSD11B gene were associated with hypertension and salt-sensitivity in hypertension.

ICAM1 (OMIM 147840) **Atherosclerosis (inflammation).** Intercellular adhesion molecule-1 protein is expressed on endothelial cells and cells of the immune system. Production of ICAM-1 ('CD54') protein is stimulated by cytokines. The protein plays an important role in leukocyte adhesion on the vessel walls at sites of inflammation. Higher level of ICAM-1 in plasma and common polymorphism G214R were associated with atherosclerosis and ischemic stroke.

IGF1 (OMIM 147440) **Atherosclerosis (inflammation).** Insulin-like growth factor-1 mediates leukocyte adhesion to endothelium. IGF-1 levels have been associated with risk of atherosclerosis, promoter polymorphism -1411 C>T was associated with MI.

IL1A (OMIM 147760) **Atherosclerosis (inflammation).** Interleukin 1a is an important proinflammatory cytokine produced by monocytes and macrophages. This cytokine is released in response to cell injury. IL-1 stimulates thymocyte and B-cell proliferation. IL-1 level in atherosclerotic plaques are increased; plasma IL-1 concentrations as well as common polymorphisms in IL1A were associated with CHD.

IRS1 (OMIM 147545) **Atherosclerosis (inflammation).** Insulin receptor substrate 1 is involved in the control of intracellular response upon the insulin stimulation. In obese subjects, insulin resistance is often associated with atherosclerosis. Several common polymorphisms in IRS1were associated with insulin resistance and CAD. In particular, variant R972 contributes to the atherosclerosis by inducing endothelial dysfunction.

ITGA2 (OMIM 192974) **Blood coagulation.** Integrin alpha-2 (platelet glycoprotein IIa) is one of the major collagen receptors on platelets. Common polymorphisms in ITGA2 were associated with CHD and MI.

ITGB3 (OMIM 173470) **Blood coagulation.** Integrin beta-3 (platelet glycoprotein IIIa) is the most abundant protein on the platelet surface. It serves as a receptor for fibrinogen-mediated aggregation of platelet. The common polymorphism L33P was associated with thrombosis, CHD and MI.

LDLR (OMIM 606945) **Atherosclerosis (lipoproteins).** Low density lipoprotein receptor is involved in receptor-mediated endocytosis of LDL. A common polymorphism A370T was associated with a higher risk of stroke.

LEP (OMIM 164160) **Adipose tissue.** Leptin is an adipocyte-specific hormone that regulates mass of the adipose tissue through hypothalamus. Leptin has a critical role in the regulation of body weight by inhibiting food intake and stimulating energy expenditure. Leptin can also promote aggregation of human platelets. CHD patients exhibit higher plasma leptin concentrations. LEP polymorphisms were associated with elevated plasma leptin and obesity, one of the risk factors in CVD.

LEPR (OMIM 601007) **Adipose tissue.** Leptin receptor. The leptin hormone acts through the leptin receptor. LEPR polymorphisms were associated with the elevated leptin levels, obesity and insulin response during glucose tolerance test.

LIPC (OMIM 151670) **Atherosclerosis (lipoproteins).** Hepatic lipase is secreted by the liver and is an important enzyme in HDL metabolism. The common promoter polymorphisms -514C/T (-480C/T) and -250G/A were associated with CHD and hypercholesterolemia.

LIPG (OMIM 603684) **Atherosclerosis (lipoproteins).** Endothelial lipase is involved in lipoprotein metabolism (in particular, metabolism of HDL). An association between 584C/T in the LIPG gene and HDL cholesterol levels was found in a Caucasian population.

LOX1 (OMIM 602601) **Atherosclerosis (lipoproteins).** Lectin-like oxidized LDL receptor ("scavenger receptor E1") protein binds, internalizes and degrades oxidized low-density lipoprotein. Activation of the endothelial cell by oxidized LDL is potentially involved in pathogenesis of atherosclerosis. Polymorphisms in LOX1 gene were associated with MI.

LPA (OMIM 152200) **Atherosclerosis (lipoproteins).** Apolipoprotein(a) is the main constituent of lipoprotein(a) ("Lp(a)"). Lp(a) is a LDL-like particle with the apo(a) covalently attached to apoB. High plasma Lp(a) is considered to be a risk factor for CAD. The variants of the kringle-IV polymorphism might account for 90% of inter-individual variation of Lp(a) concentration. Lp (a) alleles with a low kringle-IV copy number (<22) and high Lp (a) concentration were significantly more frequent in CHD groups.

LPL (OMIM 238600) **Atherosclerosis (lipoproteins).** Lipoprotein lipase is activated by apolipoprotein C2 and primarily functions as triglyceride hydrolase for the receptor-mediated lipoprotein uptake. The common polymorphisms D9N and N291S were associated with higher triglyceride, VLDL and low HDL. Polymorphisms in LPL were associated with adjustments of CVD risk in numerous studies.

LRP1 (OMIM 107770) **Atherosclerosis (lipoproteins).** Low density lipoprotein-related protein 1 is involved in clearance of chylomicron remnants from plasma. LRP1 protein is a major APOE receptor which also interacts with LPL. The protein may also function as a regulator of blood coagulation by participating in the formation of complexes between plasminogen activators and the endogenous inhibitors. Polymorphism -25 C>G in the LRP1 gene was associated with the extent of coronary atherosclerosis.

MBL2 (OMIM 154545) **Atherosclerosis (inflammation).** Mannose-binding lectin is a receptor for mannose and N-acetylglucosamine of bacterial pathogens. MBL protein activates the complement pathway. A common polymorphism Gly54Asp of MBL gene segregates with low MBL levels, recurrent infections and CVD.

MMP1 (OMIM 120353) **Vascular constitution and remodeling.** Matrix metalloproteinase 1 (fibroblast collagenase), as well as other MMPs, is involved in remodeling of the extracellular matrix. Mostly, polymorphisms in MMP1 are associated with lung function and ventricular function.

MMP12 (601046) **Vascular constitution and remodeling.** Matrix metalloproteinase 12 (macrophage metalloelastase) degrades soluble and insoluble elastins. There were a few studies associating polymorphisms in MMP12 with CHD, aortic aneurysm and arterial stiffness.

MMP3 (OMIM 185250) **Vascular constitution and remodeling.** Matrix metalloproteinase 3 (stromelysin-1) degrades fibronectin, laminin, and collagen-IV but not collagen-I. Vascular remodeling is an essential feature in the development of atherosclerosis-related changes in the arterial wall. Promoter polymorphisms of MMP3 were associated with progression of atherosclerosis in MI patients.

MMP9 (OMIM 120361) **Vascular constitution and remodeling.** The 92kD matrix metalloproteinase 9 (gelatinase B) is secreted mostly by macrophages. This enzyme degrades type IV and V collagens of extracellular matrix. Plasma levels of the MMP9 appear to correlate with severity of atherosclerosis in CHD; MMP9 polymorphisms were correlated with CHD, MI and severity of the atherosclerosis.

MTHFR (OMIM 607093) **Atherosclerosis (inflammation).** Methylenetetrahydrofolate reductase catalyzes the conversion of 5,10-methylenetetrahydrofolate to 5-methyltetrahydrofolate. MTHFR is an essential enzyme that controls folate levels in plasma. Deficiencies in folate metabolism lead to lower folate levels (folate deficiency) and higher homocystein levels (hyperhomocysteinemia) which affect as a wide range of organs and tissues including renal tissue, brain and endothelium. There is a wide spectrum of clinical manifestation of the deficiencies in folate metabolism: from birth defects and neurological disorders to CVD. The 677T variant of the common polymorphism 677 C/T corresponds to a thermolabile protein and was associated with hyperhomocysteinemia and CVD in general as well as with thrombosis and atherosclerosis, in particular.

MTP (OMIM 157147) **Atherosclerosis (lipoproteins).** Microsomal triglyceride transfer protein is involved in the transport of triglycerides, cholesteryl esters and phospholipids between phospholipid surfaces. Located in the microsomal fraction of liver and intestine, the protein also appears to have a central role in lipoprotein assembly. Polymorphisms in the promoter region were, in particular, associated with MI.

NOS3 (OMIM 163729) **Vasoconstriction and vasodilatation.** Endothelial nitric oxide synthase. Nitric oxide levels can influence vascular tone, platelet aggregation, proliferation of vascular smooth muscle cells and leukocyte adhesion. Nitric oxide is synthesized by various isoforms of NO synthase (NOS): NOS1, NOS2, NOS3. Polymorphic NOS3 were associated with higher nitric oxide levels and CVD.

NPPA (OMIM 108780) **Electrolyte balance.** Natriuretic peptides A and B are cardiac hormones that have key roles in cardiovascular homeostasis. Natriuretic peptide B is produced by the atrium and is implicated in the control of extracellular fluid volume and electrolyte balance. High concentrations of both A and B peptides in the bloodstream are indicative of heart failure. Polymorphisms in the precursor gene NPPA were associated with ischemic stroke, baseline blood pressure, hypertension, MI and the extent of CHD.

NPPB (OMIM 600295) **Electrolyte balance.** Natriuretic peptide B is a cardiac hormone produced primarily by the ventricle. Physiological activities of the peptide include natriuresis, diuresis, vasorelaxation, inhibition of renin and aldosterone secretion. Polymorphisms in the peptide precursor gene NPPB were associated with hypertension and coronary artery spasm.

PAI1 (OMIM 173360) **Blood coagulation.** Plasminogen activator inhibitor-1 is the serine protease inhibitor of fibrinolysis which is also an inflammation marker. The 4G promoter polymorphism of PAI1 gene has been associated with elevated PAI-1 level and thromboembolism.

PLAT (OMIM 173370) **Blood coagulation.** Tissue plasminogen activator (PLAT) activates proenzyme plasminogen to plasmin, the key enzyme in fibrinolysis. Accordingly, a change in the levels of PLAT will influence fibrinolysis thus affecting the rate of dissolution of thrombi. PLAT polymorphism (Glu)n is a potential risk factor for MI.

PON1 (OMIM 168820) **Atherosclerosis (inflammation).** Paraoxonase-1 is a plasma esterase associated with HDL. Apart from detoxification of exogenous organophosphates, PON1 protein also neutralizes the proinflammatory oxidized lipids present in oxidized LDLs thus protecting against atherosclerosis. In addition, PON1 polymorphisms that modulate the enzyme's activity or gene expression (55 L/M, 192 Q/R and -107 C/T) can also adjust CVD risks.

PON2 (OMIM 602447) **Atherosclerosis (inflammation).** Paraoxonase-2 is expressed mostly in the liver. It is a membrane-bound protein. As PON1, PON2 prevents or reverses LDL peroxidation as well as inhibits induction of monocyte chemotaxis. The common polymorphism 311 S/C was associated with lower risk of atherosclerosis and CHD.

PPARA (OMIM 170998) **Atherosclerosis (inflammation).** Peroxisome proliferator-activated receptor alpha. Peroxisome proliferators are substances that stimulate an increase in the size and in the number of peroxisomes. The action of peroxisome proliferators (such as hypolipidemic drugs and/or fatty acids) is mediated via specific nuclear receptors (PPARs) which affect expression of the genes involved in immune and inflammation responses. The common polymorphisms in PPARA affect the plasma levels of lipoproteins.

PPARG (OMIM 601487) **Adipose tissue/atherosclerosis (inflammation).** Peroxisome proliferator-activated receptor is a regulator of adipocyte differentiation; the common polymorphisms of PPARG were associated with obesity, atherosclerosis and attenuation of CHD risk.

PROC (OMIM 176860). **Blood coagulation.** Protein C is activated on the surface of endothelial cell by thrombin in complex with thrombomodulin. Promoter polymorphisms in PROC were associated with plasma level of protein C and an increased thrombotic risk.

PROS1 (OMIM 176880) **Blood coagulation.** Protein S inhibits blood clotting by serving as a cofactor for activated protein C. It is a vitamin K-dependent plasma protein. Polymorphisms in PROS1 were associated with thrombotic complications.

PTGIS (OMIM 601699) **Vasoconstriction and vasodilatation.** Prostacyclin synthase. Prostacylin-2 is a strong vasodilator and an endogenous inhibitor of platelet aggregation. Prostacyclin synthase catalyzes isomerization of prostaglandin H2 to prostacyclin. Variants of the 9bp repeat polymorphism in the promoter were associated with higher blood pressure in Japanese. These 9bp VNTRs are also present in European populations though no associations were reported as yet.

PTGS2 (OMIM 600262) **Vasoconstriction and vasodilatation.** Prostaglandin-endoperoxide synthase 2 (also known as cyclooxygenase 2) is the key enzyme in biosynthesis of prostaglandins. PTGS2 activity is associated with biologic events such as injury, inflammation and cell proliferation. At the same time, PTGS2 also catalyzes prostaglandin E in atherosclerotic plaques which, in turn, may be a proatherogenic factor. Antiinflammatory action of aspirin, the well-known drug, is exerted by inactivation of this enzyme through a covalent interaction. Promoter polymorphisms in PTGS2 were associated with intima media thickness, levels of inflammation markers and MI.

REN (OMIM 179820) **Renin-angiotensin system.** Renin cleaves angiotensinogen into angiotensin I. Common polymorphisms in the REN gene were associated with essential hypertension.

SAA1 (OMIM 104750) **Atherosclerosis (inflammation).** Plasma amyloid A is a major acute-phase response protein. It is also a component of the HDL particles. The plasma SAA level can increase several orders of magnitude in response to infections, inflammations and are also increased in CVDs. SAA protein has been identified in atherosclerotic lesions and SAA1 polymorphisms are perspective genetic markers of atherosclerosis.

SCARB1 (OMIM 601040) **Atherosclerosis (lipoproteins).** Scavenger receptor class B type 1 is an HDL receptor. The "exon-1" variant was associated with increased HDL-cholesterol and lower LDL-cholesterol levels.

SCNN1B (OMIM 600760) **Electrolyte balance.** Nonvoltage-gated sodium channel 1 beta subunit (amiloride-sensitive epithelial sodium channel, ENaC) mediates the diffusion of the sodium from the vessel lumen through the epithelial layer. This ion channel also controls reabsorption of the sodium in kidney. Blood pressure is to a greater extent determined by the concentration of the sodium in the vessels and polymorphisms in SCNN1B were associated with increased risk of hypertension.

SELE (OMIM 131210) **Atherosclerosis (inflammation).** E-selectin is expressed on endothelial cells in response to inflammation. The protein mediates leukocyte adhesion to the arterial endothelium. Higher levels of ICAM-1, VCAM-1 and E-selectin protein as well as common polymorphisms S128R and L554F in SELE gene were associated with CHD and atherosclerosis.

SELP (OMIM 173610). **Blood coagulation.** P-selectin is an adhesion molecule that mediates the interaction of activated endothelial cells or platelets with leukocytes. A reduced frequency of Pro-15 allele of the P-selectin gene was observed in patients with MI.

SELPLG (OMIM 600738) **Blood coagulation.** P-selectin glycoprotein ligand (PSGL1) is a high-affinity receptor for P-selectin. The protein is present in the surface of neutrophils, platelets, T-cells, monocytes and others. It plays a critical role in the tethering of the cells to platelets or endothelia expressing P-selectin. A deletion polymorphism in SELPLG was associated with a reduced risk of cerebrovascular disease.

SREBF1 (OMIM 184756) **Atherosclerosis (lipoproteins).** SRE (sterol regulatory element) are short patterns in the DNA sequence that mediate the transcriptional effects of sterols. Fllowing cleavage, the SRE-binding transcription factor 1 translocates to the nucleus and activates transcription. In particular SREBF1 binds to SRE of the LDLR gene. Polymorphisms in SREBF1 were associated with variation in the total and LDL cholesterol levels as well as with the development of atherosclerosis.

TAFI (OMIM 603101) **Blood coagulation.** Thrombin-activatable fibrinolysis inhibitor or 'carboxypeptidase B2' attenuates fibrinolysis by removing from the fibrin C-terminal lysine and arginine residues that are important for the binding and activation of plasminogen. TAFI protein is activated by thrombin and plasmin in the presence of thrombomdulin as cofactor. Elevated TAFI levels are a likely risk factor for venous thrombosis. Polymorphisms in TAFI affect the protein levels and were associated with MI and angina pectoris. Potentially, TAFI polymorphisms can explain susceptibility to bacterial agents such as *Neisseria*.

TFPI (OMIM 152310) **Blood coagulation.** Tissue factor pathway inhibitor is an important regulator of the extrinsic pathway of blood coagulation. The common polymorphism V264M was associated with lower plasma level of TFPI levels and coronary syndromes.

THBD (OMIM 188040) **Blood coagulation.** Thrombomodulin is an endothelial cell surface glycoprotein that forms a complex with thrombin. Binding of thrombin to THBD protein alters thrombin specificity and results in the activation of protein C, which degrades activated clotting factors FV and FVIII. In other words, thrombomodulin converts thrombin into an anticoagulant. Genetic defects in THBD lead to an increased predisposition to thrombosis; the common polymorphism A455V was associated with an increased risk for MI.

THBS1 (OMIM 188060) **Blood coagulation.** Thrombospondin 1 is an adhesive glycoprotein that mediates cell-to-cell and cell-to-matrix interactions. The protein binds to fibrinogen, fibronectin, laminin, type V collagen and integrins alpha-V/beta-1 and has been shown to play roles in platelet aggregation and angiogenesis. Thrombospondin-1 stimulates platelet aggregation through GPVI signaling. A common polymorphism in THBS1 was associated with CHD.

THPO (OMIM 600044) **Blood coagulation.** Megakaryocytopoiesis is the cellular development process that, eventually, leads to platelet production. Thrombopoietin, also known as megakaryocyte colony-stimulating factor, is a humoral growth factor that is necessary for megakaryocyte proliferation and maturation and thrombopoiesis. Polymorphisms in thrombopoietin were associated with adjustments of MI risk.

TNFA (OMIM 191160) **Atherosclerosis (inflammation).** Tumor necrosis factor alpha is an inflammatory cytokine produced by macrophages and monocytes. Promoter polymorphisms were associated, in particular, with CHD risk.

VDR (OMIM 601769) **Electrolyte balance.** Vitamin D receptor is the nuclear hormone receptor for vitamin D3. The skin supplies the body with 80-100% of its requirements of vitamin D. Age, latitude, exposure to sun, season of the year, pigmentation as well as VDR polymorphisms can dramatically affect the production of vitamin D in the skin. Common VDR polymorphisms were associated with changes in the bone density, decreased plasma levels of the active vitamin D, and are likely to be related to the severity of CHD. In particular, VDR is known to modify the toxicokinetics of lead and VDR genotypes were associated with hypertension in accordance with the lead status.

VEGF (OMIM 192240) **Vascular constitution and remodeling.** Vascular endothelial growth factor is the specific mitogen of the vascular endothelial cells and thus is a key regulator of angiogenesis. Potentially, VEGF can be mediator of some of the beneficial effects of dietary restriction in the brain and cardiovascular system. A common polymorphism -2578 A/C was associated with severity of atherosclerosis.

INDEX

A

B

C

D

E

F

G

H

I

M

N

O

P

S

T

U

V

W

X

Y

Z